LEARNING
MASTERCAM
X8
LATHE
Step
by
Step

James Valentino
Joseph Goldenberg

ISBN 978-0-83113-511-9

Industrial Press, Inc.
32 Haviland Street, Suite 3 South Norwalk, CT 06854
Tel: 203-956-5593, Toll-Free: 888-528-7852
Email: info@industrialpress.com
Web: books.industrialpress.com, ebooks.industrialpress.com

Sponsoring Editor: John Carleo
Cover Design: Janet Romano-Murray
Index Editor: Robert Weinstein

Cover photos supplied by Tom Lipton

DEDICATION

To my wife Barbara and to my children, Sarah and Andrew.

-James Valentino

To my students past, present and future.

-Joseph Goldenberg

ACKNOWLEDGEMENTS

The authors would like to express their thanks to CNC Software, Inc, especially Mr Ben Mund for his continued support.

We would also like to express our appreciation to Mr Bernard Hunter, the laboratory technician in the manufacturing processes laboratory at Queensborough Community College for proofing many portions of the manuscript.

PREFACE

The CNC programmer now has a powerful tool to assist in the job of creating and verifying part programs. *MasterCam X8* CNC software provides the programmer with a full array of easy to use features. The benefits of using *MasterCAM X8* include: automatic calculation of toolpath coordinates, determination of speeds and feeds, animation of the machining process and postprocessing the part program off-line, without tying up the CNC machine.

MasterCAM X7 is a robust PC based package. Its many cababilities must be presented in a clear and logical sequence. This text was written to provide a thorough introduction to *MasterCAM X8*'s LATHE package for students with little or no prior experience. Past users of *MasterCAM* will find release *X8* has been upgraded to run on Windows7or 8. Several enhancements such as the new verifier and Code Expert are also included.LEARNING *MasterCAM X8* LATHE Step by Step includes chapters on basic turning, grooving and threading and executing C-Axis machining operations.

Several learning aids have been designed throughout.

- Good graphical displays rather than long text and definitions are emphasised.

- An overview of the process of generating a word address program is presented.

- Key definitions are boxed in.

- Examples provide step-by-step instructions with excellent graphical displays.

- Needless cross-referencing has been eliminated. Each example is presented with all explainations appearing on the same page.

- Exercises are presented at the ends of chapters.

- A process plan is provided for many machining exercises to indicate the machining operations to be performed and the tools to be used.

- A CD provided with the text contains:

  ▶ *MasterCAM X8,* DEMO version. Students can use the DEMO to practice interactively on their own PC's.

  ▶ Files now keyed in sequence to the selected examples. Students can follow interactively when learning the procedure with the concepts presented in the text.

  ▶ Files containing CAD parts for machining exercises.

LEARNING *MasterCAM X8* LATHE can be used for many different types of training applications; these include:

- Undergraduate one-semester or two semester CNC programming courses.
- Computer assisted component of a CNC programming course.
- Indistrial training courses.
- Trade school courses on computer assisted CNC programming.
- Seminar on computer assisted CNC programming.
- Adult education courses.
- Reference text for self-study.

This text is designed to be used in many types of educational institutions such as:

- Four-year engineering schools.
- Four-year technology schools.
- Community colleges.
- Trade schools.
- Industrial training centers.

CONTENTS

CHAPTER-3　BASIC LATHE OPERATIONS　　3-1

CHAPTER-4　GROOVING AND THREADING OPERATIONS 4-1

CHAPTER - 1

INTRODUCTION TO *Mastercam X8*

1-1 Chapter Objectives

After completing this chapter you will be able to:

1. State the system requirements for installing *Mastercam X8*
2. Describe the general process of generating a word address program via *Mastercam X8*
3. Know the types of files created by *Mastercam X8*
4. Understand how to start *Mastercam X8*
5. Describe the nine general elements of *Mastercam X8*'s interface window
6. State how to set the system's working parameters
7. Understand the basic concepts of Gview, View, WCS, Cplane and Tplane
8. Know how to use the **Help**, **Save**, and **Exit** commands

1-2 *Mastercam X8* CNC Software

One of the most popular CNC software packages available today is *Mastercam X8* from CNC Software, Inc located at 671 Old Post Road, Tolland, CT 06084. CNC Software can also be reached at 860-875-5006. For information on the latest product developments and downloads of enhancements and patches, visit their web site at *www.mastercam.com* .

This software has a short learning curve. It presents the user with an easy to follow menu system that works fully with the Windows 8 operating system. Part geometry can be easily created with *Mastercam*'s CAD package. The CAM package enables the operator to quickly select the part material, machining operations and cutting tools. The software allows the operator to identify the CAD geometry to be selected for a machining operation then quickly generates the required tool path. The operator uses the machine and control definitions managers to specify the features of the CNC machine to be used. The appropriate postprocessor is also selected from the system's library. *Mastercam* is then directed to generate the corresponding word address part program. A very powerful feature of the software is its ability to verify the part program by animating the entire machining process. *Mastercam* automatically checks for any tool collisions when verification is running.

1-3 System Requirements for Version X8

Mastercam X8 is a 64-bit CNC software package.
The following minimum system hardware and software must be installed.

◆ Windows 7 or 8
◆ Intel or AMD 64 bit processor
◆ 1024 x 1280 resolution (minimum)
◆ System memory, 4GB
◆ 250GB drive, 20GB free hard disk space

1-4 Conventions Used Throughout the Text

The following conventions are used throughout this text.

DISPLAY	MEANING	PICTORIAL
[Enter ⏎]	Directs the operator to *press* the [Enter] key on the *keyboard*.	
[Space Bar]	Directs the operator to *press* the [Space Bar] key on the *keyboard*.	
Click Ⓝ	Means to move the mouse cursor to position Ⓝ and press the *LEFT* mouse button	
Bold	Commands to be ***typed at the keyboard*** appear in **bold**	

(ON CD icon)	When placed next to examples and exercises, this icon indicates a ***file by the same name is on the enclosed CD***. The student can ***get the file and follow the work interactively***.
	◇ Place the DEMO+EXERCISES CD in the drive
	◇ When the *Mastercam X8* graphic auto-displays:
	      Click the [Exit] button.
	◇ ***Double*** Click on the Mastercam X8 Demo icon
	◇ Click File ; Click 📂 Open
	◇ Click DVD RW Drive(E:) MCAMX8-EXERCISES
	◇ the chapter's file folders.
	                    📁 CHAPTER2
	                    📁 CHAPTER3

1-5 Installation of *Mastercam* X8 Demo CD Software for Student Use

⯈ Place the DEMO X8 CD in the CD drive

The *Mastercam X8* graphic shown below will auto-display

⯈ Click ① the Mastercam X7 Installs

⯈ Follow the prompts *automatically triggered* by the software on the CD to complete the installation.

1-6 An Overview of Generating a Word Address Program Via *Mastercam X8*

Any part to be machined using *Mastercam X* 8 software must first be drawn using either the *Mastercam X8* computer aided drafting (CAD) package or imported from another CAD package such as AutoCAD or SolidWorks. The part geometry created by the CAD package is used directly by the computer aided machining (CAM) package in specifying the location of machining cycles and in the determination of the corresponding tool paths. Ultimately, the CAM package produces a complete word address program for machining a specified part on a particular CNC machine tool (to learn more about word address programming refer to Introduction to Computer Numerical Control 5th Ed by J.Valentino and J.Goldenberg, published by Prentice Hall.
    The sequence of steps to be followed to direct *Mastercam X8* to generate a word address program for a milling CNC machine tool are shown in Figure 1-1.

CAM OPERATIONS

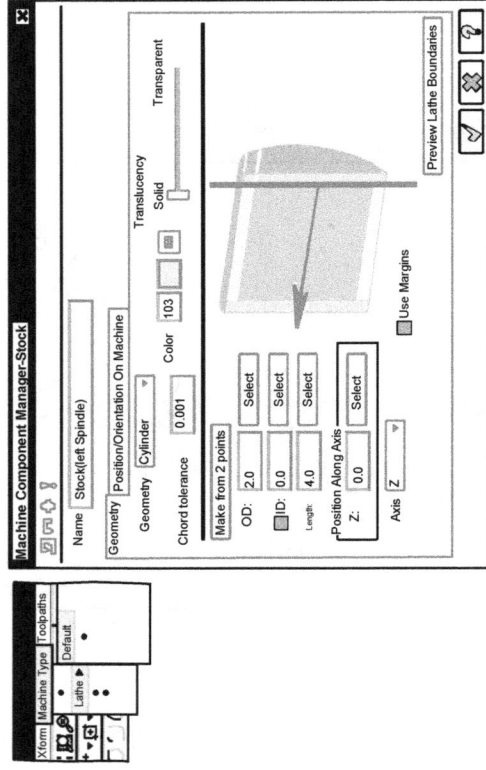

Step-2

Machine Component Manager-Stock

Name   Stock(left Spindle)

Geometry   Position/Orientation On Machine

Geometry   Cylinder

Chord tolerance   0.001

Color   103

Make from 2 points

OD: 2.0   Select

ID: 0.0   Select

Length: 4.0   Select

Position Along Axis

Z: 0.0   Select

Axis   Z

Translucency

Solid   Transparent

Use Margins

Preview Lathe Boundaries

The Machine Type(Turn) is specified and the stock is set up.

Step-1

CAD OPERATION

Mastercam Design X8

File Edit View Analyze Create Solids Model Prep Xform Machine Type Toolpaths Screen Settings Help

Gview:TOP   WCS:TOP   Cplane:+D+Z [TOP]

For help, press Alt+H.

A CAD model of the part is created or imported from another.
CAD package such as AutoCAD, Inventor or Solidworks

Step-3

C:\users\public\documents\mcam8\Materials

ALUMINUM inch - 2024
ALUMINUM inch - 5050
ALUMINUM inch - 6061
ALUMINUM inch - 6066
ALUMINUM inch - 7075
ALUMINUM inch - CAST - 65BHN
ALUMINUM inch - WROUGHT - 55BHN
COPPER inch - CAST - 65BHN
COPPER inch - WROUGHT - 40BHN
GRAPHITE inch
HASTELLOY inch
INCONEL inch

Display options
   Show all   Millimeters
   Inch   Meters

Source   Mill - library

Compress

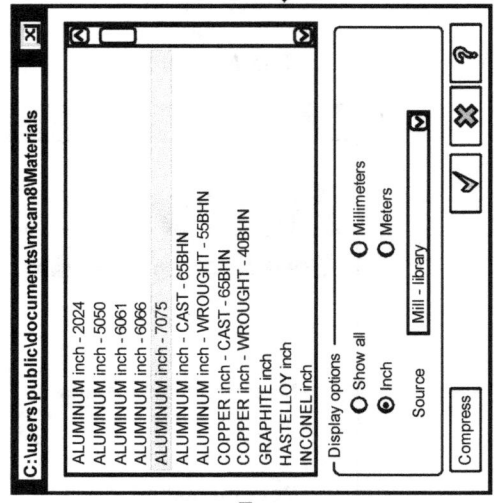

The material to be machined is selected.

Step-4

Mastercam Lathe X8

File Edit View Analyze Create Solids Model Prep Xform Machine Type Screen Settings Help

Toolpaths

Rough

Finish

Gview:TOP   WCS:TOP   Cplane:+D+Z [TOP]

For help, press Alt+H.

The operator selects the operation(rough) and clicks   geometry to be turned.

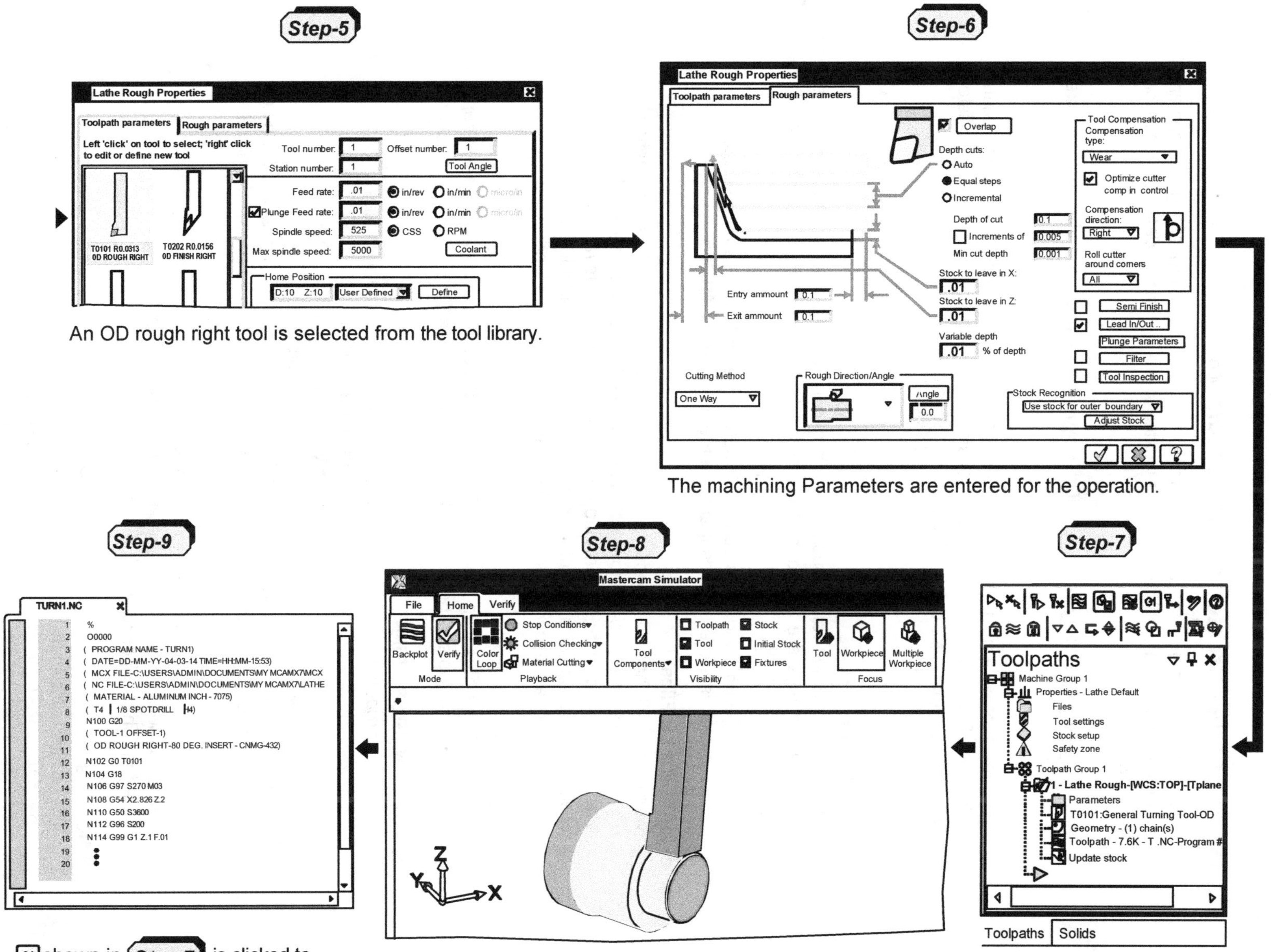

Step-5

An OD rough right tool is selected from the tool library.

Step-6

The machining Parameters are entered for the operation.

Step-7

The rough operation is clicked then the verify button is clicked.

Step-8

Mastercam animates the rough operation and checks for any collisions during machining.

Step-9

is clicked to direct Mastercam to generate the word address program.

Figure 1-1 The sequence of steps for generating a part program with *Mastercam's* lathe package.

1-7 Types of Files Created by *Mastercam X8*

Mastercam X8 is designed to be more compatible with Microsoft's guidelines and security measures. To this end the it is located in the following *default installation* directory:

64-bit operating systems   **C:\Program Files\mcamx8\Mastercam.exe**

The operator will encounter various file extension names in the course of working with *Mastercam X8* software. The extension names and their meanings are listed below in one place for quick reference.

a) CONFIG File

The system default values such as *units,allocations, tolerances, NC settings, screen* and *CAD settings* are stored in the configuration file that has the *extension* **.CONFIG**

b) MCX-8 Files

All the *part geometry* generated in DESIGN mode and *tool path information* created in MILL *mode* is stored in files that have the *extension* **.MCX-8**

c) MATERIALS Files

Mastercam X8 has an extensive library of various materials. The information on a particular *material* is stored in the materials file with extension **.MATERIALS** *Mastercam X8* uses this information to *automatically set the recommended speeds and feeds* for a particular cutting tool used in a machining operation. The operator can also *manually* enter desired speeds and feeds.

d) TOOLS Files

The *tool library* files with the extension **.TOOLS** contain a comprehensive set of the most common *tools* to be selected in order to execute the machining of a part.

e) PST or MCPOST Files

The *post processors* for various CNC control units are stored in files that have the *extension* **.PST** or the newer extension **.MCPOST** . The change has been made because Winsows ocasionally confuses **.PST** with the same extension used with Microsoft Outlook.
Note: For each post renamed to **.MCPOST** the operator must also build a new control definition and import the settings from the old control definition.

f) NCI Files

NCI(Numerical Control Intermediate) files contain the tool path coordinate values as well as speeds, feeds and other important machining information for a job. These files have the extension **. NCI**

f) NC Files

These files contain the *word address part programs* . Mastercam X5 uses a particular **.PST** file and a selected **.NCI** file to generate the corresponding **.NC** file. The **.NC** file is sent to the CNC machine tool for producing the part.

1-8 Starting *Mastercam X*8

Mastercam X8 is started by double clicking ① on the *MastercamX8* icon appearing on the Windows desktop. See Figure 1-2.

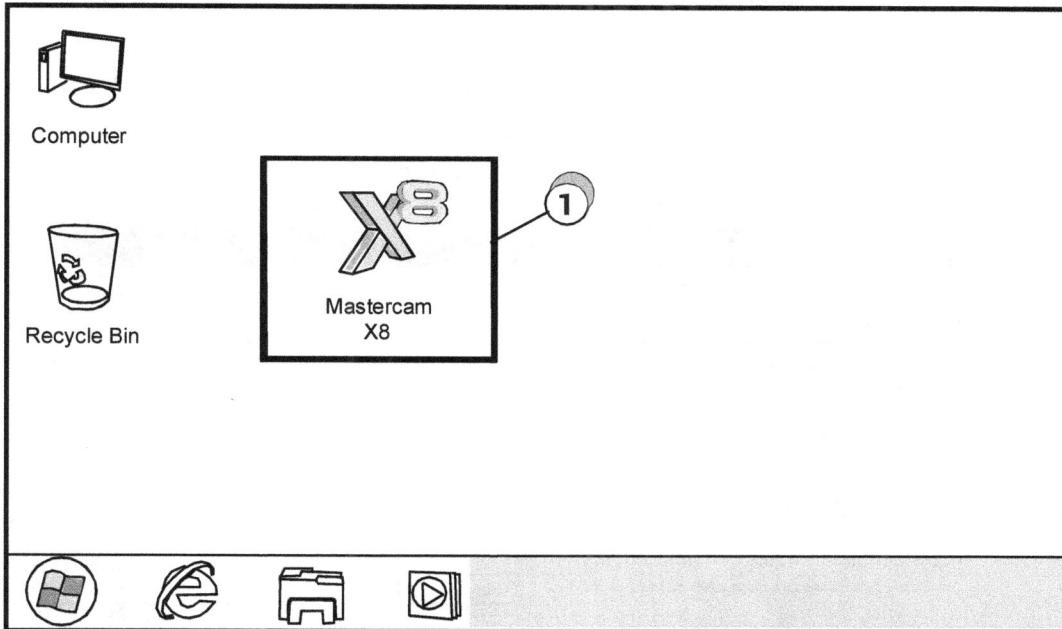

Figure 1-2 Starting *MastercamX8* from the Windows Desktop

1-9 Entering the *Mastercam X8* Lathe Package

To enter the Mill package the operator must select a milling machine from the machine type drop down menu.

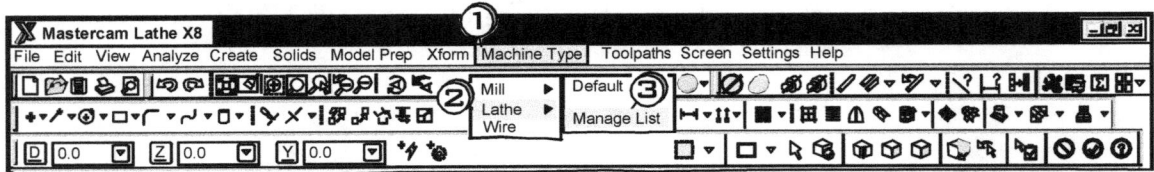

➤ Click ① the   Machine Type   drop down menu   ➤ Click ③ Manage List

➤ Click ② the   Lathe   package

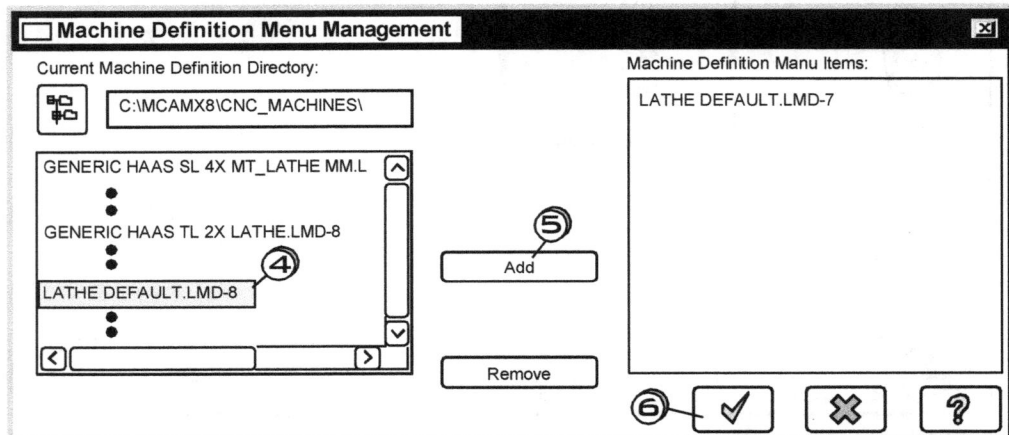

➤ Click ④ LATHE DEFAULT.LMD-7   ➤ Click ⑥ the OK button ✓

➤ Click ⑤ the [Add] button.

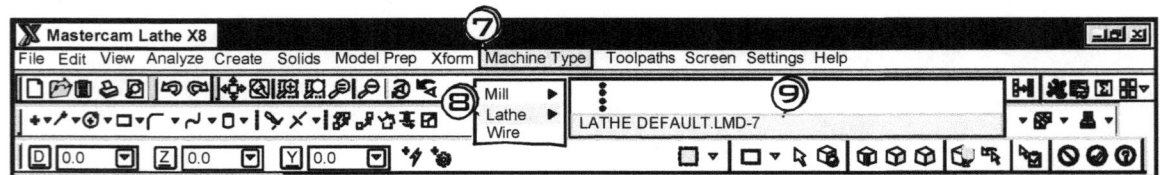

➤ Click ⑦ the   Machine Type   drop down menu   ➤ Click ⑨
                                                       LATHE DEFAULT.LMD-8
➤ Click ⑧ the Lathe package

Mastercam will then *change* the main interface window listing from **Design** to **Lathe**

1-10 A Description of the *Mastercam X8* Lathe Main Interface Window

The Mastercam X8 Lathe main interface window is similar to any other window that is used in the Windows operating system. The interface window consists of **nine** general elements:

- menu bar
- tool bars
- autocursor ribbon bar
- general selection ribbon bar
- current function ribbon bar
- operations manager
- graphics window
- status bar
- MRU(most recently used) function bar

These elements are shown in Figure 1-3.

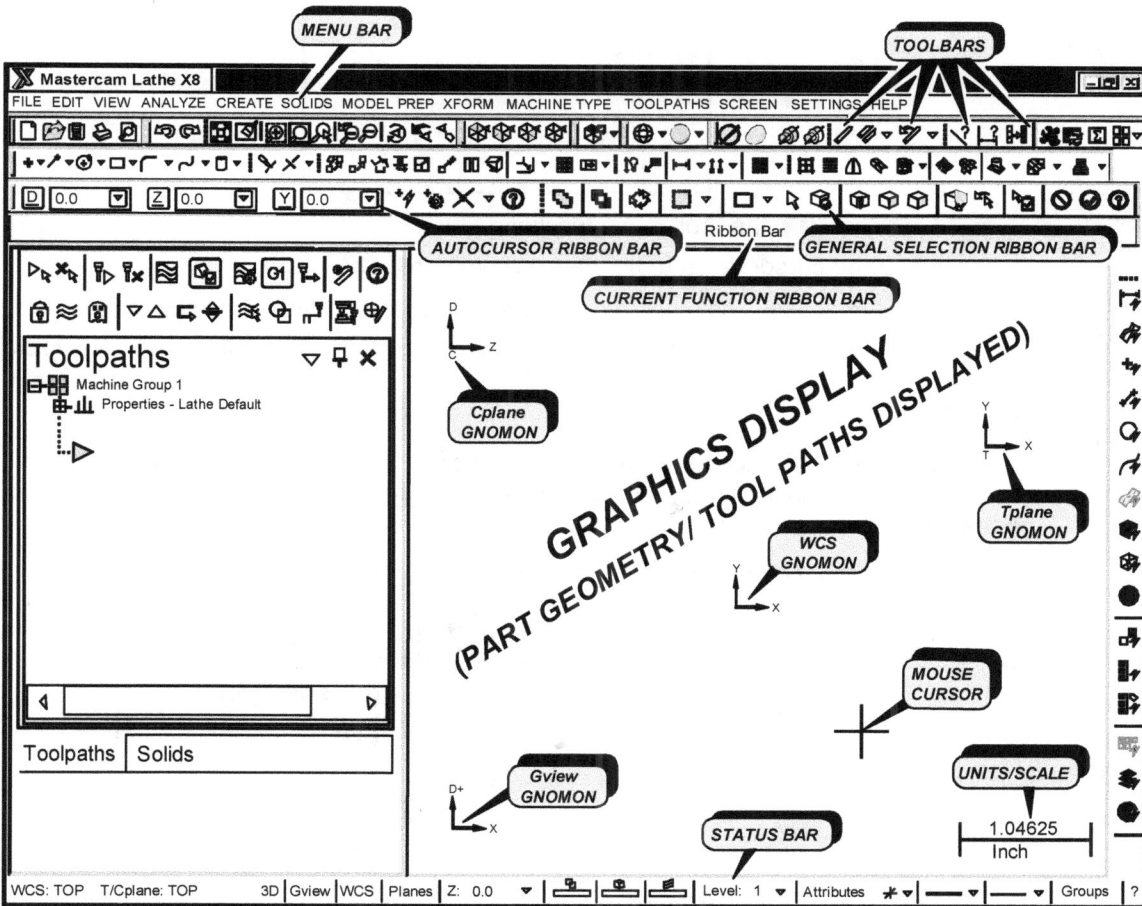

Figure 1-3 *Mastercam* X8 design/lathe main interface window.

MENU BAR

This element contains **twenty five** drop down menus. Several of the functions have icons next to them so the operator gets a graphic picture of what the command accomplishes. A function is **executed** by moving the mouse cursor **over** it and pressing the **left** mouse button. The operator also has the ability to create customized drop down menus. Refer to Chapter 11.

FILE   EDIT   VIEW   ANALYZE   CREATE   SOLIDS   MODEL PREP   XFORM   MACHINE TYPE   TOOLPATHS

Point   ▶
Line   Endpoint
Arc   ▶   Closest
Fillet   ▶   Bisect
Chamfer   ▶   Perpendicular
Spline   ▶   Parallel
Curve   ▶   Tangent Through Point
Surface   ▶
Drafting   ▶

MOUSE CURSOR
CASCADING MENU

⊞ Rectangle
⊡ Rectangular
⬡ Polygon
◯ Ellipse
◎ Spiral
❀ Helix

➤ **Left** Click ① the drop down function

➤ move the cursor down to the desired command ②

➤ The symbol ▶ indicates a cascading sub-menu relating to the command exists

➤ move the cursor across to the desired command in the sub-menu and **left** click ③
to **execute** it

TOOLBARS

Toolbars are sets of buttons that execute Mastercam functions. Toolbars are normally docked but they can be undocked and relocated by clicking on their "grab" handle. Toolabrs can also be reshaped and customized.

EXECUTING A FUNCTION IN A TOOLBAR

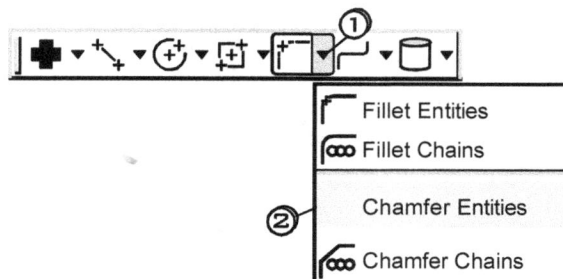

Fillet Entities
Fillet Chains
Chamfer Entities
Chamfer Chains

➤ **Left** Click ① the down arrow

➤ move the cursor down to the desired command; **Left** click ②

RELOCATING A TOOLBAR

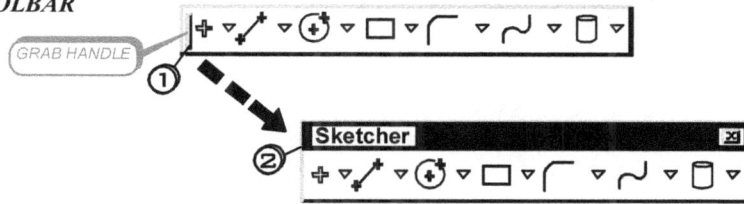

➤ Move the mouse cursor on the grab bar and press the left mouse button ①

➤ **Keeping the left button depressed,** move the toolbar to its new location ②, release.

RESHAPING AN UNDOCKED TOOLBAR

➤ Move the cursor *near* the edge of the toolbar until the double headed arrow appears.

➤ **Depress the left** mouse button and **keeping it depressed** drag it to position② ; release.

REMOVING or ADDING TOOLBARS TO THE INTERFACE DISPLAY

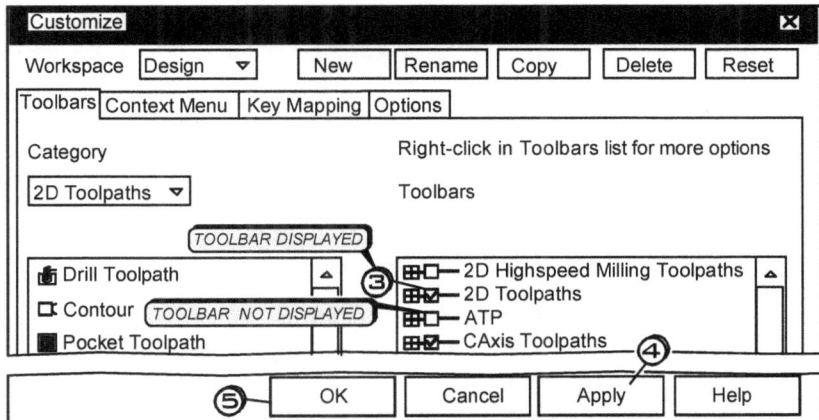

➤ **Left** Click ① the ⬛ Settings ⬛ function

➤ **Left** Click ② the 🔲 Customize dialog box

➤ **Left** Click ③ the

    added check on ✓
        or
    removed check off

➤ **Left** Click ④ the

    Apply

➤ **Left** Click ⑤ the

    OK

    when done

AUTOCURSOR RIBBON BAR

Autocursor is used to **automatically detect and snap to** certian point locations on graphic entities displayed on the screen. This is done as the operator moves the mouse cursor over the entities. Snap locations include endpoints, centers of arcs, intersections, midpoints, etc. Autocursor is activated every time Mastercam prompts the operator to select a location on the screen. Upon detecting a snapped point location autocursor displays a "visual cue" to the right of the cursor. Cues include:

- ⁺↘ endpoint snap
- ✕ intersection point snap
- ⊕ center of arc snap

If Autocursor does not detect any points to snap to it automatically defaults to the **Sketcher** function. The operator then enters the point location using coordinates.

Autocursor

| D | 0.0 | ▼ | Z | 0.0 | ▼ | Y | 0.0 | ▼ | ⚡ ⚙ ✕ ▾ ⓘ |

SETTING AND UNSETTING AUTOCURSOR SNAP LOCATIONS

⌖ Zoom Window	F1
⊖ Un-Zoom	F2
Dynamic Rotation	
Fit	
Repaint	
Autocursor	

Autocursor Settings

☑ Origin ☐ Angular 150
☑ Arc Center ☒ Tangent
☑ Endpoint ☐ Perpendicular
☑ Intersection ☐ Nearest
☑ Midpoint ☑ Horizontal/Vertical
☑ Quadrant
☑ Point

[Enable All] [Disable All]

☑ Default to Fast Point mode
☑ Enable power keys

[✓] [✗] [?]

➤ **Right** Click ① in the display screen area

➤ **Left** Click ② **Autocursor**

➤ **Left** Click ③
added snap check on ✓
or
removed snap check off.

Autocursor *two* default settings:

☑ Default to Fast Point mode

Sets fastpoint mode as the default method of inputting XYZ corrdinates .

☑ Enable power keys

Enables the operator to specify point locations to snap to *by using single keystrokes* .

FUNCTION KEY   COMMAND ENTERED

- O → Snap to origin
- E → Snap to endpoint
- M → Snap to midpoint
- P → Snap to point
- C → Snap to arc center
- I → Snap to intersection
- Q → Snap to quadrant

GENERAL SELECTION RIBBON BAR

The General Selection ribbon bar is used to select entities displayed in the graphics window. General Selection *is active any time the operator is not in an active function* such as Sketcher, Analyze or View Manipulation. General Selection is activated by *Mastercam X8* functons that prompt the operator to select entities.

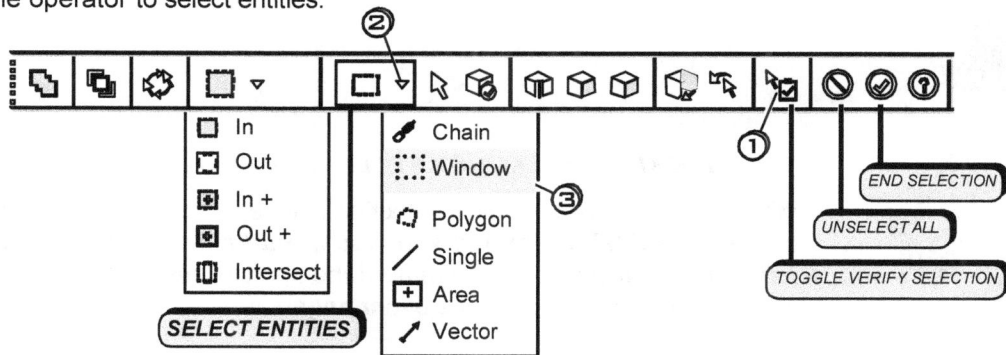

☐ In
☐ Out
☐ In +
☐ Out +
☐ Intersect

SELECT ENTITIES

- Chain
- Window ③
- Polygon
- Single
- Area
- Vector

END SELECTION
UNSELECT ALL
TOGGLE VERIFY SELECTION

ACTIVATING THE SELECT ENTITIES MENU

➤ *Left* Click ① Toggle Verify Selection button

➤ *Left* Click ② the down arrow

➤ *Left* Click ③ the entity selection method

CURRENT FUNCTION RIBBON BAR

The current function ribbon bar is used to create and edit CAD geometry. Ribbon bars have a close appearance to toolbars and work very much like dialog boxes. They can be docked or undocked and relocated in a convenient area of the graphics screen. If a ribbon bar for a function exists it is automatically activated and displayed when the function is selected. A blank ribbon bar is displayed by *Mastercam X8* above the graphics window when no ribbon bar is activated. When a function with a ribbon bar is selected *Mastercam X8* replaces the blank with the ribbon bar for the function.

ACTIVATING THE CURRENT FUNCTION RIBBON BAR

➤ *Left* Click ① The Circle Center pt Button.

LOCKING A SELECTION BY LEFT CLICKING ON IT

Mastercam X8 has a left click selection feature. If the operator *left* clicks on the **SELECTION** it *remains locked in that mode until the operator left clicks on the selection again.* If the operator *left* clicks in the **VALUE** box *it is in effect for a single event only.*

RELOCATING THE CURRENT FUNCTION RIBBON BAR

➤ move the mouse cursor on the boundary of the ribbon bar and press the left mouse button ①

➤ ***keeping the left button depressed*** move the ribbon bar to its new location ② and release.

THE OPERATIONS MANAGER

The Operations Manager is displayed to the left of the graphics window. It contains both the Toolpath Manager , the Solids Manager and Art. The Operations Manager is *Mastercam X8*'s control center for a job. Some of its functions include:

- *listing* all machining operations in the order which they will occur

- *backplotting and verifying* toolpaths for listed machining operations

- *editing* any of the machining parameters of listed operations

- *re-sequencing* the order of listed machining operations

- *deleting* listed machining operations

- *creating new* machining operations by copying existing ones and editing

- *postprocessing* a word address part program

RESIZING THE PANE WIDTH OF THE OPERATIONS MANAGER

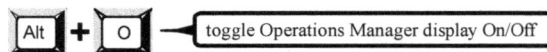

➣ Move the cursor **near** the edge of the toolbar until the double headed arrow appears

➣ **Depress the left** mouse button and **keeping it depressed** drag it to position②; release

TURNING THE OPERATIONS MANAGER DISPLAY ON/OFF

The Operations Manager is display is turned on, by default. The operator can toggle the display *off* or *on* by using the short-cut keys on the keyboard:

Alt + O — toggle Operations Manager display On/Off

THE GRAPHICS WINDOW

Part geometry is created, edited and viewed in the working area called the Graphics window. Backplotted toolpaths and verification of machining operation is also displayed in this area. The Graphics window can be resized. Doing so also resizes the width of the Operations Manager window. When the Graphics Window is resized *Mastercam X8* remembers the setting and will maintain it when the software is opened again.

STATUS INDICATORS DISPLAYED IN THE GRAPHICS WINDOW

Mastercam X8 displays the following indicators in the graphics window to keep the operator updated with important parameters relating to the job at hand. These include:

THE CURRENT WORKING UNITS AND SCALE
inch or metric and the number of units displayed per inch or millimeter. This type of scaling provides a feel for the *actual* size of the part.

|— 1.04625 —|
Inch

THE COORDINATE AXIS ICON("GNOMON")

Gview Gnomon

*Mastercam X7 displays the orientation
of the current Gview in the lower left
corner of the Graphics Window. The
X axis is shown red, D+ axis green.
Below the gnomon Mastercam X8
indicates that the Gview is aligned
with TOP and views WCS, Cplane
and Tplane are also aligned TOP
or with the current Gview.*

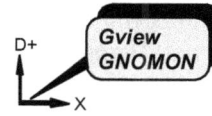

Gview:TOP   WCS:TOP   T/Cplane:TOP

Cplane Gnomon

Mastercam X8 displays the Cplane gnomon
in the *upper left* corner of the Graphics Window.
This indicates the orientation of the Cplane.

Tplane Gnomon

Mastercam X8 displays the Tplane gnomon
in the *upper right corner* of the Graphics Window
to show the Tplane's orientation. This indicates
the orientation of the tool.

WCS Gnomon

The WCS gnomon is displayed at the *origin
of the current WCS* and indicates its current
orientation. The color of the WCS gnomon is
set in the View Manager dialog box.

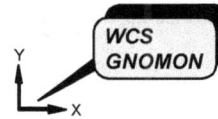

Refer to Section 1-12 for the steps to be
taken to direct *Mastercam X7* to display these
gnomons at all times.

When the operator works in *Mastercam X7*
Design the Tplane gnomon *is not displayed*
in the Graphics Window.

MOUSE CURSOR

The Graphics cursor is used as a *pointing/identifying*
device. The *location* of a point can be inputted
by moving the cursor to a location in the
Graphics Window and pressing the mouse button.
The cursor is also used to *identify* which objects
in the Graphics Window are to be operated on by
the current command.

STATUS BAR

Mastercam X8's status bar is positioned at the bottom of the main interface window. The *left* side of the status bar displays *status messages* and the *right* side enables the operator to specify *important parameter settings* . These include entity creation types(2D,3D) , orientations (Gview and Planes), depth (Z), entity color, levels on which information is placed, line types/widths and point types(attributes), view settings(WCS) and Groups.

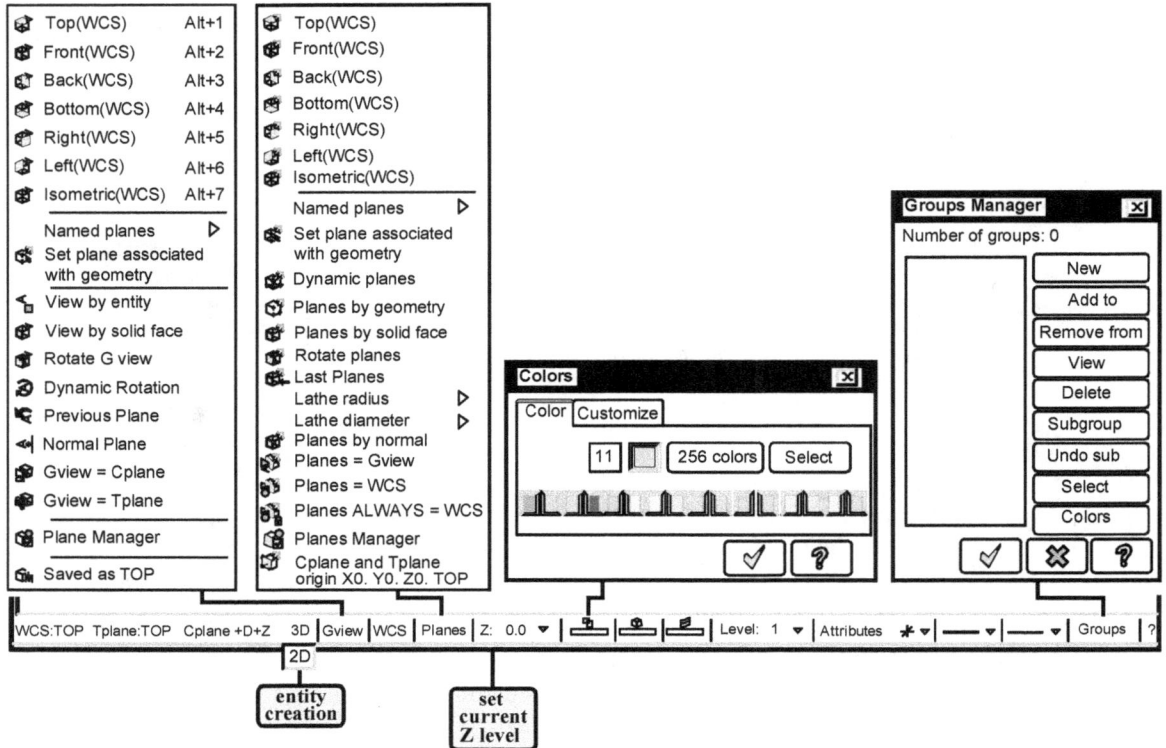

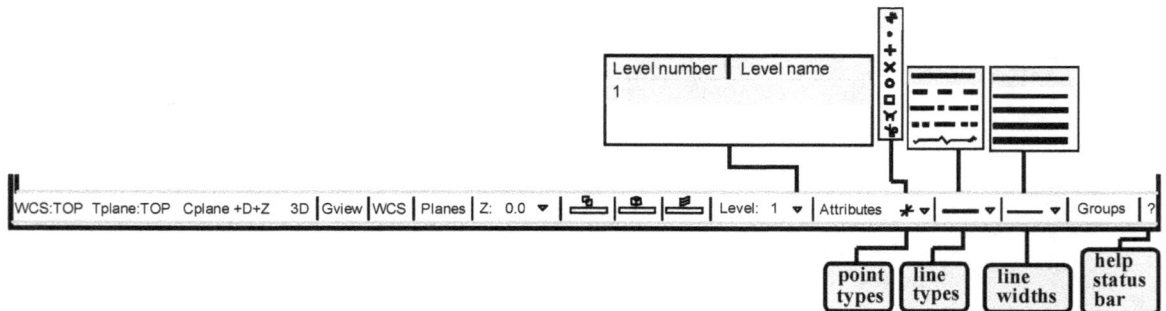

Top(WCS)	Alt+1
Front(WCS)	Alt+2
Back(WCS)	Alt+3
Bottom(WCS)	Alt+4
Right(WCS)	Alt+5
Left(WCS)	Alt+6
Isometric(WCS)	Alt+7
Named planes	▷
Set plane associated with geometry	
View by entity	
View by solid face	
Rotate G view	
Dynamic Rotation	
Previous Plane	
Normal Plane	
Gview = Cplane	
Gview = Tplane	
Plane Manager	
Saved as TOP	

Top(WCS)	
Front(WCS)	
Back(WCS)	
Bottom(WCS)	
Right(WCS)	
Left(WCS)	
Isometric(WCS)	
Named planes	▷
Set plane associated with geometry	
Dynamic planes	
Planes by geometry	
Planes by solid face	
Rotate planes	
Last Planes	
Lathe radius	▷
Lathe diameter	▷
Planes by normal	
Planes = Gview	
Planes = WCS	
Planes ALWAYS = WCS	
Planes Manager	
Cplane and Tplane origin X0. Y0. Z0. TOP	

Colors

Color | Customize

11 [ ] | 256 colors | Select

Groups Manager

Number of groups: 0

New
Add to
Remove from
View
Delete
Subgroup
Undo sub
Select
Colors

WCS:TOP Tplane:TOP Cplane +D+Z   3D │Gview│WCS │ Planes │ Z: 0.0 ▼ │ ▢ │ ▢ │ ▢ │ Level: 1 ▼ │ Attributes ✱ ▼│ ── ▼│ ── ▼│ Groups │ ?

2D

entity creation

set current Z level

Level number | Level name
1

WCS:TOP Tplane:TOP Cplane +D+Z   3D │Gview│WCS │ Planes │ Z: 0.0 ▼ │ ▢ │ ▢ │ ▢ │ Level: 1 ▼ │ Attributes ✱ ▼│ ── ▼│ ── ▼│ Groups │ ?

point types   **line types**   **line widths**   **help status bar**

MRU FUNCTION BAR

The MRU(Most Recently Used) function bar is located to the right of the main interface window. *Mastercam X8* keeps track of the *last function the operator selected and creates a button for it in the MRU toolbar* . The MRU makes the most recently used functions easily accessible for use again in one location.

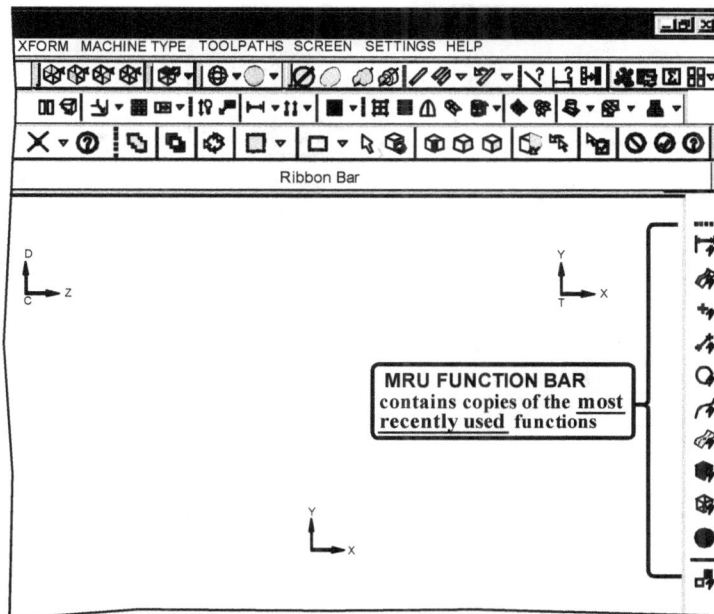

MRU FUNCTION BAR
contains copies of the most recently used functions

1-11 *Mastercam X8*'s Short-Cut Keys for Entering Commands

In addition to using toolbar and drop down menus, the operator can also use short-cut keys for quickly entering functions. Key use reduces the need for clicking into sub-menus thereby dramatically lowering the time it takes to complete operations in *Mastercam X8* Keys can provide additional flexibility. For example, the keys that cause pan and zoom can be used *while* a line function is being executed.

FUNCTION KEYS   FUNCTION ENTERED

Keys	Function Entered
Alt + 1	Top view
Alt + 2	Front view
Alt + 5	Right side view
Alt + 7	Isometric view
Alt + A	Save file
Alt + A	Autosave
Alt + C	Access C-Hooks
Alt + D	Open Drafting Options dialog
Alt + E	Hide/Show entities
Alt + F1	Fit all geometry to screen
Alt + F2	Unzoom
Alt + F4	Exit file
Alt + F8	Configure
Alt + G	Display screen grid
Alt + H	Open online help
Alt + O	Hide/Show Operations Manager
Alt + P	Revert to previous view
Alt + S	Shade On/Off
Alt + T	Toggle toolpath display

FUNCTION KEYS   FUNCTION ENTERED

Keys	Function Entered
Alt + U	Undo the function just performed
Alt + V	Display the *Mastercam* version and SIM numbers
Alt + X	Set the attributes from a selected entity
Ctrl + A	Select all entities
Ctrl + C	Copy
Ctrl + U	Undo
Ctrl + X	Cut
Ctrl + Z	Undo
F1	Zoom window
F2	Unzoom current display by .5
F3	Repaint the graphics area
F4	Analyze selected entity
F5	Delete selected entities
F9	Display coordinate info
Esc	*Cancel* current command
Pg Up	Zoom graphics display *up*
Pg Dn	Zoom graphics display *down*
→	Pan graphics display *right*
←	Pan graphics display *left*
↑	Pan graphics display *up*
↓	Pan graphics display *down*

1-12 Setting Working Parameters Via the System Configuration Dialog Box

Important working parameters in *Mastercam X8* are specified in the System Configuration dialog box. These include tolerance setting, file subdirectory creation, screen settings, NC settings, CAD settings, etc. The configuration file is saved as **MCAMX.CONFIG** for English units or **MCAMXM.CONFIG** for metric units.

➤ Click ① SETTINGS

➤ Click ② [⚙ Configuration]

SPECIFYING WORKING UNITS(ENGLISH OR METRIC)

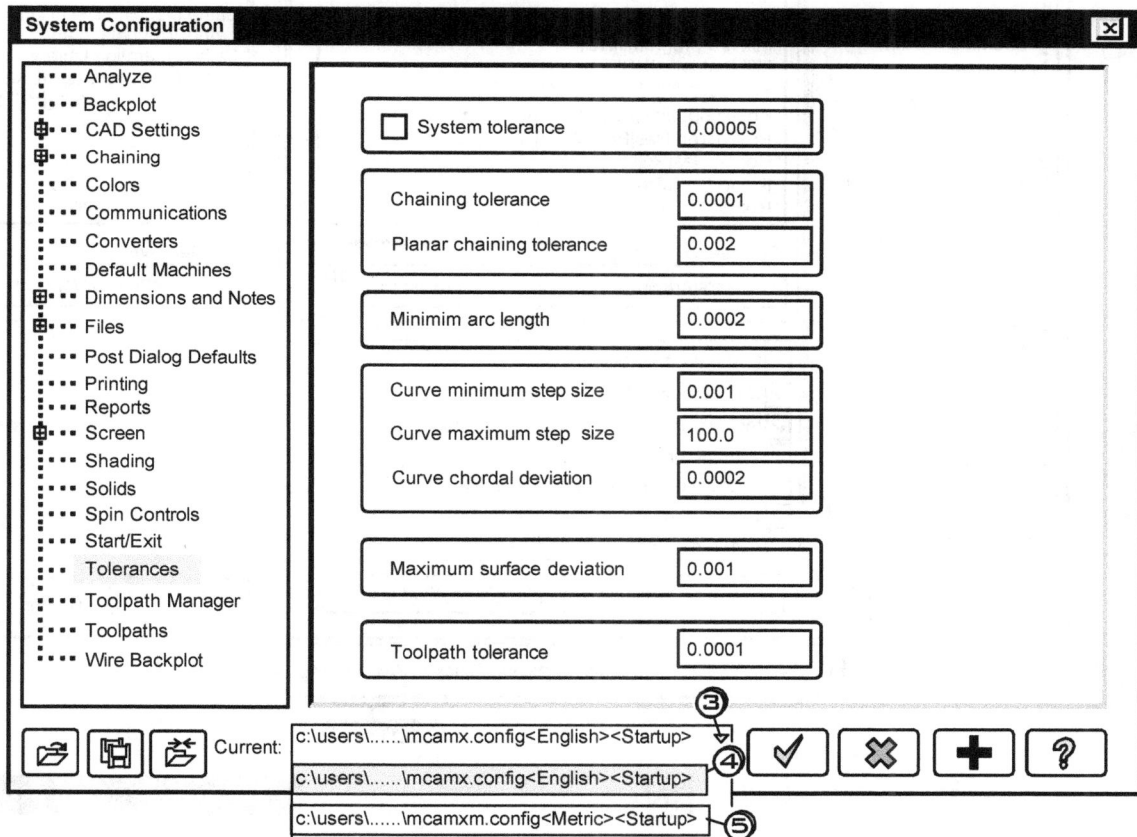

➤ Click ③ the Current configuration file down button [▼]

➤ Click ④ for *Lathe* part programs where *English units* are used
  or
➤ Click ⑤ for *Lathe* part programs where *metric units* are used

CREATING A DIRECTORY FOR STORING AND GETTING .MCX-8 STUDENT EXERCISE FILES

➤ Click ⑥   Files

➤ Click ⑦   Mastercam Current Version Parts[MCX-5,EMCX-5]

➤ Click ⑧   in the Selected Item box to scroll forward to the *end* of the directory name

➤ Click ⑨   in the Selected Item box; enter   YOUR INITIALS

C:\DOCUMENTS AND SETTINGS\JAMES\MY D OCUMENTS\MY MCAMX7**JVAL-MILL**

System Configuration

Data paths ⑦

- Mastercam Current Version Parts[MCX-8,EMCX-8]
- Mastercam Previous X Version Parts[MCX-6,MCX
- MC8 Files [MC8]
- MC9 Files[MC9]
- Mill NC programs[NC]
- Mill toolpaths[NCI]
- Parasolid files[X,T,X_B,XMT_TXT]
- Postscript files[EPS, AI, PSI]
- ProE/Creo files[PRT, ASM]
- Regen files[RGN]
- Rhino files[3DM]
- Router NC programs[NC]

Left panel:
- • • Analyze
- • • Backplot
- ⊞ • • CAD Settings
- ⊞ • • Chaining
- • • Colors
- • • Communications
- • • Converters
- • • Default Machines
- ⊞ • • Dimensions and Notes
- ⊞ • • Files ⑥
- • • Post Dialog De
- • • Printing
- ⊞ • • Screen
- • • Shading
- • • Solids
- • • Spin Controls
- • • Start/Exit
- • • Tolerances
- • • Toolpath Manager
- • • Toolpaths
- • • Wire Backplot

Selected item:
C:\Users|Admin\Documents\my mcamx8\mcx ⑧ ⑨

- ☐ Use default Data paths
- ☑ Use Windows Temp Directory
- ☑ Include bitmap in the file when saving
- ☑ Prompt for the file descriptor when saving
- ☑ Restore entire toolpath data in File, Open
- ⑪ ☑ Delete duplicate entities in File, Open
- ☐ Apply last machine definition

Current: c:\users\......\mcamx.config<English><Startup>

File usage
- Default Component Library
- Default Control Definition
- Lathe Defaults Library
- Lathe Machine Definition
- Lathe Material Library
- Lathe Operation Library
- Lathe Post Processor[EXE]
- Lathe Post Processor[PST]
- Lathe Setup Sheet
- Lathe Tool Library
- Mill Defaults Library
- Mill Machine Definition
- Mill Material Library

Selected item:
ALL COMPONENTS.GMD-5

Number of files/folders to show in MRU fields: 10

C:\Users\Admin\Documents\my mcamx8\mcx **JVAL-LATHE**

operator's entry

➤ Click ⑨   the Select button

➤ Click ⑩   the [Yes] button

Data path does not exist-create it?
⑩ [Yes]   [No]

Your mill jobs will now be saved and retrieved from the directory named

C:\Users\Admin\Documents\my mcamx8\mcx **JVAL-LATHE** when the edited configuration file is saved.

Configure *Mastercam* to *automatically find and delete duplicate entities* every time a file is opened.

➤ Click ⑪   check on for   ☑ Delete duplicate entities in File, Open

DISPLAYING THE WCS GNOMON AND LEARNING MODE PROMPTS

By default *Mastercam X8* has the *continuous display* for the WCS gnomon set to *off*. The operator can change this setting to *on* so the location of the **working origin** is *displayed at all times.* Beginners also find it helpful to have the Learning Mode prompts turned **on** when executing functions.

➤ Click (1 2)   Screen

➤ Click (1 3)   the check **on** ☑ for Display WCS XYZ axes

➤ Click (1 4)   the check **off** ☑   for Enable Ribbon modality

Note: If Ribbon Bar modality is *on* ☑ , all settings previously entered *will remain* when it is opened again. This feature is intended to save time for entering the same function in production work but should be **turned off** ☐ when learning different features of the same function.

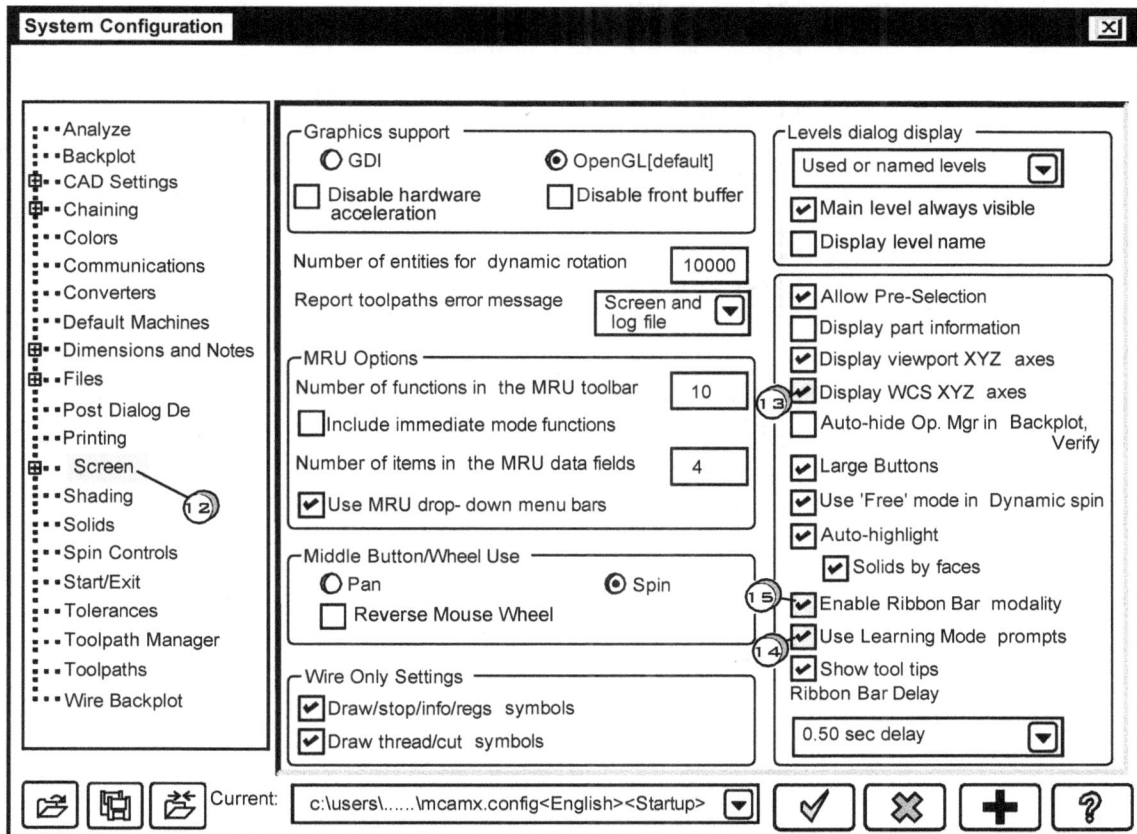

System Configuration   ☒

- • • Analyze
- • • Backplot
- ⊞• • CAD Settings
- ⊞• • Chaining
- • • Colors
- • • Communications
- • • Converters
- • • Default Machines
- ⊞• • Dimensions and Notes
- ⊞• • Files
- • • Post Dialog De
- • • Printing
- ⊞• • Screen
- • • Shading   (1 2)
- • • Solids
- • • Spin Controls
- • • Start/Exit
- • • Tolerances
- • • Toolpath Manager
- • • Toolpaths
- • • Wire Backplot

┌─ Graphics support ──────────
  ○ GDI   ◉ OpenGL[default]
  ☐ Disable hardware   ☐ Disable front buffer
      acceleration

Number of entities for dynamic rotation   [10000]
Report toolpaths error message   [Screen and log file ▼]

┌─ MRU Options ──────────
Number of functions in the MRU toolbar   [10]
☐ Include immediate mode functions
Number of items in the MRU data fields   [4]
☑ Use MRU drop- down menu bars

┌─ Middle Button/Wheel Use ──────────
  ○ Pan   ◉ Spin
  ☐ Reverse Mouse Wheel

┌─ Wire Only Settings ──────────
☑ Draw/stop/info/regs symbols
☑ Draw thread/cut symbols

┌─ Levels dialog display ──────────
[Used or named levels ▼]
☑ Main level always visible
☐ Display level name

☑ Allow Pre-Selection
☐ Display part information
☑ Display viewport XYZ axes
(1 3) ☑ Display WCS XYZ axes
☐ Auto-hide Op. Mgr in Backplot, Verify
☑ Large Buttons
☑ Use 'Free' mode in Dynamic spin
☑ Auto-highlight
   ☑ Solids by faces
(1 5) ☑ Enable Ribbon Bar modality
(1 4) ☑ Use Learning Mode prompts
☑ Show tool tips
Ribbon Bar Delay
[0.50 sec delay   ▼]

Current: [c:\users\......\mcamx.config<English><Startup> ▼]   ✓   ✖   ✚   ?

SETTING THE SYSTEM DEFAULT COLORS

➤ Click ⟨16⟩   Colors

➤ Click ⟨17⟩ the *Mastercam X8* interface element   Graphics backround color

➤ Click ⟨18⟩ the color setting for the element

System Configuration   ☒

Color [9] [ ]   ☑ Show 256 colors   ⟨18⟩   [Customize]

- • • Analyze
- • • Backplot
- ⊞ • • CAD Settings
- ⊞ • • Chaining ⟨16⟩
- • • Colors
- • • Communications
- • • Converters
- • • Default Machines
- ⊞ • • Dimensions and Notes
- ⊞ • • Files
- • • Post Dialog De
- • • Printing
- ⊞ • • Screen
- • • Shading
- • • Solids
- • • Spin Controls
- • • Start/Exit
- • • Tolerances
- • • Toolpaths
- • • Toolpath Manager
- • • Wire Backplot

Calculate field color
Construction origin color
Default Groups color
Draft dirty color
Draft phantom color
Edge motion color
General Selection backround color
Geometry color
Gradient backround end color
Gradient backround start color
Graphics backround color
Grid color ⟨17⟩

☑ Use Group and Result color in Xform

Group colors
◉ Use entity colors
○ Use group's colors

Gradient backround direction
None[use Graphics backround color] ▼

Current: c:\users\......\mcamx.config<English><Startup> ▼   ✓   ✗   ✚   ?

SPECIFYING DEFAULT SETTINGS FOR POINT AND LINE STYLES AND LINE WIDTHS

➤ Click ⑲ Cad Settings

➤ Click ⑳ the Point Style down arrow ▾

➤ Click ㉑ the desired point style

➤ Click ㉒ the Line Width down arrow ▾

➤ Click ㉓ the desired line width

➤ Click ㉔ the Line Style down arrow ▾

➤ Click ㉕ the desired line style

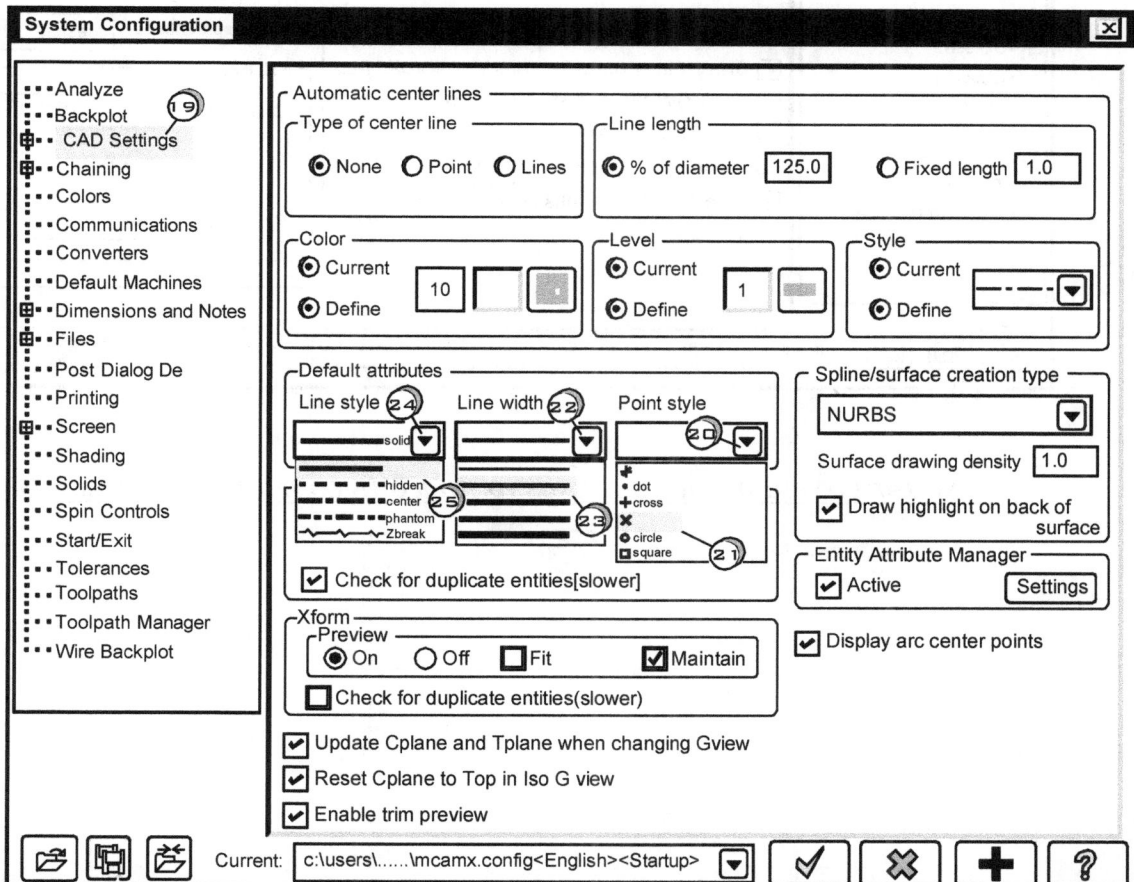

SPECIFYING TO IMPORT THE SOLIDS HISTORY TREE OF A SOLIDWORKS FILE

➢ Click ㉖ Converters

➢ Click ㉗ the check **on** ☑ for Import SolidWorks Solids History

---

System Configuration   ☒

- Analyze
- Backplot
- CAD Settings
- Chaining
- Colors
- Communications
- Converters ㉖
- Default Machines ㉗
- Dimensions and Notes
- Files
- Post Dialog De
- Printing
- Screen
- Shading
- Solids
- Start/Exit
- Tolerances
- Toolpaths
- Toolpath Manager
- Wire Backplot

Solid import

◉ Solids   ☑ Edge curves

○ Trimmed surfaces

☐ Attempt to heal solids during import

☐ Use surface stiching to import solids

☑ Import SolidWorks Solids History

ASCII file entry creation   [Points ▼]

☑ Use System Color for imported Solids

☑ Blank "Paper Space" entities in DWG/DXF files

☐ Break DWG/DXF Drafting Entities

☐ Use IGES files tolerance values

☐ Import Datum entities from Pro/E files

☑ Import MCX Toolpaths from SolidWorks files

Untrimmable surface level: [10000]

Export versions

SaveAs Parasolid   [26.0 ▼]

SaveAs ACIS   [24.0 ▼]

SaveAs AutoCAD   [2013-2014 ▼]

Unit conversion

○ Scale data

◉ Override units

STL files

STL Import entities   [Meshes ▼]

STL Export

Coordinates   [Current WCS ▼]

◉ Binary   ○ ASCII

Resolution   [0.001]

Current: [c:\users\......\mcamx.config<English><Startup> ▼]   [✓] [✗] [+] [?]

㉘

SAVING THE CONFIGURATION SETTINGS TO YOUR OWN CONFIGURATION FILE   NAME

➢ Click ㉘ the Save As button 🖬

Save As   X

(←) (→) ▽ ▯ ▷ Admin ▷ My Documents ▷ my mcamx8 ▷ CONFIG ▷ | ↯ | Search CONFIG 🔍

Organize ▽   New folder   ▤≡ ▽   ⑦

	Name	Date modified	Type	Size
✗ Mastercam X8 ㉙	▯ batch	8/20/2014 1:23 PM	File folder	
▯ my mcamx8	•			
▯ shared mcamx8	•			
▯ MCX	•			

File name | mcamx   ▽

Save as type: | Mastercam Configuration File(*.config)   ▽

(▲) Hide Folders   ㉚ [Save]   [Cancel]

myconfig1
⎵
[operator's entry]

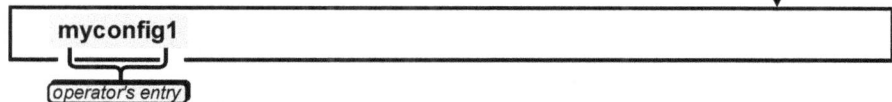

➤ Click ㉙ ▯ my mcamx8   ; enter the name of the new configuration file **myconfig1**

➤ Click ㉚ the [Save)] button

➤ To open a customized configuration file, **myconfig1,** open the System Configuration dialog box.

➤ Click on the file open button [📂]

➤ Click ▯ my mcamx8

➤ Click 📄 **myconfig1**

➤ Click the [Open] button

Note: it is advisable to save your *customized* configuration file(s) with extension (**.CONFIG**)
to a FLASH drive that way they are readily avalable if restoration is needed.

1-13 Using On Line Help

Mastercam provides quick access to information regarding basic concepts, commands, tools,and information about the latest release of the software.

General Use Of Help

➤ Press the   Alt   +   H   keys

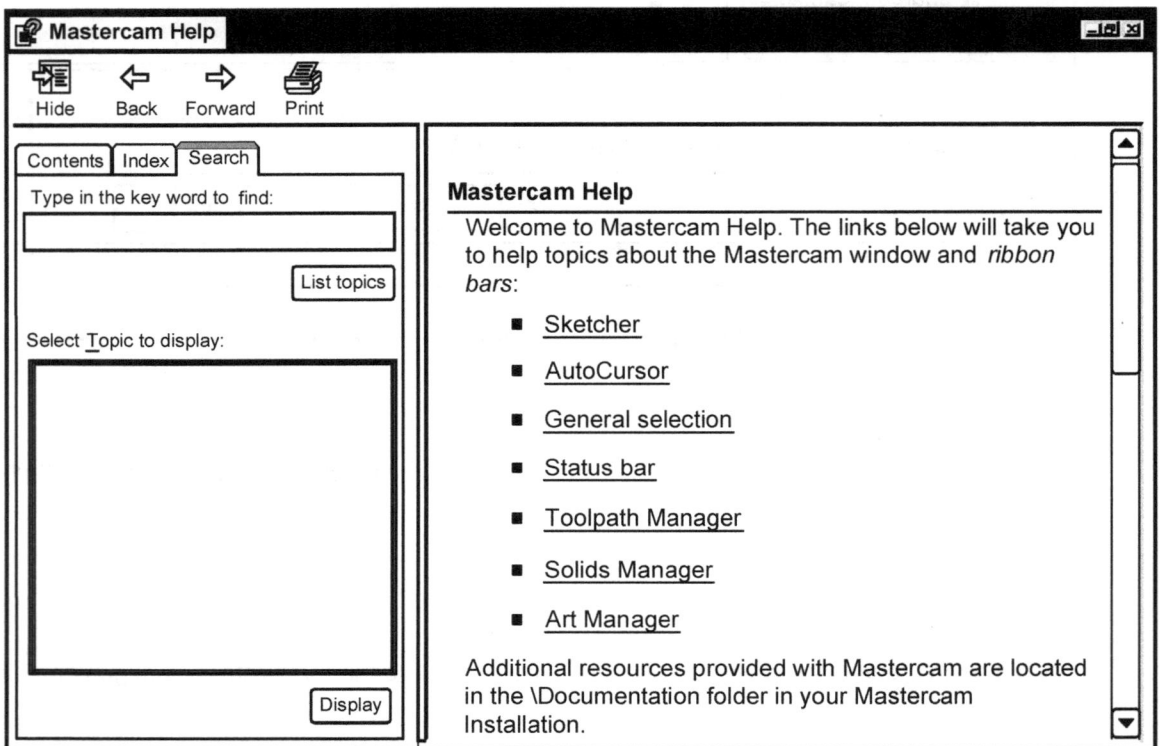

Mastercam Help

Hide   Back   Forward   Print

Contents | Index | Search

Type in the key word to find:

List topics

Select Topic to display:

Display

Mastercam Help

Welcome to Mastercam Help. The links below will take you to help topics about the Mastercam window and *ribbon bars*:

- Sketcher
- AutoCursor
- General selection
- Status bar
- Toolpath Manager
- Solids Manager
- Art Manager

Additional resources provided with Mastercam are located in the \Documentation folder in your Mastercam Installation.

USING THE CONTENTS TAB TO NAVIGATE HELP

➤ Click ① the [Contents] tab

➤ Click ② open the contents of Mastercam Help

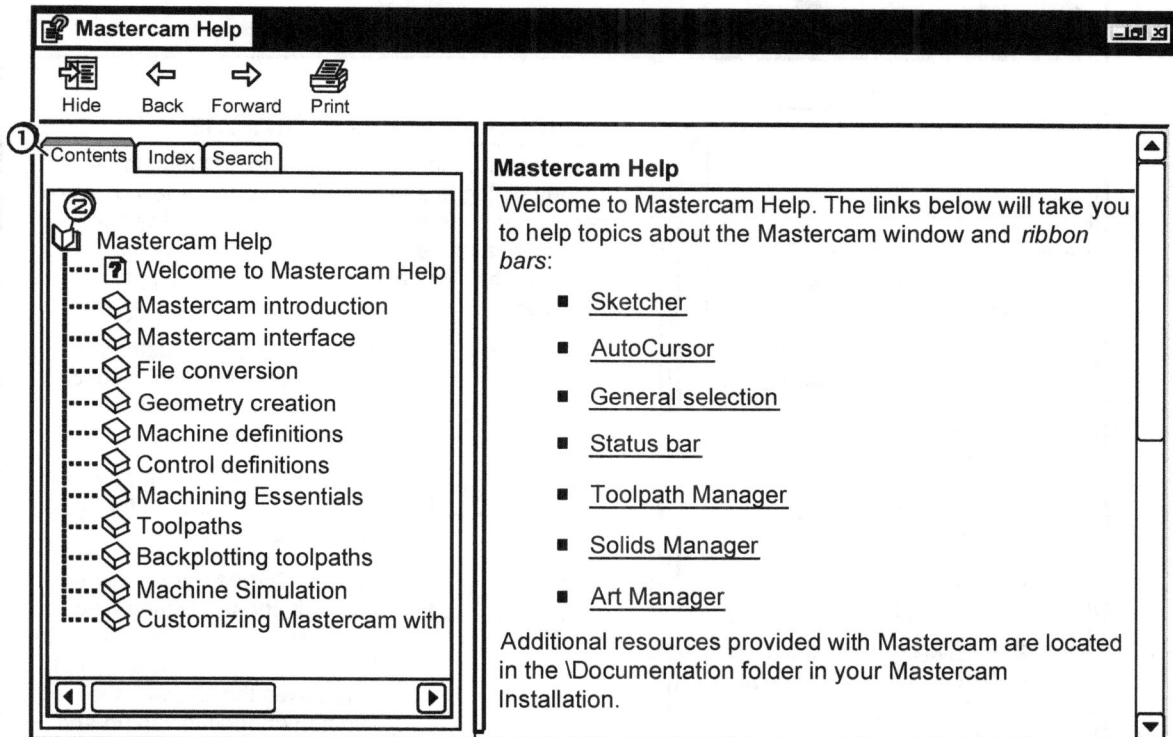

➤ **Double** Click ③ open the contents of Mastercam Interface 📖

➤ **Double** Click ④ open the contents of Interface Overview 📖

➤ Click ⑤ Common Mastercam buttons

Mastercam Help   _□×

🗐	⇐	⇒	🖨
Hide	Back	Forward	Print

Contents | Index | Search

📖 Mastercam Help
····🔞 Welcome
····◇ Mastercam introdction
····📖 Mastercam interface
③····📖 Interface overview
④····🔞 Mastercam interface
······🔞 Graphics window ove
······🔞 Dynamic Gnomon
······🔞 Mastercam shortcut
······🔞 Common Mastercam
······🔞 Toolbars ⑤
······🔞 Prompts
······🔞 Right-click menus
······🔞 Learning mode and
······🔞 Data entry shortcuts
······🔞 Mastercam's built in
······🔞 Root and Immediate
····◇ File conversion
····◇ Geometry creation
····◇ Machine definitions
····◇ Control definitions
····◇ Machining Essentials
····◇ Toolpaths
····◇ Backplotting toolpaths
····◇ Machine Simulation
····◇ Customizing Masterca

Common Mastercam buttons

In Mastercam, there are buttons that appear in most or many dialog boxes and bibbon bars. A few of these may also appear in toolbars or menus. These common buttons have universal functions throughout the interface. Their images and functional descriptions are as follows:

[?] or [?]   **Help** - Opens the Mastercam help system to the topic specific to the active dialog box or ribbon bar.

[✔] or [✔]   **OK** - *Fixes the live entry* (where applicable) and *closes* the dialog box or ribbon bar.

[⊕] or [+]   **Apply** - *Carries out the current task* with chosen settings. For example. If you create a line from 2 endpoints, choosing the Apply button fixes the line with the selected endpoints, length and position and style. Pressing the Enter key performs the same function.
Note: When you you use the Apply button the dialog box or ribbon bar does not close and that function remains active until you exit using the OK button.

[✖]   **Cancel** - *Closes dialog box* and *cancels* any task currently in progress. Pressing ESC performs the same function.

Note:
📖 *double* Clicking *closes* the contents on the topic
⬇
◇

📖 ⬆ *double* Clicking *opens* the contents on the topic
◇

right Click
Click ⌐

| Open all |
| Close all |
| Print |

USING THE SEARCH TAB TO NAVIGATE HELP

▷ Click ⑥ the [Search] tab

▷ Click ⑦ in the key word box and enter the search word: **LATHE TOOLPATHS**

▷ Click ⑧ the [List topics] button

▷ Click ⑨ on   Adding a tailstock

▷ Click ⑩ the [Display] button

Mastercam Help   _ ⻆ ⊠

| Hide | Back | Forward | Print |

⑥

| Contents | Index | Search |

Type in word(s) to search for ⑦

LATHE TOOLPATHS ▼

[List topics] ⑧

Select Topic to display:

Absolute and incremental values
Adding a center to lathe tailstock ⑨
Adding a tailstock
Adding chuck jaws
Adding geometry to a machine definition
Adding lathe stock, chucks, and periphe
Adjust stock dialog box
Arc page
Assigned axis combinations dialog box
Automatically calculating entry and exit
Axis Combination page

⑩ [Display]

Adding a tailstock

A tailstock typically consists of three parts: the tailstock block
, the quill, and a center.Setting this up in Mastercam is a two-
stage process:

■ Follow the steps in this procedure to add the tailstock
body in the Machine Definition Manager. This component
includes the tsailstock block and quill.

■ Then, follow these steps to add a center component to the
tailstock in the Toolpath Manager as part of the machine
group properties.

•
•
•
•

1-14 Saving a File

TO SAVE A FILE

➤ Click ① FILE pull down menu

➤ Click ② 📑 Save As

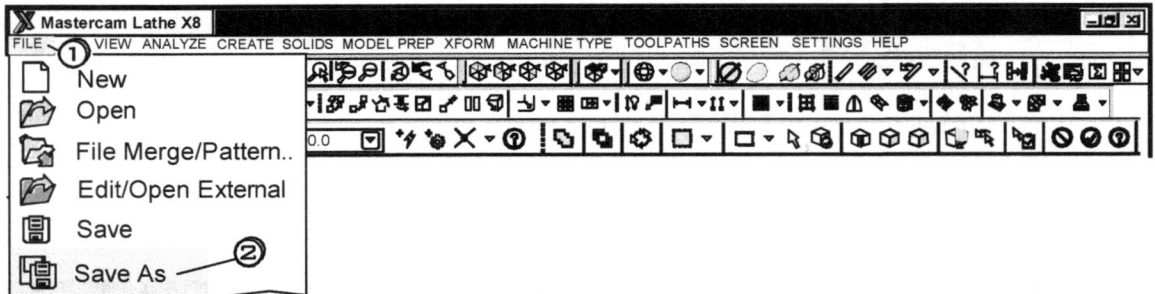

➤ Click ③ 📁 my mcamx8

➤ Click ④ New folder ; enter **JVAL-LATHE**

➤ **Double** Click ⑤ on the file folder

➤ Click ⑥ in the file name box; enter the name of the file to be saved **EX1-1JV**

➤ Click ⑦ the [Save   ] button

1-15 Opening a File

TO OPEN A FILE

Click ① the file open icon ⊞ in the **File** toolbar

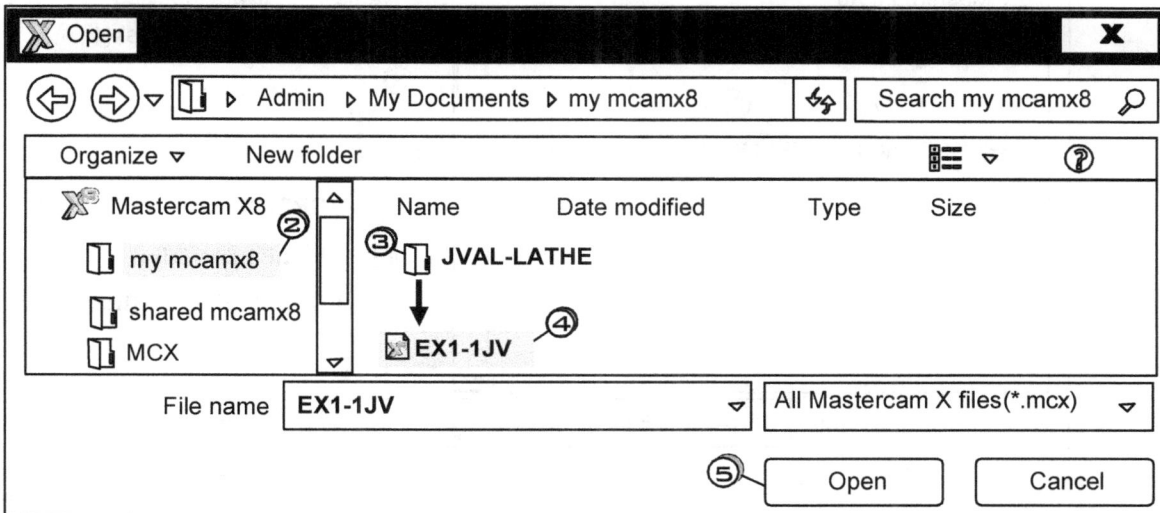

Click ② ⊡ my mcamx8

Double Click ③ on the file folder ⊡ **JVAL-LATHE**

Click ④ on the file to open ▦ **EX1-1JV**

Click ⑤ the [Open] button

TO OPEN A PREVIOUSLY STORED FILE ON THE ENCLOSED CD

➤ Click ① the file open icon 📂 in the FILE   toolbar

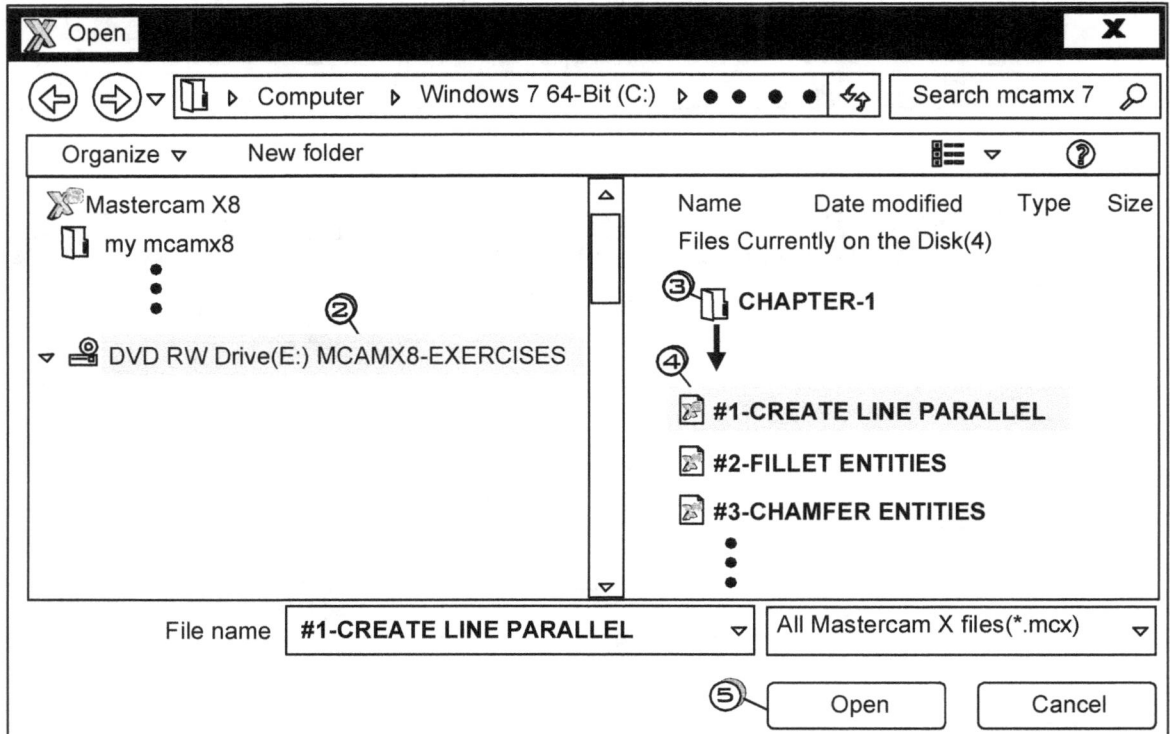

Open   **X**

← → ▽ │ 📁 ▷ Computer ▷ Windows 7 64-Bit (C:) ▷ ● ● ● ● ● ⚡   Search mcamx 7 🔍

Organize ▽   New folder   ▦ ▽   ❓

	Name Date modified Type Size
✖ Mastercam X8	Files Currently on the Disk(4)
📁 my mcamx8	③ 📁 CHAPTER-1
⁝	
⁝	④
▽ 💿 DVD RW Drive(E:) MCAMX8-EXERCISES	↓
	📄 #1-CREATE LINE PARALLEL
	📄 #2-FILLET ENTITIES
	📄 #3-CHAMFER ENTITIES
	⁝

File name │ **#1-CREATE LINE PARALLEL**   ▽ │ All Mastercam X files(*.mcx) ▽

⑤ [Open   ]   [Cancel]

➤ Click ② DVD RW Drive(E:) MCAMX8-EXERCISES

➤ ***Double*** Click ③   📁 CHAPTER-1

➤ Click ④   📄 #1-CREATE LINE PARALLEL

➤ Click ⑤ the [Open   ] button

1-16 Using the Zip2Go Utility

The Zip2Go utility is used to gather and compress the current Mastercam part data and other files into a .Z2G file. This is especially useful for sending files over the internet .Other utlities such as WinZip can then be used to unzip all the files.

➤ Click ① the Help pull down menu

➤ Click ② Zip2Go Utility

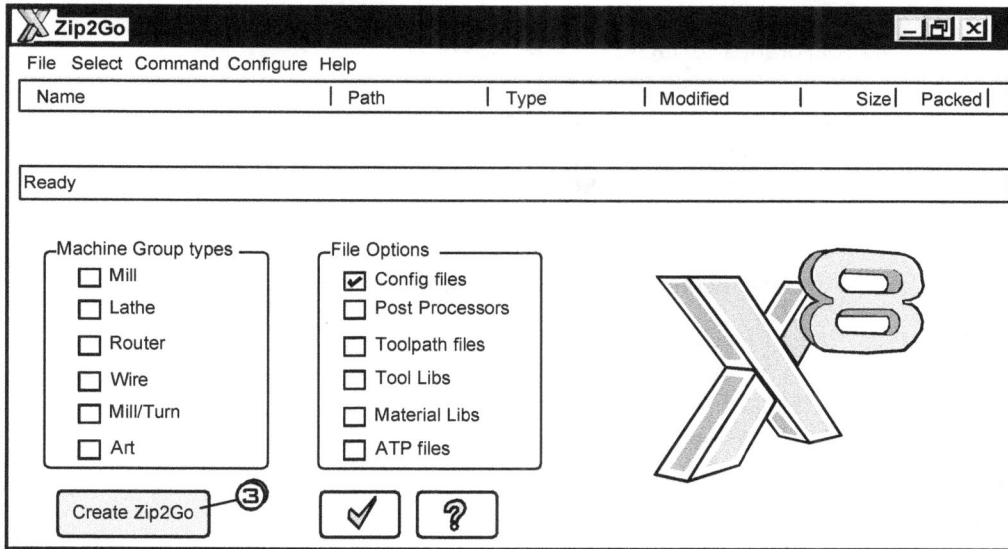

➤ Click ③ the Create Zip2Go button

1-17 Creating a New File

TO CREATE A NEW FILE WHILE IN Mastercam X8

➤Click ① the file new icon ☐ in the **File** toolbar

1-18 Converting Files from Previous Releases of *Mastercam X* to *X8*

Users of previous versions of *Mastercam X* should convert the following files before working with them in *Mastercam X8*

- Tool libraries ***.TOOLS**
- Material libraries ***.MATERIALS**
- Toolpath and operation defaults ***.DEFAULTS**
- Operation libraries ***.OPERATIONS**
- Machine definitions ***.MMD**
- Component libraries ***.GMD**
- Part files ***.MCX**
- Some posts(those that *check for the version number*)***.PST**

To *convert* files using *Mastercam X8* select **File** pull down menu then click **Open.** Open the individual file and *save it* using the proper extension such as ***.TOOLS** or ***.PST** etc.

The reader is encouraged to consult the *MastercamX8* Transition Guide that comes with the installation software for further information.

1-19 Exiting the *Mastercam X8* Design/Lathe package

TO EXIT Mastercam X8

➤ Click ① the FILE drop down menu

➤ Click ② Exit

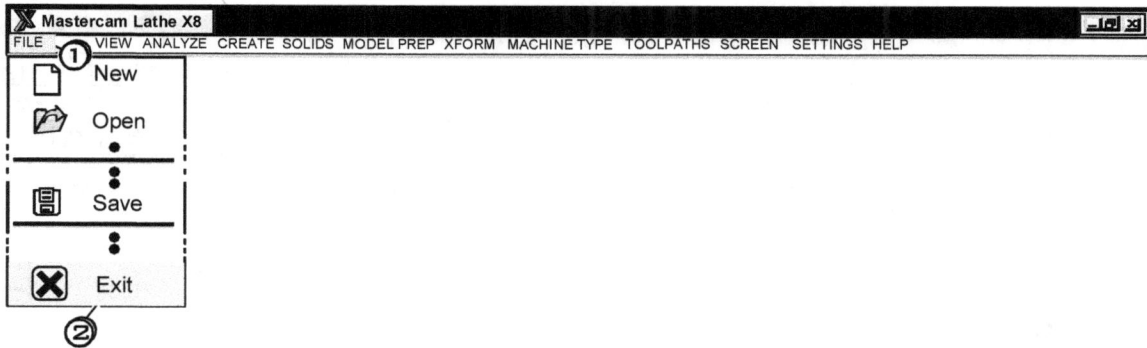

If any *changes* to the current **.MCX** file were *not saved* the system
will display the dialog box below

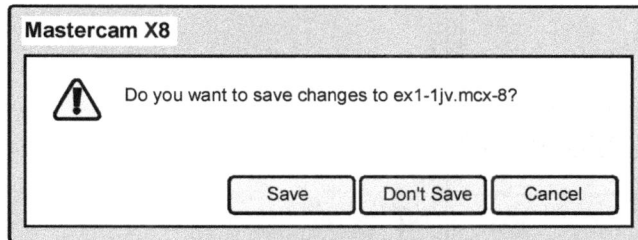

➤ Click ④ [Save] to **save** the changes and exit *Mastercam X8*.

➤ Click ⑤ [Don't Save] to **discard** the changes and exit *Mastercam X8*.

EXERCISES

1-1) Execute the steps necessary to turn on the computer and start *Mastercam X8* Lathe.

1-2) What information is contained in the following files: a) **CONFIG** b) **MCX** c) **NC**

1-3) Identify and describe the use of each of the areas *A, B, C, D,E, F, G.... O* of *Mastercam X8's* main interface window shown in Figure 1p-1.

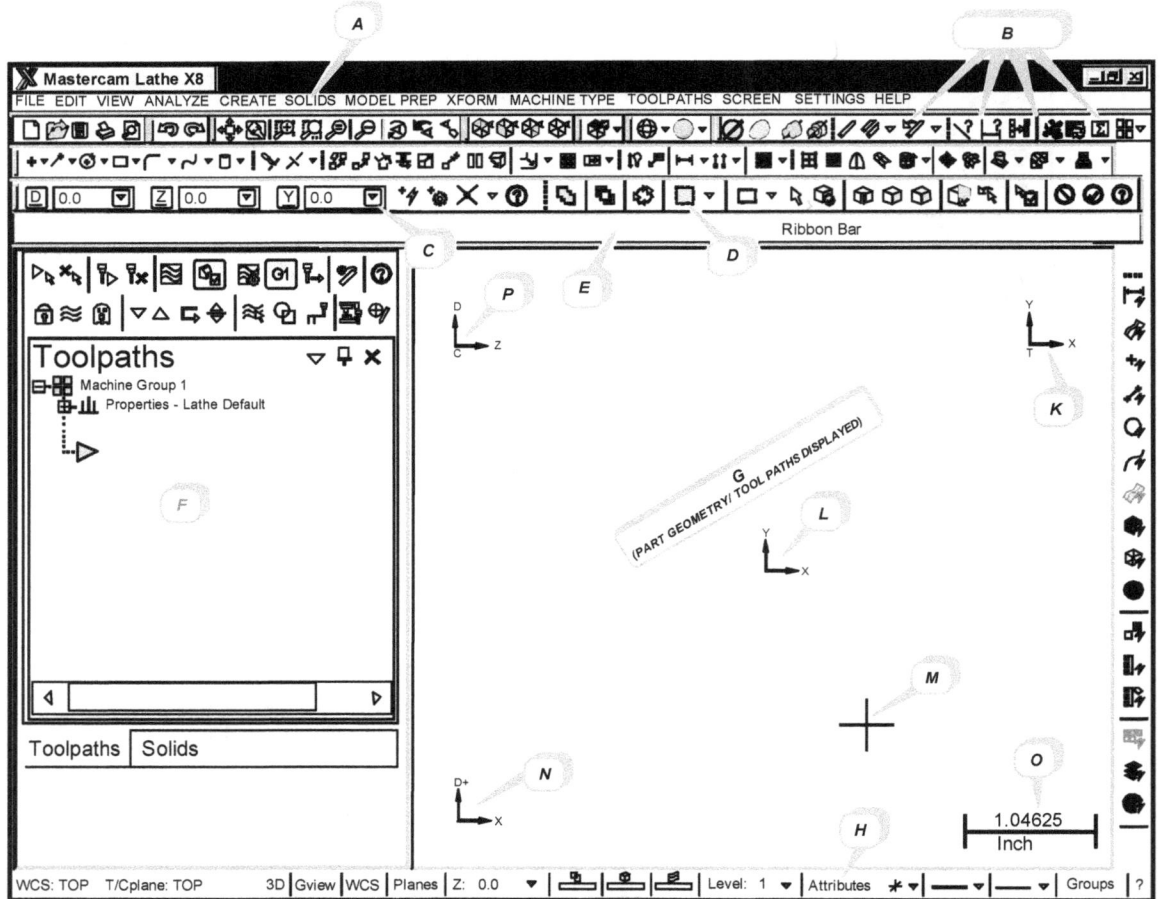

Figure 1-p1

1-4) What is the significance of using the [Esc] key

1-5) What are some advantages of using the short-cut keys

1-6) What functions are executed by the following short-cut keys

    a) [Alt] + [H]

    b) [Alt] + [O]

    c) [Pg Up]

    d) [→]

1-7) Describe the steps required to set the system to *metric* part programming.

1-8) Describe the steps required to create the file folder **SPEEDY-LATHE** for saving and getting *MastercamX8* **.MCX-8** files.

1-9) A Gview _____

    A. defines the active working coordinate system   C. is a line of sight on the part

    B. is a orientation plus an origin   D. controls 2D arc/line orientations

1-10) A Cplane _____

    A. defines the active working coordinate system   C. is a line of sight on the part

    B. is a orientation plus an origin   D. controls 2D arc/line orientations

1-11) A View _____

    A. defines the active working coordinate system   C. is a line of sight on the part

    B. is a orientation plus an origin   D. controls 2D arc/line orientations

1-12) A Tplane _____

    A. defines the active working coordinate system   C. is a line of sight on the part

    B. sets the orientation of the cutting tool   D. controls 2D arc/line orientations

1-13) Use the Contents tab in on line help to find and print information about the topic *ribbon bars*

1-14) Use the Search tab in on line help to find and print information about the topic *Configuration files.*

1-15) Explain the steps for saving the file **EX1-2JV** in the file folder **JVAL-LATHE** and exiting *Mastercam X8.*

CHAPTER - 2

BASIC CAD OPERATIONS

2-1 Chapter Objectives

After completing this chapter you will be able to:

1. Construct basic 2D wireframe geometric entities such as lines, arcs, fillets and chamfers.

2. Execute delete and trim operations in 2D space.

3. Pan and zoom screen displays.

4. Repaint and regenerate screen displays.

2-2 Generating and Editing a Wireframe CAD Model of a Part

This chapter as well as Chapter 3 presents the commands and techniques for generating 2D wireframe CAD models using *Mastercam X7* Design. A Wireframe model is the most basic method of representing a real 2D part as a collection of lines, arcs, points and splines. A wireframe defines the *boundaries of the part only* . The space enclosed by the boundaries is void of points and undefined. Only parts containing *no surface intersections or flat surface planes between the boundaries* can be modelled by wireframe.

interior is *empty* space with *flat* planes and *no surfaces*

only *boundary* or *edge* of part defined

Wireframe model of a part

2-3 Setting the Construction Plane To Lathe Diameter

For creating wireframe geometry for lathes the operator works in terms of diameter or *twice* the distance from the spindle centerline.

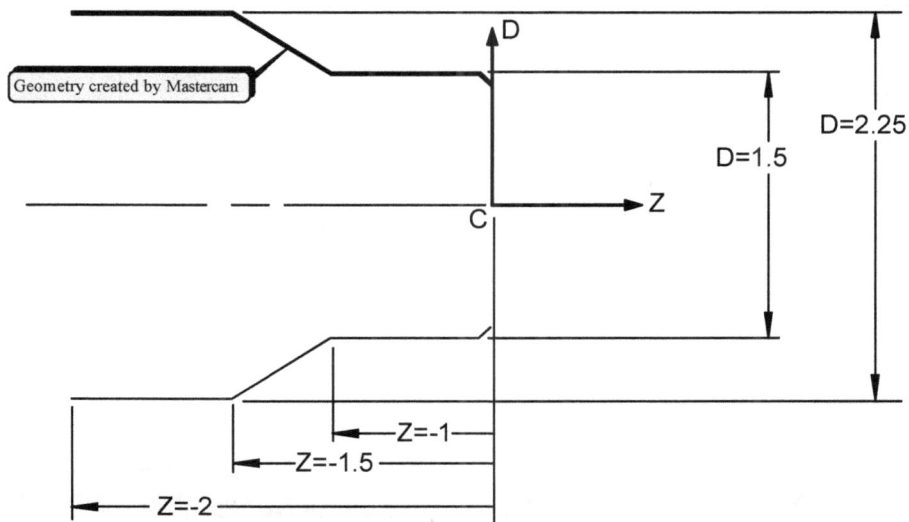

Geometry created by Mastercam

D

D=2.25

D=1.5

C

Z

Z=-1

Z=-1.5

Z=-2

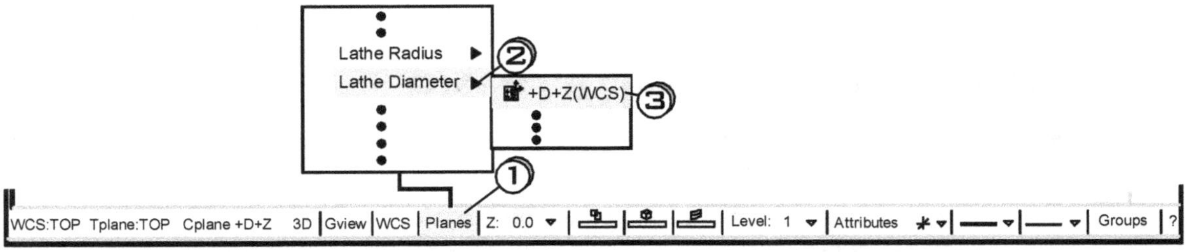

Lathe Radius
Lathe Diameter
+D+Z(WCS)
WCS:TOP Tplane:TOP Cplane +D+Z 3D Gview WCS Planes Z: 0.0 Level: 1 Attributes Groups ?

- Click ① the Planes button
- Click ② the Lathe Diameter arrow ▸
- Click ③ +D+Z(WCS)

2-4 Creating Lines in 2D Space

Create Line Endpoint ▸ ***MULTIPLE*** or M ◂ ☑ Default to Fast Point mode

Multiple lines are created between inputted *start* and *end points.*
Absolute coordinates are used: each new *D,Z is measured from the WCS* ***origin***

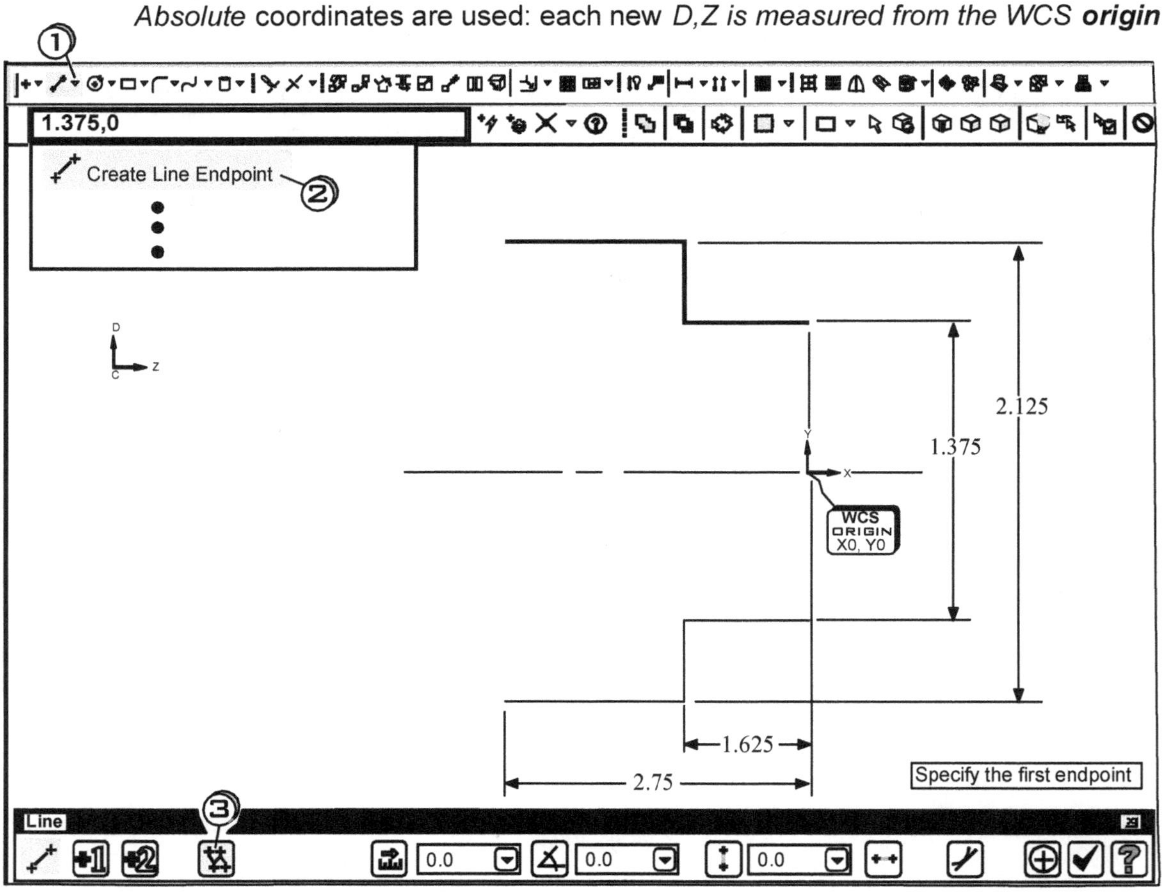

1.375,0
Create Line Endpoint
D
C
Z
2.125
1.375
WCS ORIGIN X0, Y0
1.625
2.75
Specify the first endpoint
Line
0.0
0.0
0.0

➤ Tap the `Pg Dn` `Pg Up` keys to zoom fit the display

➤ Click ① the line down arrow ▼

➤ Click ② ✧ Create Line Endpoint

  | Specify the first endpoint |

➤ Click ③ the multi line button 🔲
        or tap the `M` key

➤ Enter | 1.375,0 | | Enter |

  | Specify the second endpoint |

➤ Enter | 1.375,-1.625 | | Enter |

➤ Enter | 2.125,1.625 | | Enter |

➤ Enter | 2.125,-2.75 | | Enter |

> Note: Use the Undo 🔁 and Redo 🔁
> buttons to *remove incorrect* line
> entities created or *restore correct*
> ones.

➤ Press `Esc` to cancel the function

✗ **Create Line Endpoint** ▸ ; *LENGTH* or ; *ANGLE* or ◂ ☑ Default to Fast Point mode
L A

Lines are created using *Length, Angle polar* coordinates

Create Line Endpoint ②

angle is measured *positive counterclockwise(+ccw) from the +X-axis direction*

3.5 -120° ④

WCS ORIGIN X0, Y0

Specify the first endpoint

Line ③ 3 30 0.0

▰ Tap the Pg Dn Pg Up keys to zoom fit the display

▰ Click ① the line down arrow ▾

▰ Click ② ✗ Create Line Endpoint

Specify the first endpoint

▰ Click ③ the length button or tap the L key

▰ Enter 3 Tab to ⤬ button

▰ Enter 120 ; Enter↵

▰ Click ④ near the end of the line

\\ Create Line Parallel ► *SIDE/DIST* or ◄

A *line* is created *parallel* to an existing *line* at an *offset distance* inputted by the operator.

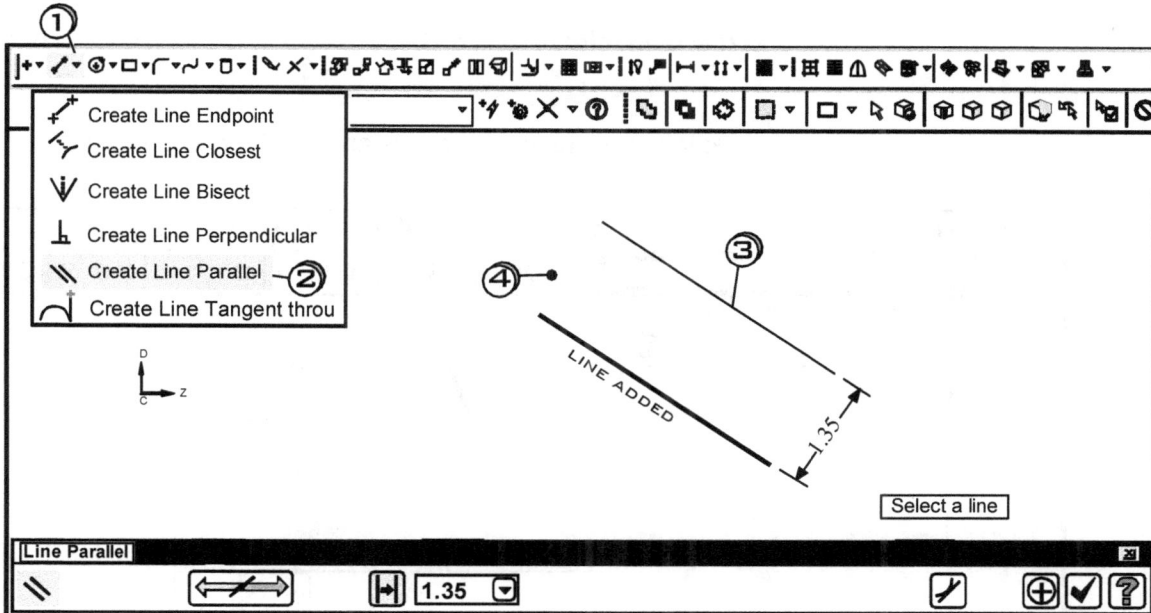

Create Line Endpoint
Create Line Closest
Create Line Bisect
Create Line Perpendicular
Create Line Parallel
Create Line Tangent throu

LINE ADDED

Select a line

Line Parallel

1.35

➤ Click ① the down arrow ▼

➤ Click ② \\ Create Line Parallel

➤ Tap the [D] key for *offset distance*

➤ Enter [1.35]

Select a line

➤ Click ③ near the line entity

Indicate iffset direction

➤ Click ④ the side on which the parallel line is to be created

➤ Press [Esc] to cancel the function

2-5 Creating Circles in 2D Space

Create Circle Center Point   ► *RADIUS* or ◄   ☑ Default to Fast Point mode
                                    R

A *circle* is created by entering its *radius* and *center point* values

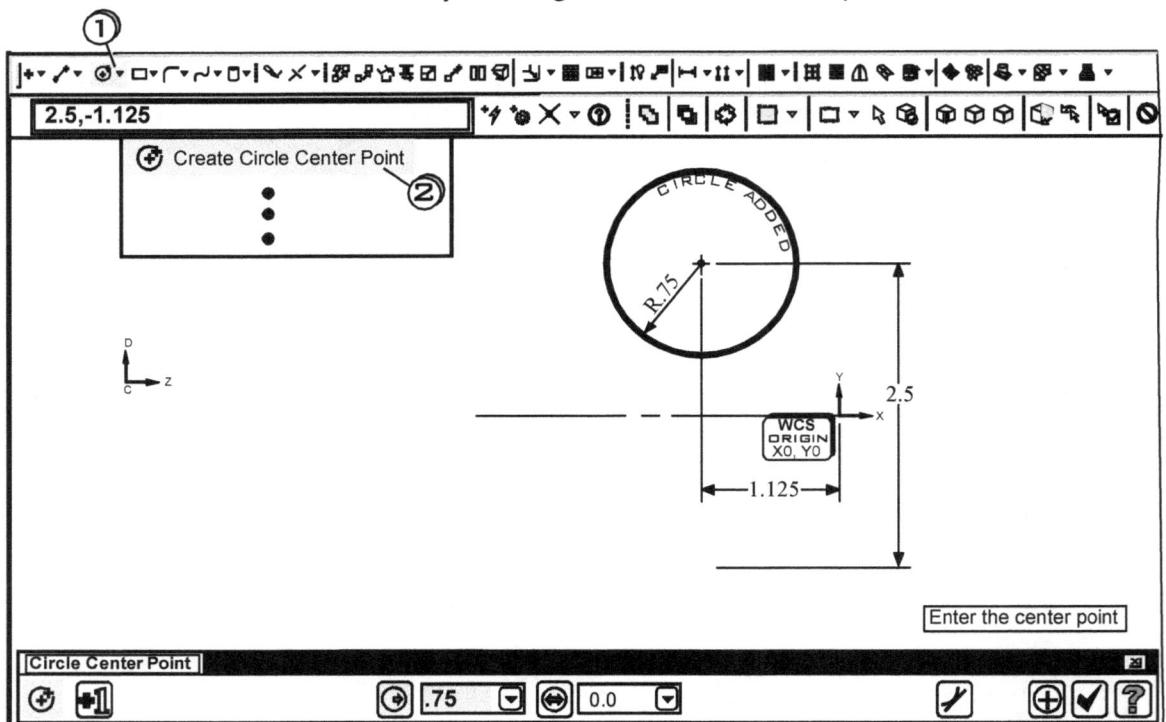

➣ Click ① the circle down arrow ▼   ➣ Tap the [R] key for *Radius*

➣ Click ② ⊕ Create Circle Center Point   ➣ Enter [.75]

Enter the center point   ➣ Press [Enter] to create the circle

➣ Enter [2.5,-1.125]   [Enter]   ➣ Press [Esc] to cancel the function

2-6 Creating Arcs in 2D Space

Create Arc Polar ► *RADIUS* or ; *START ANGLE* or ; *END ANGLE* or ◄

R   S   A

☑ Default to Fast Point mode

An *arc* is created at inputted *center point, radius, starting angle* and *ending angle* values.

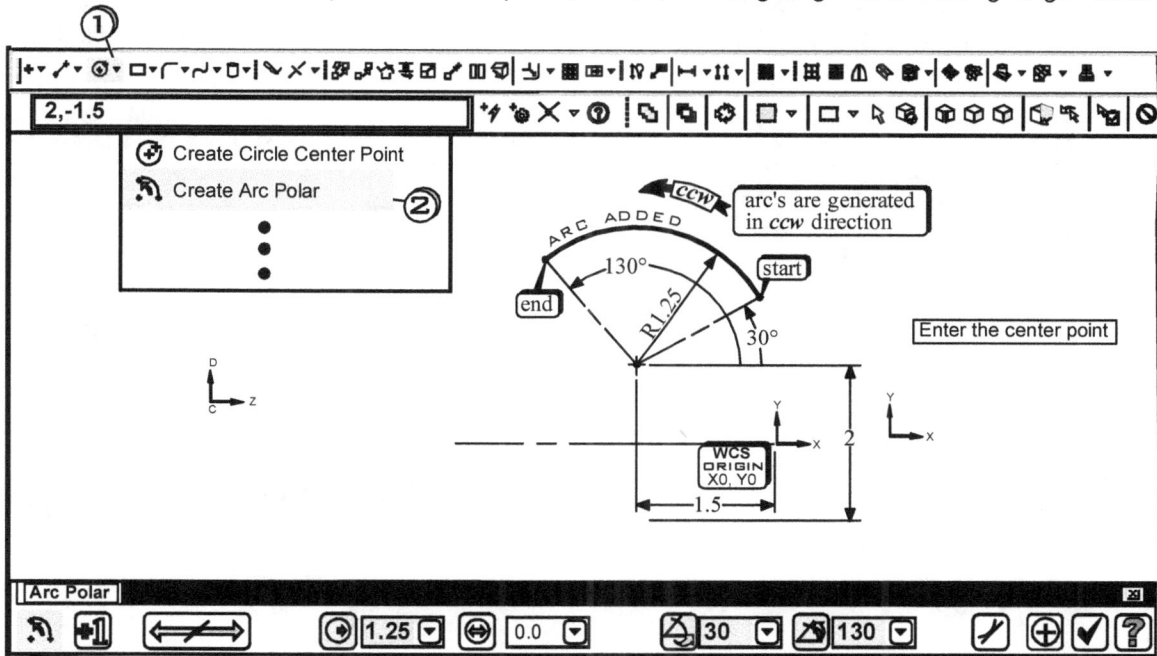

> Click ① the circle down arrow ▾

> Click ② 🏹 Create Arc Polar

Enter the center point

> Enter 2,-1.5   Enter

> Tap the R key for *Radius*

> Enter 1.25   Tab

> Tab into the *Start Angle* box

Sketch the initial angle

> Enter 30

> Tab into the *Final Angle* box

Sketch the final angle

> Enter 130   Enter

> Press Esc to cancel the function

2-7 Creating Fillets in 2D Space

Fillet Entities ▸ *RADIUS* or ;*STYLE* or ;*TRIM ON* or ◂

R   S   T

The operator *sets* the fillet *Radius* for *all subsequent* fillets created. The fillet arc *Style* is set to *Normal*(an arc *less than 180°*). *Trim* is set to *On.*The two entities filleted *will be automatically trimmed* to their *tangency* point with the fillet.

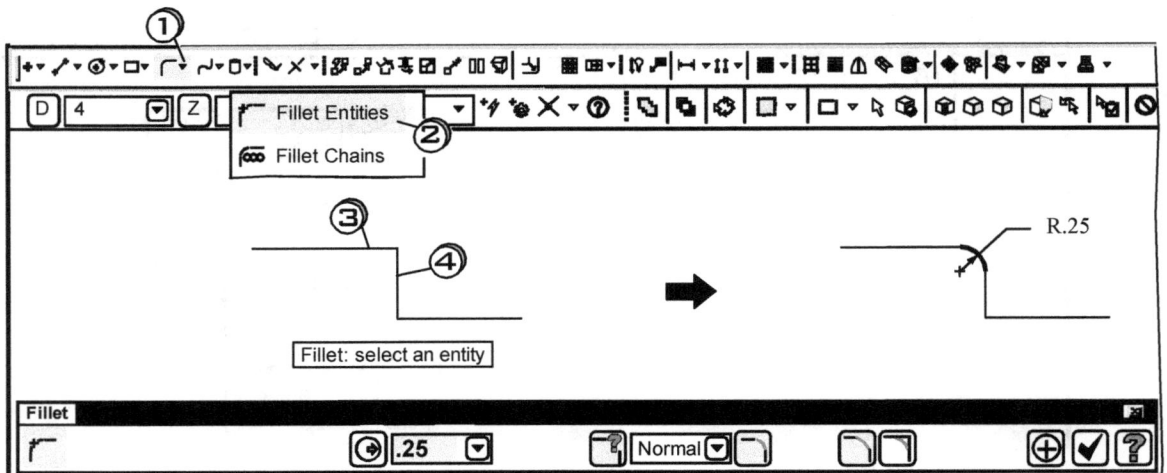

➤ Click ① the down arrow ▾

➤ Click ② 〔 Fillet Entities

➤ Tap the [R] key for *Radius*; enter **.25**

➤ accept the default settings: [Normal▾]〔 ; Trim *On* 〔

Fillet: select an entity

➤ Click ③ the first line entity

Fillet: select another entity

➤ Click ④ the second line entity

➤ Tap [Esc] for function cancel

2-8 Creating Chamfers in 2D Space

Chamfer Entities ▸ *DIST 1* or ;*Angle* or ;*STYLE* or ;*TRIM ON* or ◂

A *chamfer* is created. The operator specifies the *first* chamfer distance *and angle* then clicks the *first* and *second lines or arcs* to be chamfered.

- Click ① the down arrow ▾
- Click ② Chamfer Entities
- Click ③ the down arrow
- Click ④ the *Distance/Angle* style
- Tap the 1 key for *Dist 1*; enter .06
- Tab into *Angle*; enter 45
- accept the default setting: Trim *On*

Select line or arc
- Click ⑤ the first line entity

Select line or arc
- Click ⑥ the second line entity

- Tap Esc for function cancel

2-9 Deleting Entities in 2D Space

/ **Single**

The operator *clicks on each* entity to be *deleted*

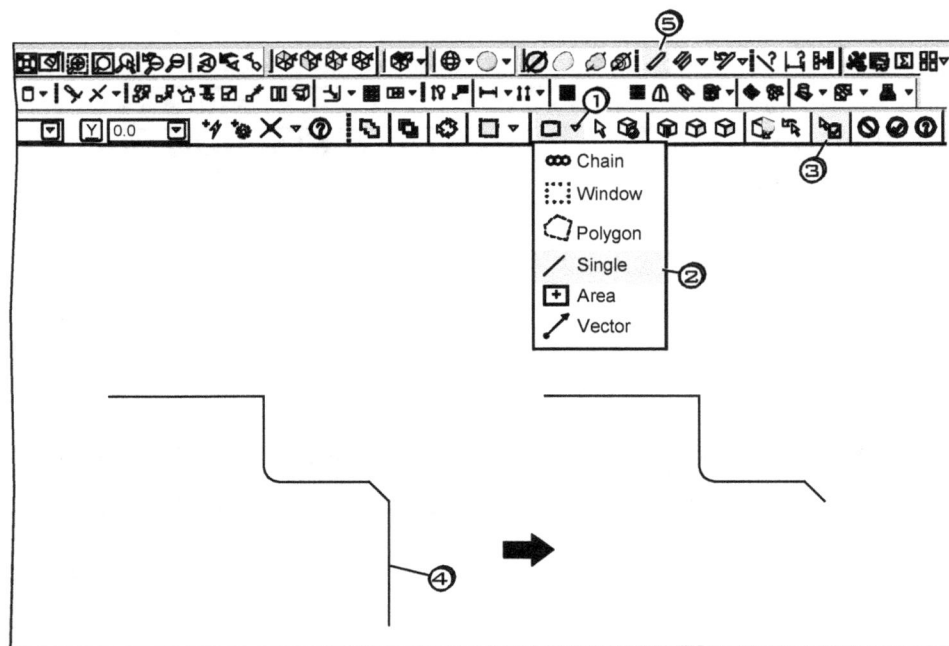

➤ Click ① the down arrow ▾

➤ Click ② / Single   single click

➤ Click ③ the toggle verify selection button 🖱

➤ Click ④ the line entity

➤ Click ⑤ the Delete key ✎ or tap Del

2-10 Triming Entities in 2D Space

✎ Trim   ▸ *1 ENTITY* or ◂

Trims(shortens or extends) an existing line, arc, fillet, ellipse or spline entity to its *intersection* with another entity. The operator *first* clicks the entity *to be trimmed* on the **portion to remain.** The entity to be *trimmed to* is *then* clicked.

➤ Click ① the Trim icon ✎

   Trim 1 Entity function will be selected by *default*

Select the entity to trim/extend

➤ Click ② the portion to **remain**

Select the entity to trim/extend to

➤ Click ③ the entity to be trimed to

➤ Press Esc to cancel the function

2-11 Zooming Graphics Window Displays

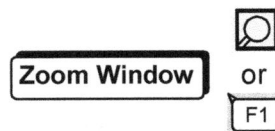

Mastercam X8 provides the user with the ability to control the size of the graphics window geometry displayed. Zooming can be executed while the operator is currently in an active function such as Sketcher. The various zoom functions provided are described in this section.

Zooming Graphics Window Displays

Zoom Window   or   F1

Magnifies the geometry contained within a *rectangular* window clicked by the operator.

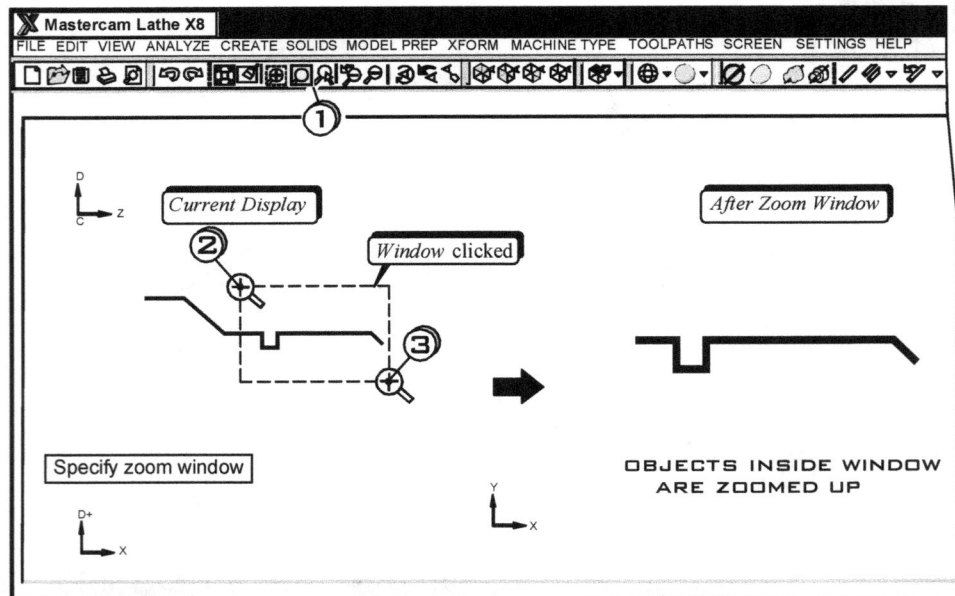

➤ Click ① the Zoom window button 🔍 or tap the F1 key

Specify zoom window

➤ Click ② ③ corners of the zoom window

➤ Press Esc to cancel the function

Unzoom   or

F2

Displays the geometry at the *scale of the previous zoom*. If *no* previous zoom *exists* the system will *reduce* the geometric display *by half its original size*. Unzoom can be used a *maximum* of *eight times* to reduce the current geometric display.

FILE EDIT VIEW ANALYZE CREATE SOLIDS MODEL PREP XFORM MACHINE TYPE TOOLPATHS SCREEN SETTINGS HELP

Mastercam Lathe X8

FILE EDIT VIEW ANALYZE CREATE SOLIDS MODEL PREP XFORM MACHINE TYPE TOOLPATHS SCREEN SETTINGS HELP

①

After Unzoom

Current Display

If previous zoom was entered

PREVIOUS ZOOM SCALE

If no previous zoom was entered

After Unzoom

.5 CURRENT SCALE

Click ① the Unzoom button   or tap the F2 key

Zoom Target ⊕

Magnifies the geometry contained within a *rectangular* window or target boundary clicked by the operator. The operator is prompted *first* to click the target point where the *center* of the window is to be placed. Next, the operator is prompted to drag the mouse cursor and click the location of the *corner* of the window to define its size.

➤ Click ① the Zoom Target button ⊕

Pick point to zoom from

➤ Click ② near the center of the circle

Choose a second corner of your zooming box

➤ Click ③ the location of the corner of the target window

➤ Press Esc to cancel the function

Zoom Selected

The operator first *identifies the entities* to be considered *in the scaling set*. When Zoom Selected is clicked Mastercam *scales the set* such that it *fits* the graphics window display.

Mastercam Lathe X8

FILE EDIT VIEW ANALYZE CREATE SOLIDS MODEL PREP XFORM MACHINE TYPE TOOLPATHS SCREEN SETTINGS HELP

Current Display

window used to identify scaling set

➤ Click ② ③ corners of the zoom window to *identify* the *entities in the scaling set*

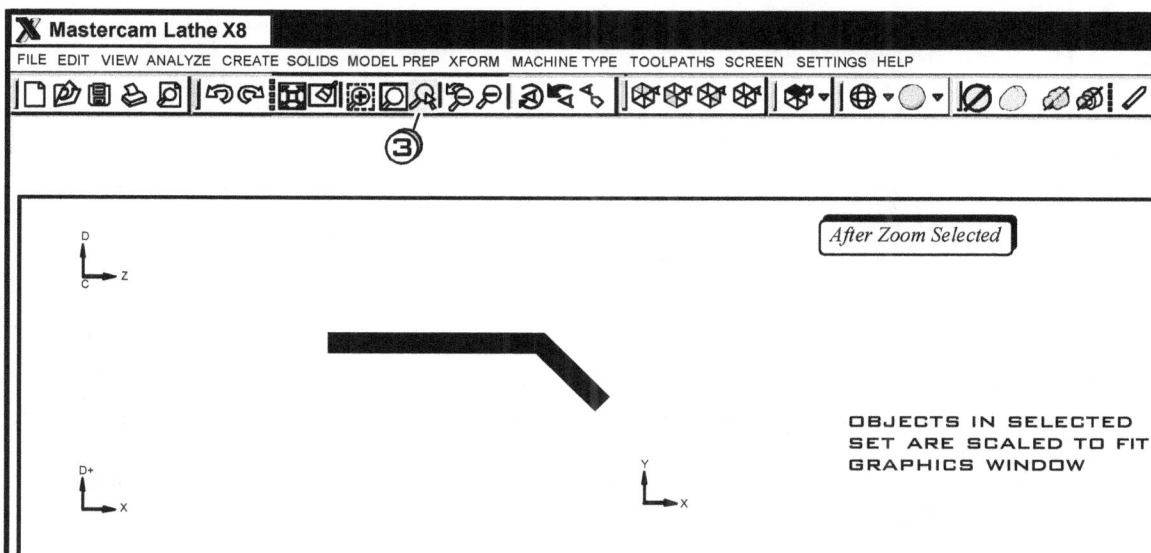

Mastercam Lathe X8

FILE EDIT VIEW ANALYZE CREATE SOLIDS MODEL PREP XFORM MACHINE TYPE TOOLPATHS SCREEN SETTINGS HELP

③

After Zoom Selected

OBJECTS IN SELECTED
SET ARE SCALED TO FIT
GRAPHICS WINDOW

➤ Click ③ the Zoom Selected button

➤ Press [Esc] to cancel the function

Zoom in/out

Enables the operator to *dynamically zoom in or out* from a point selected. After clicking the focal point the function features any one of the following methods for dynamically zooming:

▸ Moving the mouse up(*zoom up*) or down(*zoom down*)

▸ Spin the mouse wheel up(*zoom up*) or down(*zoom down*)

▸ Tap [Pg Up] key(*zoom up*) or tap [Pg Dn] key(*zoom down*)

➤ Click ① VIEW

➤ Click ② Zoom In/Out

Note:
The operator can use the mouse wheel and the [Pg Up] [Pg Dn] keys independent of the Zoom in/out function. They can be applied when any function such as Sketcher is currently active.

2-12 Panning Screen displays

The operator can *pan* the geometric screen display in any of *four directions*: *up, down, left* or *right*. This is accomplished by one of two methods.

▶ Tapping the directional arrow keys on the keypad 🔼up 🔽down ◀left ▶right

▶ Press the [Shift] key and ***keeping it depressed*** ;

press the mouse wheel and drag the mouse in the desired pan direction

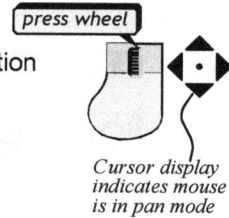

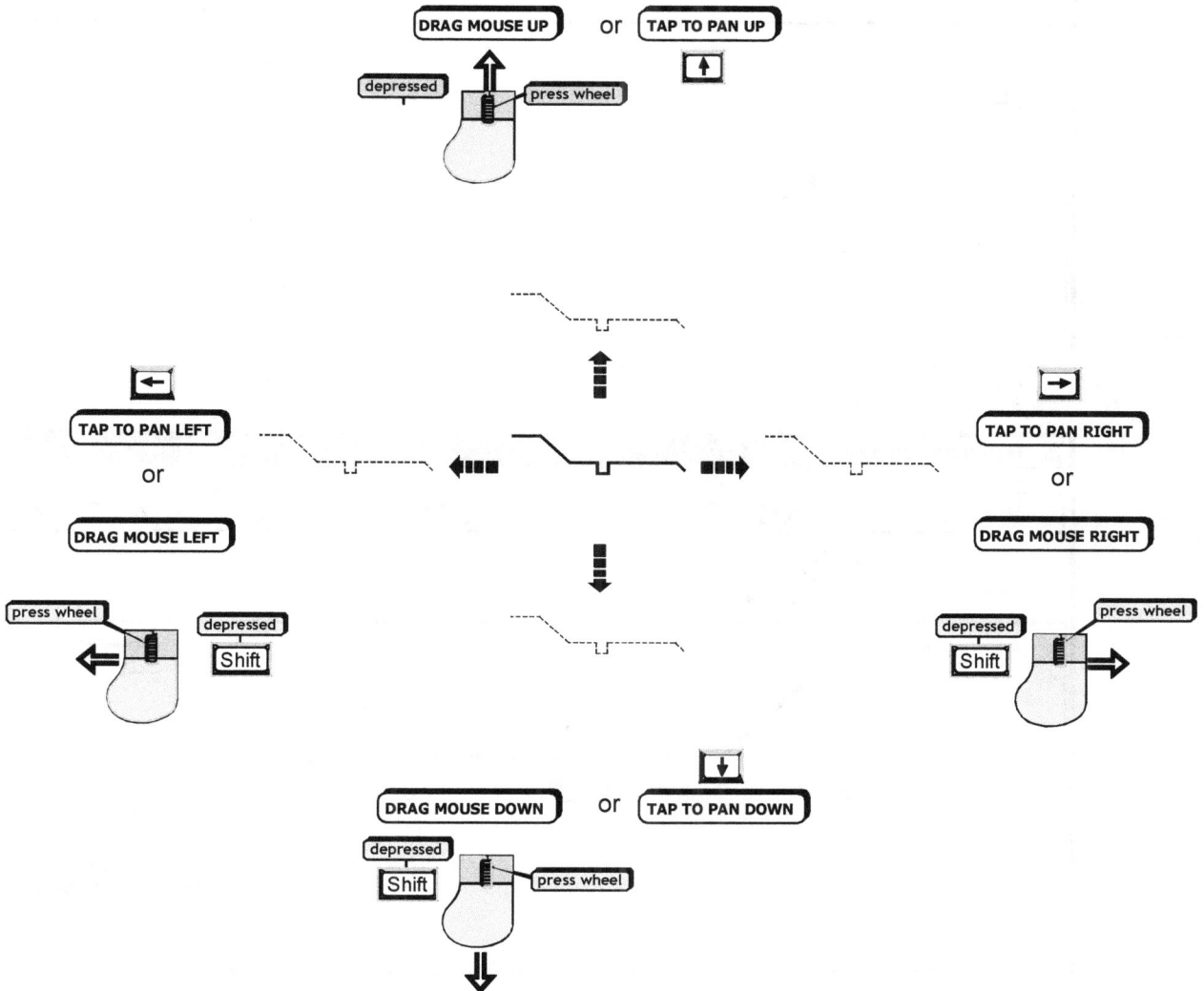

press wheel

Cursor display indicates mouse is in pan mode

[DRAG MOUSE UP] or [TAP TO PAN UP]

depressed   press wheel

[TAP TO PAN LEFT]

or

[DRAG MOUSE LEFT]

press wheel   depressed
             [Shift]

[TAP TO PAN RIGHT]

or

[DRAG MOUSE RIGHT]

depressed   press wheel
[Shift]

[DRAG MOUSE DOWN] or [TAP TO PAN DOWN]

depressed
[Shift]   press wheel

2-13 Fitting the Existing Geometry to the Screen Display Area

[Fit]   or

[Alt] [F1]

Scales the *shape* formed by all the *visible* graphic entities such that it *fits* the *graphics window display* area.

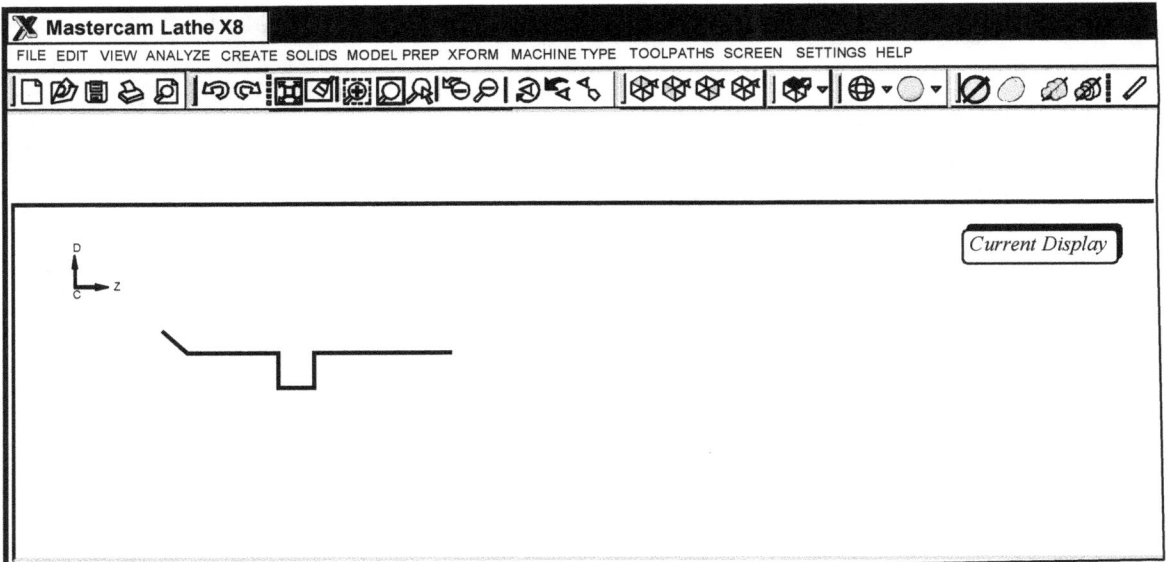

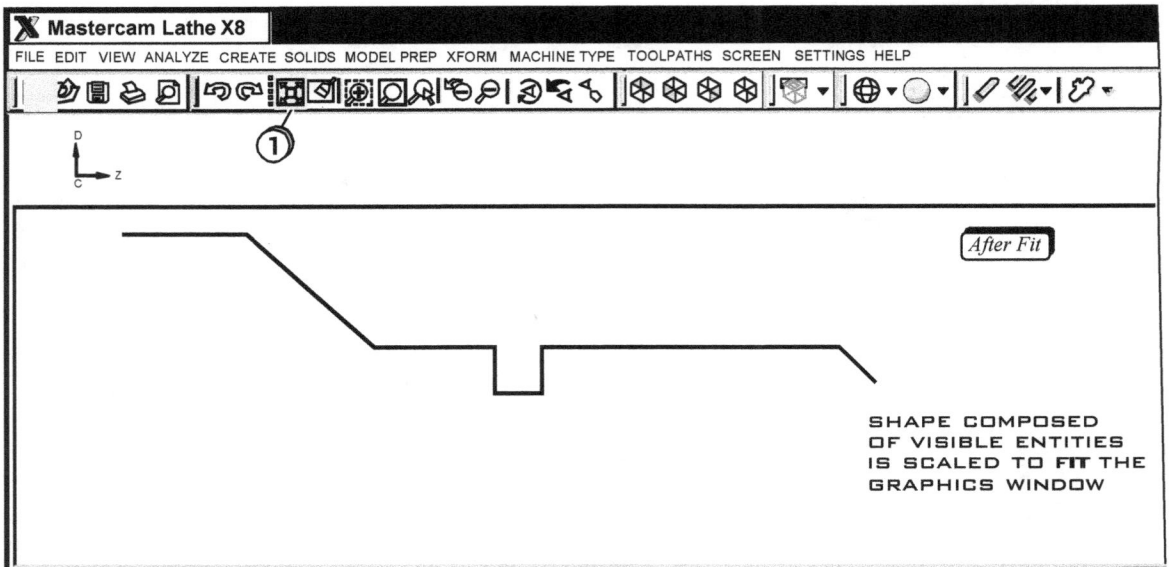

Mastercam Lathe X8

FILE EDIT VIEW ANALYZE CREATE SOLIDS MODEL PREP XFORM MACHINE TYPE TOOLPATHS SCREEN SETTINGS HELP

Current Display

Mastercam Lathe X8

FILE EDIT VIEW ANALYZE CREATE SOLIDS MODEL PREP XFORM MACHINE TYPE TOOLPATHS SCREEN SETTINGS HELP

①

After Fit

SHAPE COMPOSED
OF VISIBLE ENTITIES
IS SCALED TO **FIT** THE
GRAPHICS WINDOW

➤ Click ① the Fit button 🔲 or tap the [Alt] [F1] keys.

2-14 Repainting the Screen

| Repaint Screen |

Ocassionally, *Mastercam* may display incomplete or distorted images of the entities in the graphics window. This depends upon the graphic capabilities of your PC, the size the part file and amount of available memory. Repaint is the *first* remedy the operator uses to correct these display problems in the graphics window.

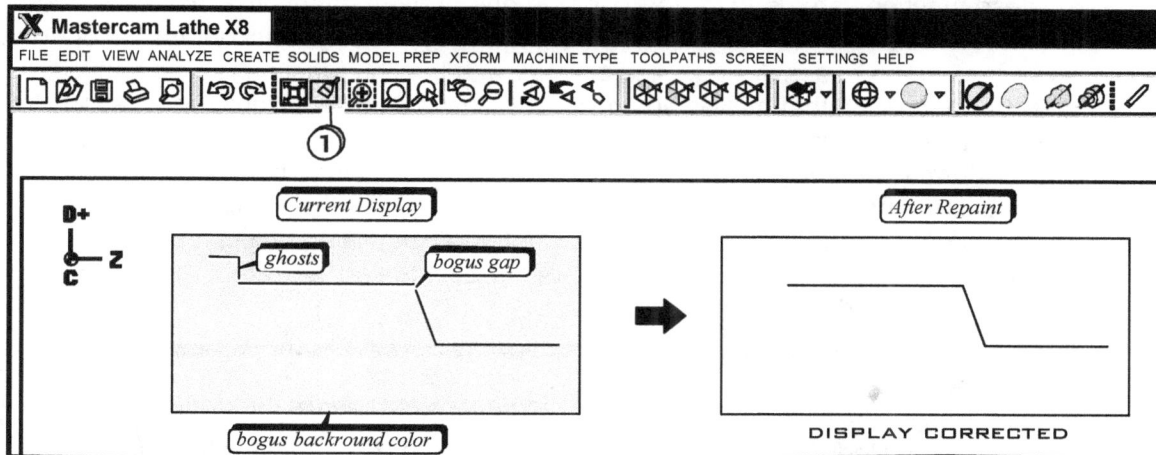

➤ Click ① the Repaint button ⟨⟩

2-7 Regenerating the Screen

| Regenerate Screen | 📇 or | Shift | Ctrl | R |

Regenerates all the graphic entities and *displays all* the *current existing geometry.* This command is used if Repaint *fails* to restore all the existing graphics.

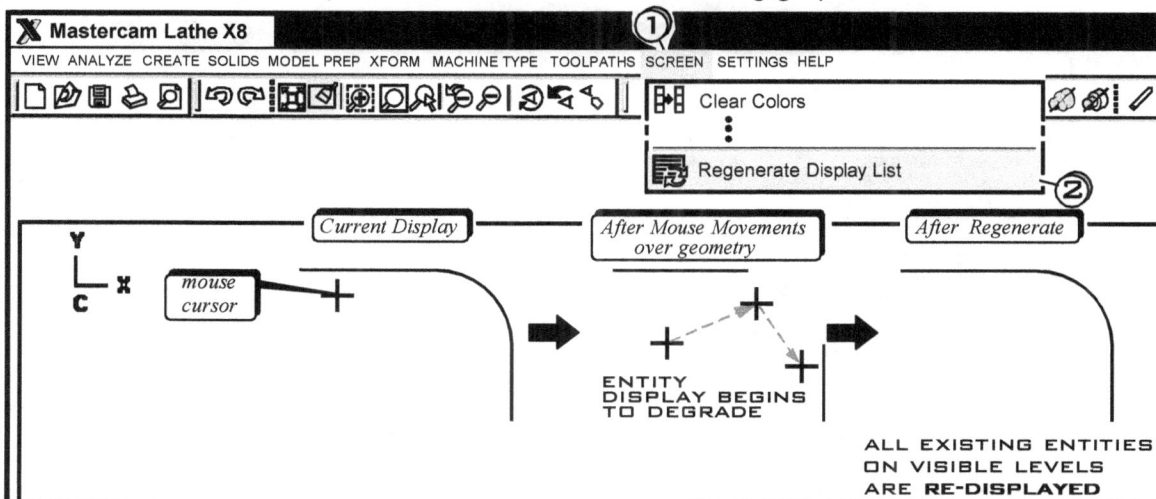

➤ Click ① the SCREEN drop down menu   ➤ Click ② Regenerate Display List 📇

2-15 The Undo/Redo Functions

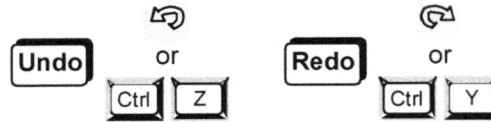

Undo	or	Redo	or
	Ctrl Z		Ctrl Y

These functions are used to *undo or redo one or more events that have just been executed in sequential order in the current w*orking file. Any *function -based* operation is defined as an *event*. A single line created with the line function is an event. Similarly, the Xform function that makes 50 copies of a line is also considered as an event. When applied to Xform, the Undo and redo functions will undo and redo all 50 copies at once since Xform is treated as a *single event*. The default setting enables Mastercam to save up to 2 billion undo/redo events. This is restricted only by the availability of ramdom access memory(RAM) in your PC. The System Configuration Dialog box can be used to direct *Mastercam* to store only a specific number of events and allocate a maximum amount of RAM to the undo/redo functions.

> Note:
> - ► Opening a part file or creating a new file *clears* the undo/redo list
> - ► Undo/redo *cannot* be applied to *toolpath* or *CAM* related functions
> - ► Undo/redo *can* only be applied to *CAD* related functions including:
>   - entity creation/edit
>   - drafting entities(dimensions)
>   - file annotations
>   - modify entity attributes

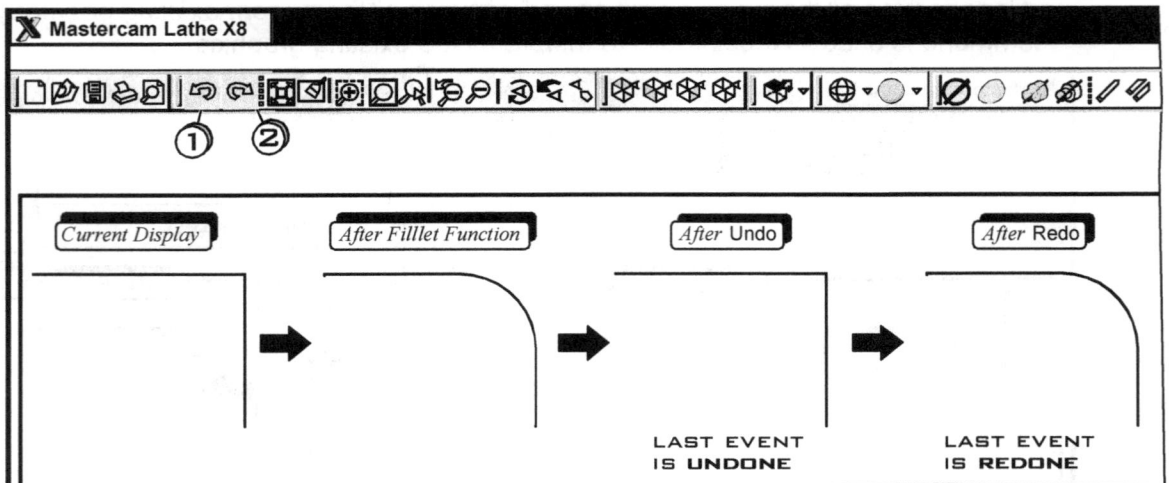

X Mastercam Lathe X8

① ②

| Current Display | After Filllet Function | After Undo | After Redo |

LAST EVENT IS UNDONE

LAST EVENT IS REDONE

➤ Click ① the Undo button ↶ to *undo* the last event(fillet creation)

➤ Click ② the Redo button ↷ to *redo* the last event(fillet creation)

EXERCISES

In each case create the 2D profile of the part as indicated in the **PICTORIAL** by using *Mastercam's* CAD package.

2-1) File Name: **EX2-1JV** ← YOUR INITIALS

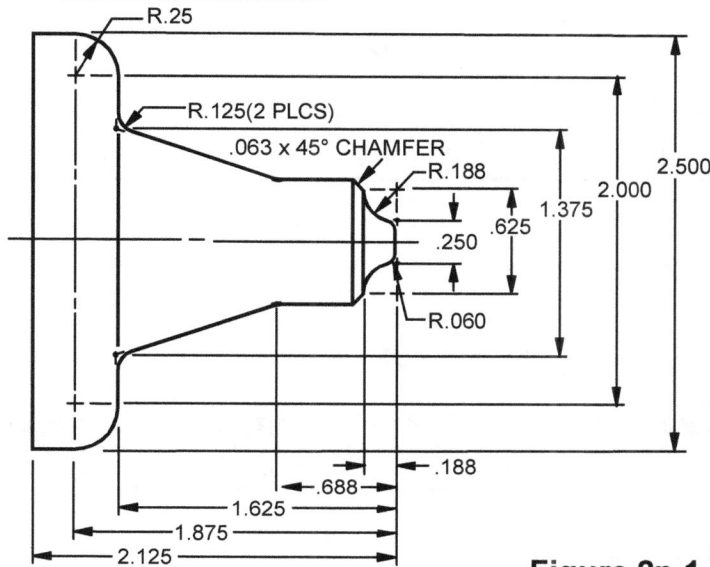

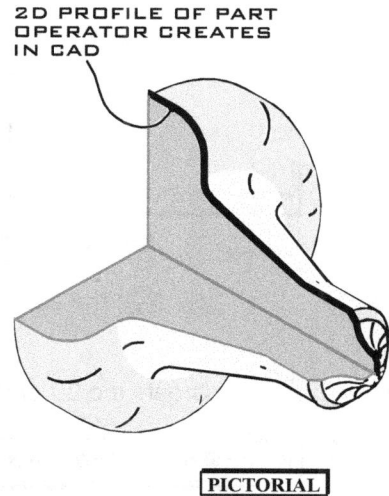

2D PROFILE OF PART
OPERATOR CREATES
IN CAD

R.25

R.125(2 PLCS)

.063 x 45° CHAMFER

R.188

.250

.625

1.375

2.000

2.500

R.060

.188

.688

1.625

1.875

2.125

PICTORIAL

Figure 2p-1

Enter *Mastercam*, open a new **.MCX-8** file and set the Machine Type to Lathe

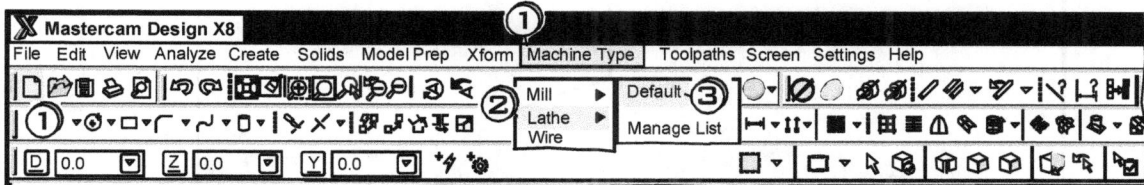

➤ Click ① the new file icon 🗋

➤ Click ② the   Machine Type   drop down menu

➤ Click ③ the   Lathe   package

➤ Click ④   Default

Assign *Mastercam* to **diameter** programming mode with the working construction plane set to **+DZ.**

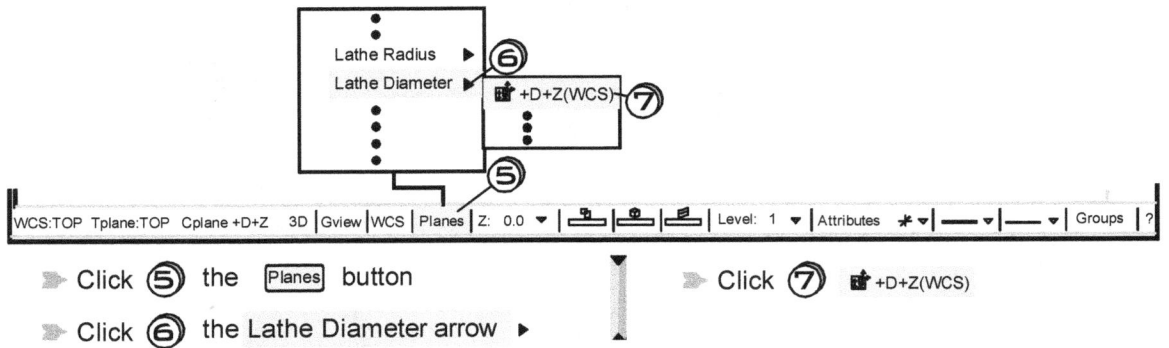

Lathe Radius ▶ ⑥

Lathe Diameter ▶   ⚙ +D+Z(WCS) ⑦

⑤

| WCS:TOP Tplane:TOP Cplane +D+Z   3D │Gview│WCS │Planes │ Z: 0.0 ▼│ ⚏ ⚏ ⚏ │ Level: 1 ▼ │Attributes ✳ ▼│ ── ▼│── ▼│ Groups │ ? |

➤ Click ⑤ the [Planes] button   ➤ Click ⑦ ⚙ +D+Z(WCS)

➤ Click ⑥ the Lathe Diameter arrow ▶

Create the 2D profile illustrated in the **PICTORIAL** .

①

⚞ Create Line Endpoint ②

ⓘ   ⓗ

D
└─ Z
C

Y
T └─ X

ⓖ

ⓕ

ⓔ   ⓓ

ⓒ

ⓐ ⓑ

X

D+
└─ X

WCS
ORIGIN
X0, Y0

Specify the first endpoint

Line   ⊠

⚞ ▪1 ▪2 [🎌] ③   [📊][0.0][▽] [📐][0.0][▽] [↕][0.0][▽] [↔] [✏] ⊕ ✔ ?

➤ Click ② ⚞ Create Line Endpoint

[Specify the first endpoint]

➤ Click ③ the multi line button [🎌]
       or tap the [M] key

≫ Enter `0,0` `Enter` ◄---ⓐ

`Specify the second endpoint`

≫ Enter `.25,0` `Enter` ◄---ⓑ

≫ Press `Esc` to cancel the operation

`Specify the first endpoint`

≫ Enter `.625,-.188` `Enter` ◄---ⓒ

`Specify the second endpoint`

≫ Enter `.75,-.188` `Enter` ◄---ⓓ

`Specify the second endpoint`

≫ Enter `.75,-.688` `Enter` ◄---ⓔ

`Specify the second endpoint`

≫ Enter `1.375,-1.625` `Enter` ◄---ⓕ

`Specify the second endpoint`

≫ Enter `2.0,-1.875` `Enter` ◄---ⓖ

`Specify the second endpoint`

≫ Press `Esc` to cancel the operation

`Specify the first endpoint`

≫ Enter `2.5,-1.875` `Enter` ◄---ⓗ

`Specify the second endpoint`

≫ Enter `2.5,-2.125` `Enter` ◄---ⓘ

Add the .188 radius arc

④

2,-1.5

(B) Create Circle Center Point

⑤ Create Arc Polar

D
C ⌐ Z

180°

Enter the center point

270°

Enter the center point

Arc Polar

1.25   0.0   30   130

➤ Click ④ the circle down arrow ▾

➤ Click ⑤ Create Arc Polar

Enter the center point

➤ Enter .625,0   Enter

➤ Tap the R key for *Radius*

➤ Enter .188 Tab

➤ Tab into the *Start Angle* box

Sketch the initial angle

➤ Enter **180**

➤ Tab into the *Final Angle* box

Sketch the final angle

➤ Enter **270** Enter

➤ Press Esc to cancel the function

Create the R.25, R.125 and R.063 fillets

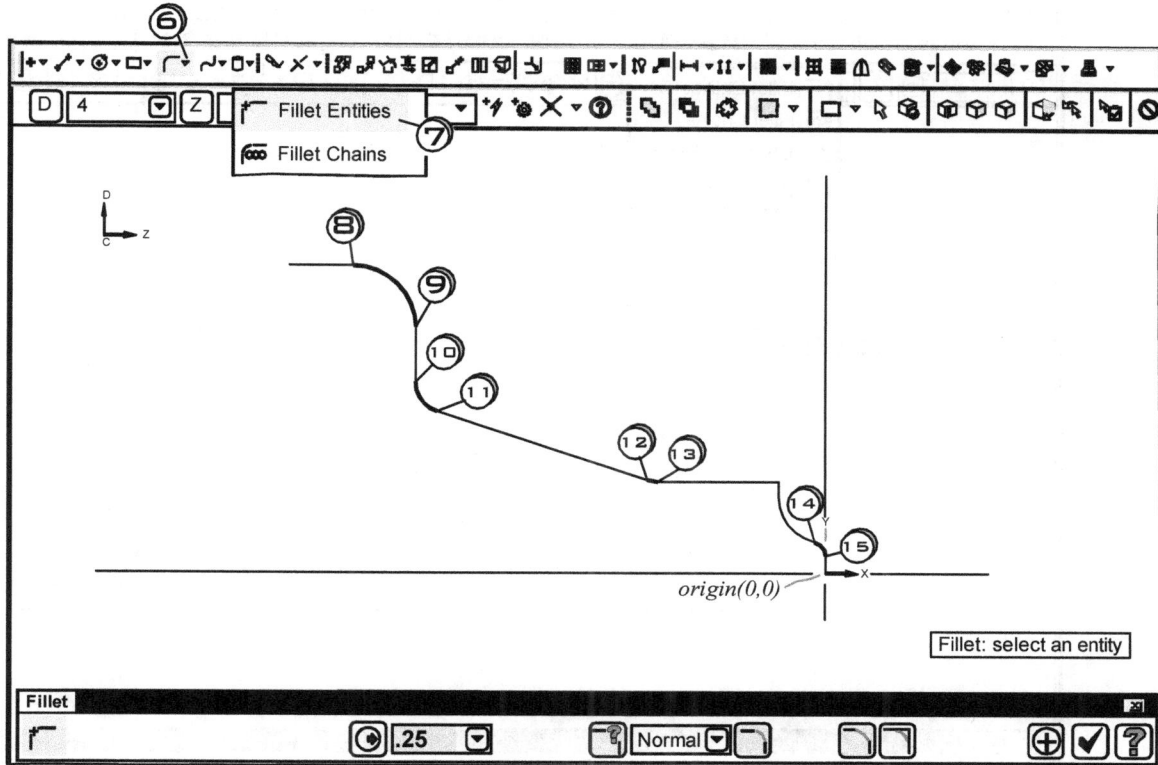

- Click ⑥ the down arrow ▾

- Click ⑦ ⌐ Fillet Entities

- Tap the [R] key for *Radius*; enter **.25**

  Fillet: select an entity

- Click ⑧ the first line entity

  Fillet: select another entity

- Click ⑨ the second line entity

- Tap [Esc] for function cancel

  Fillet: select an entity

- Tap the [R] key for *Radius*; enter **.06**

- Click ⑩ the first line entity

  Fillet: select another entity

- Click ⑪ the second line entity

  Fillet: select an entity

- Click ⑫ the first line entity

  Fillet: select another entity

- Click ⑬ the second line entity

  Fillet: select an entity

- Click ⑭ the first line entity

  Fillet: select another entity

- Click ⑮ the second line entity

Complete the 2D CAD profile by constructing the .063 x 45° chamfer.

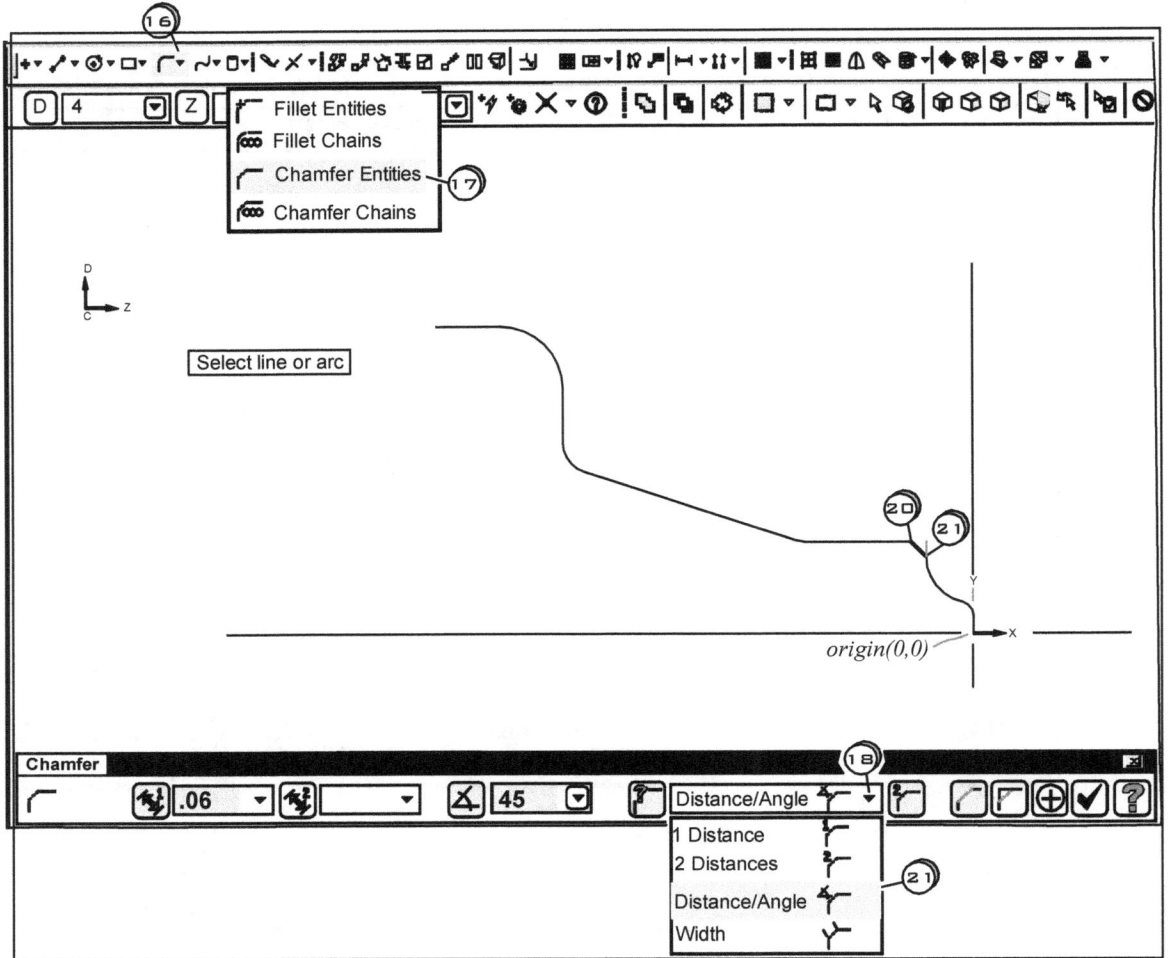

▸ Click ⑯ the down arrow ▾

▸ Click ⑰ Chamfer Entities

▸ Click ⑱ the down arrow ▽

▸ Click ⑲ the *Distance/Angle* style ⬚ Enter

▸ Tap the 1 key for *Dist 1*; enter **.06**

▸ Tab into *Angle*; enter **45**

Select line or arc

▸ Click ⑳ the first line entity

Select line or arc

▸ Click ㉑ the second line entity

Tap Esc for function cancel

Delete the line entities.

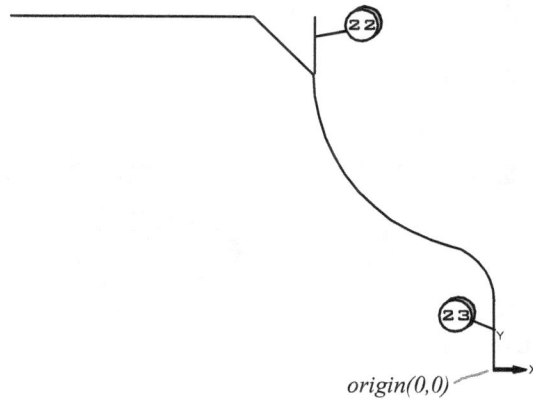

➤ Click ㉒

➤ Tap the [Del] key

➤ Click ㉓

➤ Tap the [Del] key

Save the *Mastercam* file **EX1-2JV.MCX-8** which contains the CAD profile.

— YOUR INITIALS

Click (24)   File

Click (25)   📑 Save As

Click (26) 🔲 my mcamx8

Click (27) New folder; enter **JVAL-LATHE**

Double Click (28) on the file folder

Click (29) in the file name box; enter the name of the file to be saved **EX1-2JV**

Click (30) the [Save] button

In each case create the 2D profile of the part as indicated in the pictorial

YOUR INITIALS

2-2) File Name: **EX2-2JV**

.094 x 45° CHAMFER

R.063(2PL)

.125R

.060 x 45° CHAMFER

Ø.500

Ø1.125

Ø1.375

Ø1.875

.375

.625

.875

1.188

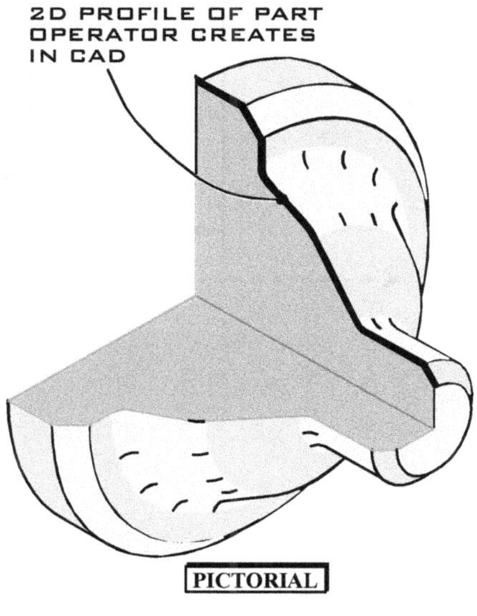

2D PROFILE OF PART
OPERATOR CREATES
IN CAD

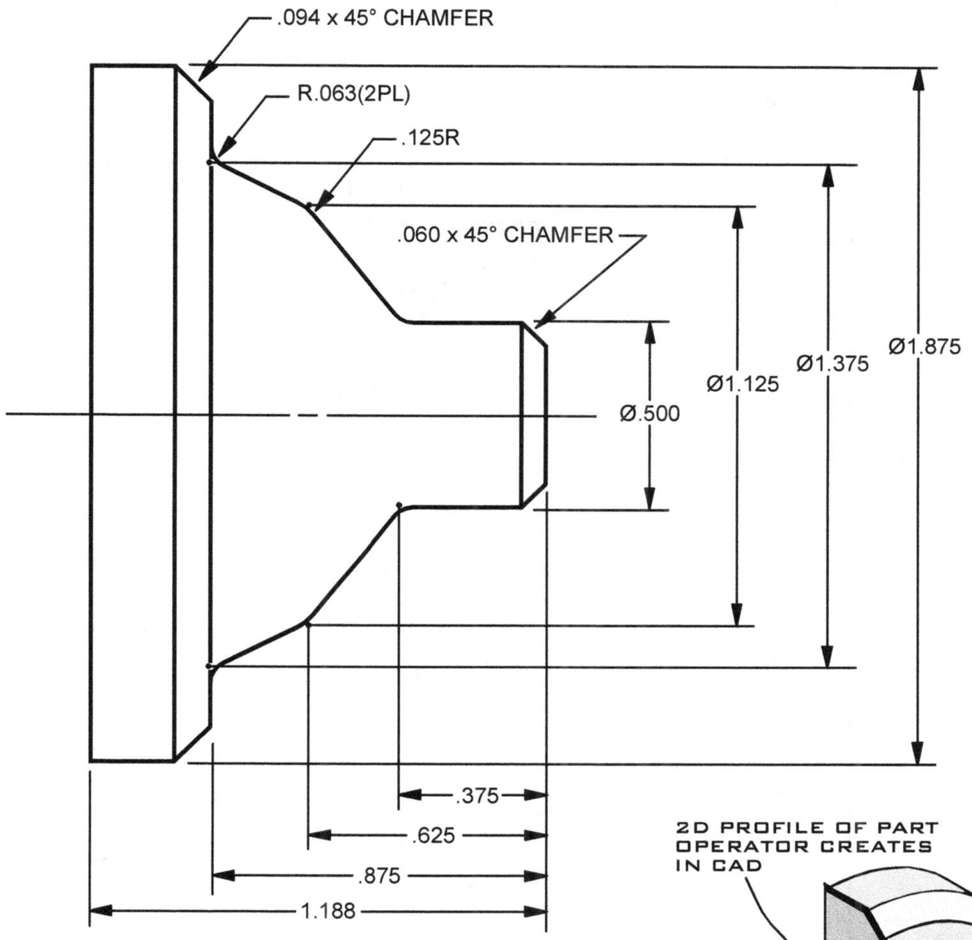

Figure 2p-2

PICTORIAL

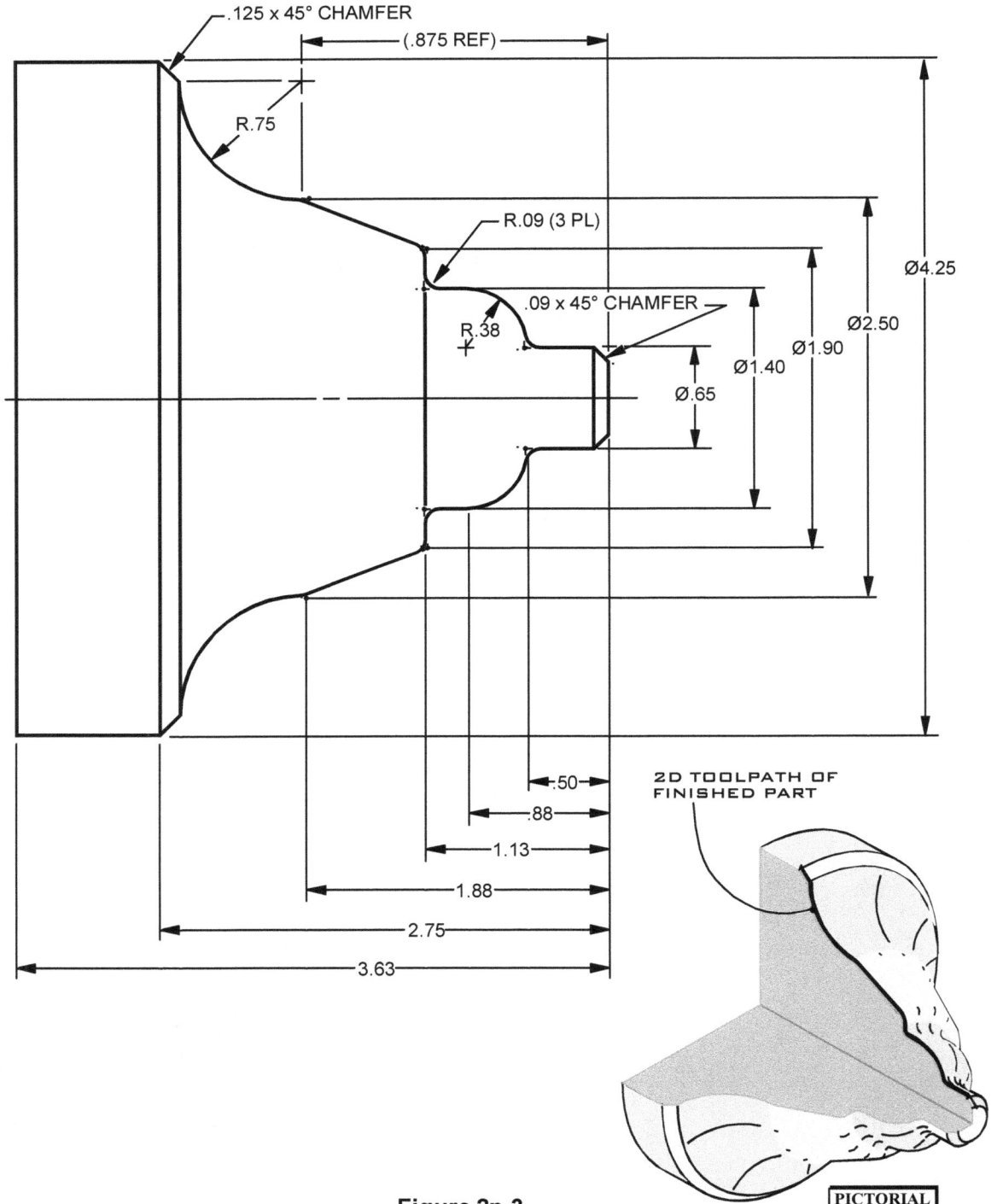

YOUR INITIALS

2-3) File Name: **EX2-3JV**

.125 x 45° CHAMFER

(.875 REF)

R.75

R.09 (3 PL)

.09 x 45° CHAMFER

R.38

Ø.65

Ø1.40

Ø1.90

Ø2.50

Ø4.25

.50

.88

1.13

1.88

2.75

3.63

2D TOOLPATH OF
FINISHED PART

PICTORIAL

Figure 2p-3

YOUR INITIALS

2-4) File Name: **EX2-4JV**

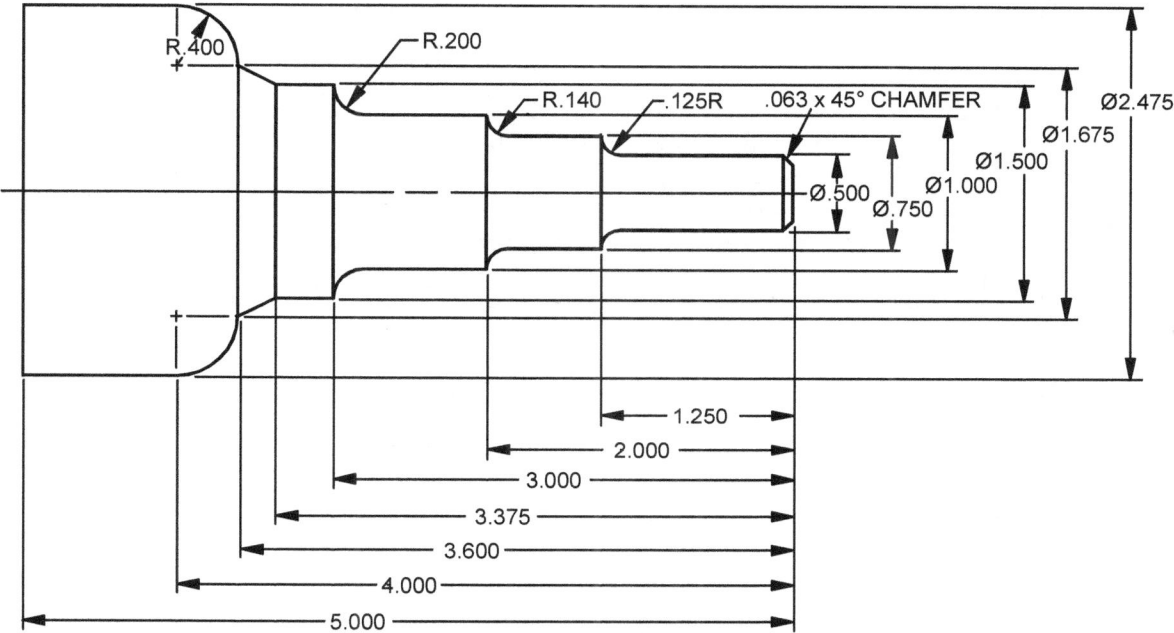

R.400

R.200

R.140 .125R .063 x 45° CHAMFER

Ø2.475

Ø1.675

Ø1.500

Ø1.000

Ø.500

Ø.750

1.250

2.000

3.000

3.375

3.600

4.000

5.000

2D PROFILE OF PART
OPERATOR CREATES
IN CAD

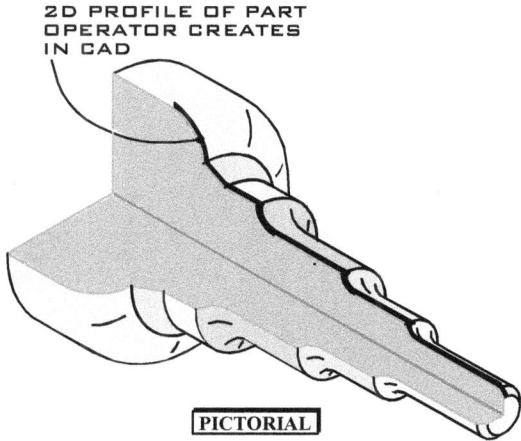

PICTORIAL

Figure 2p-4

YOUR INITIALS

2-5) File Name: **EX2-5JV**

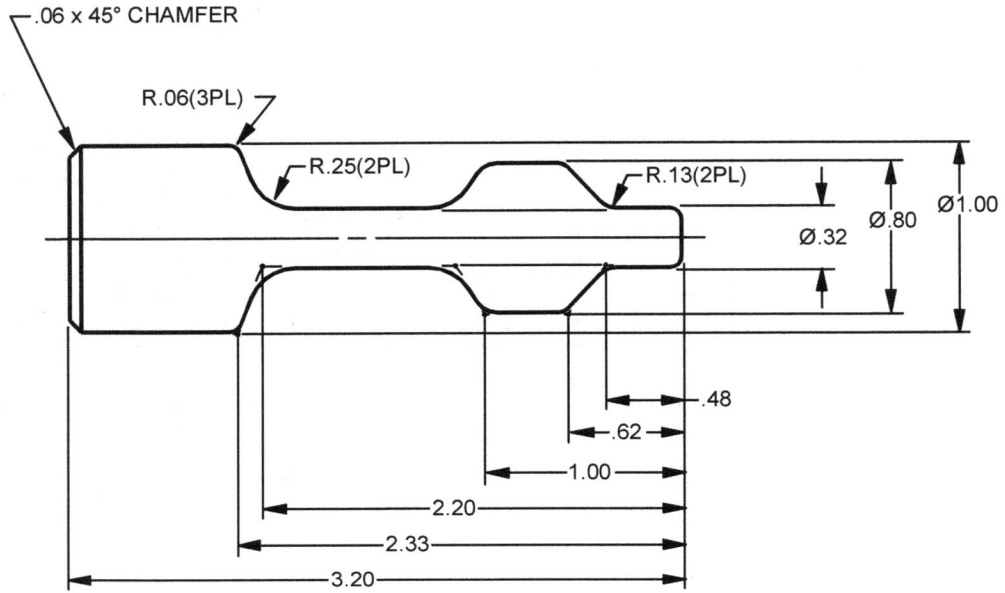

.06 x 45° CHAMFER

R.06(3PL)

R.25(2PL)

R.13(2PL)

Ø1.00

Ø.80

Ø.32

.48

.62

1.00

2.20

2.33

3.20

2D PROFILE OF PART
OPERATOR CREATES
IN CAD

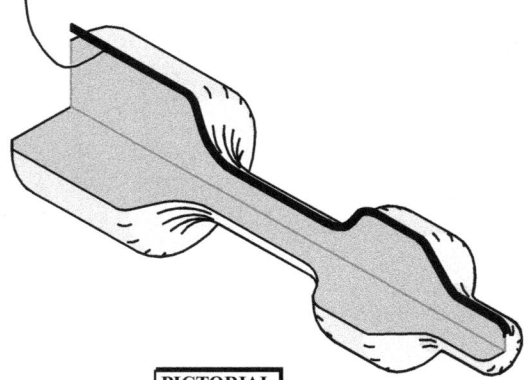

PICTORIAL

Figure 2p-5

YOUR INITIALS

2-6) File Name: **EX2-6JV**

R.25

.13 x 45° CHAMFER

R.13(2PL)

Ø4.75

Ø3.75

Ø2.75

Ø1.50

R.5

.38

.50

1.50

2.50

2D PROFILE OF PART
OPERATOR CREATES
IN CAD

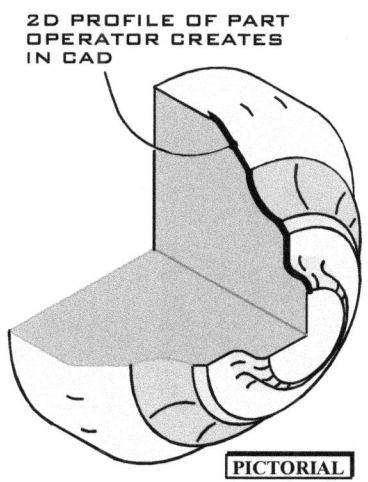

PICTORIAL

Figure 2p-6

YOUR INITIALS

2-7) File Name: **EX2-7JV**

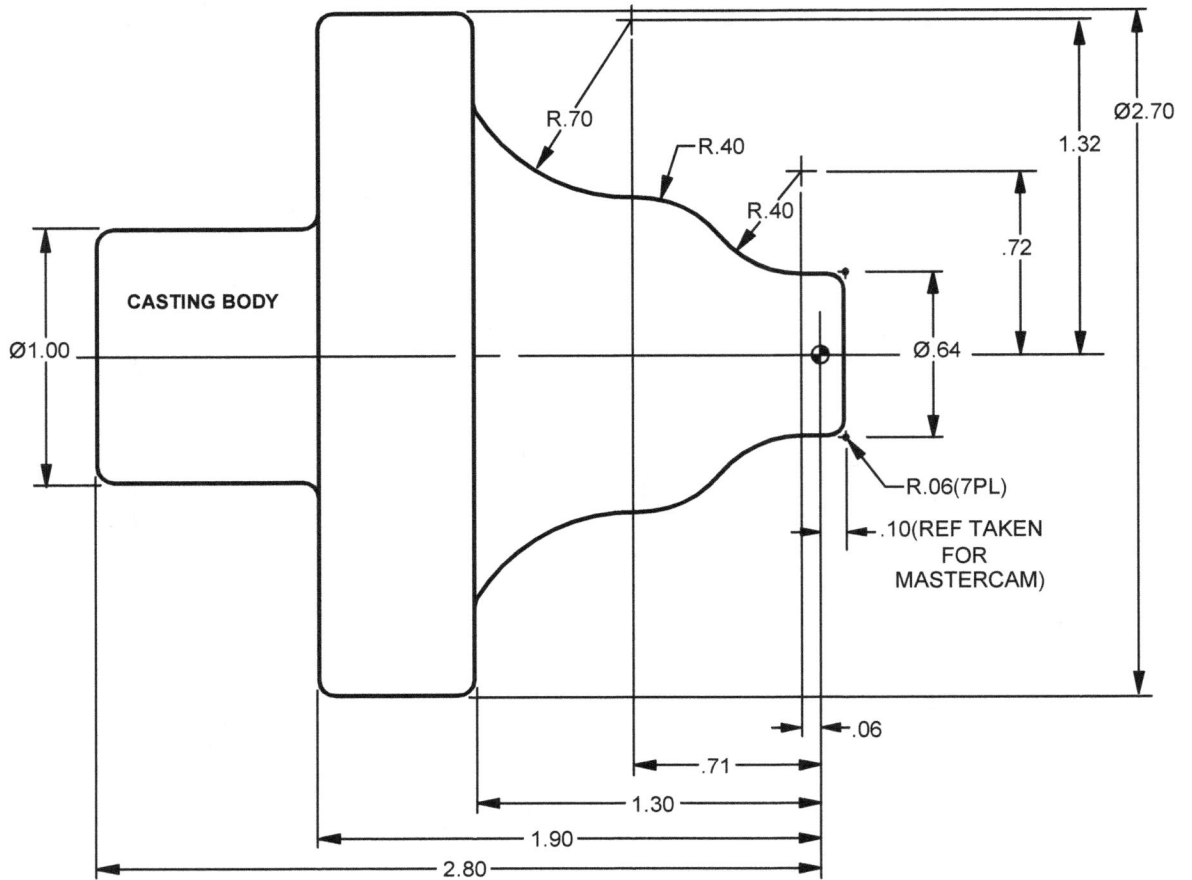

CASTING BODY

R.70

R.40

R.40

Ø2.70

1.32

.72

Ø.64

Ø1.00

R.06(7PL)

.10(REF TAKEN FOR MASTERCAM)

.06

.71

1.30

1.90

2.80

2D PROFILE OF CASTING OPERATOR CREATES IN CAD

CASTING BODY

PICTORIAL

Figure 2p-7

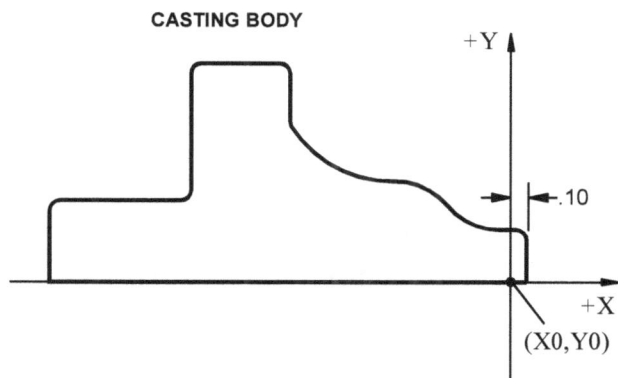

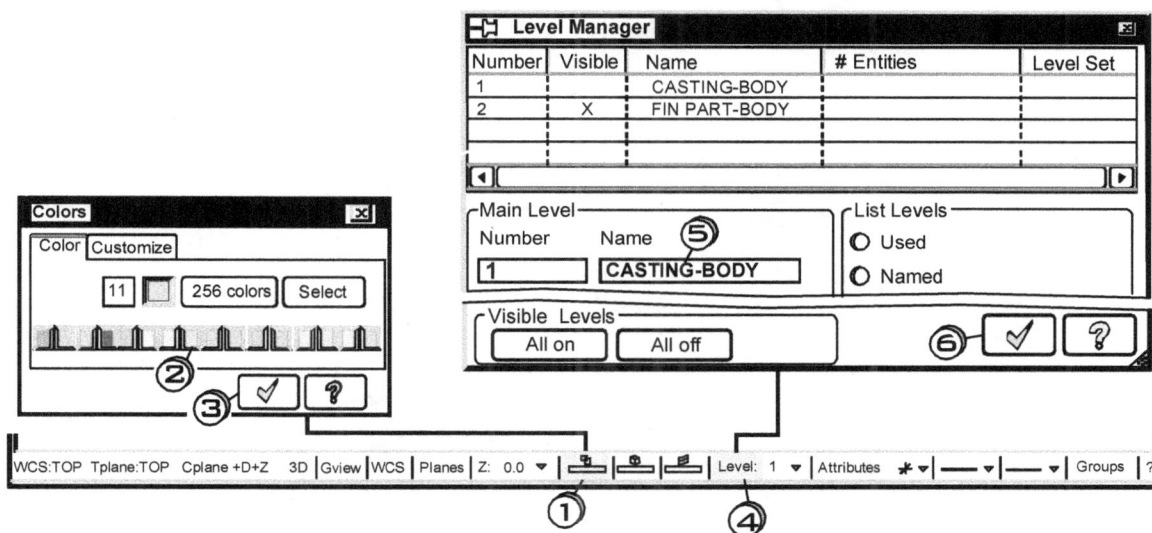

CASTING BODY

Level Manager

Number	Visible	Name	# Entities	Level Set
1		CASTING-BODY		
2	X	FIN PART-BODY		

Main Level
Number   Name   ⑤
1   **CASTING-BODY**

List Levels
○ Used
○ Named

Visible Levels
All on   All off   ⑥ ✓   ?

Colors

Color | Customize

11 [ ] 256 colors Select

②

③ ✓ ?

WCS:TOP Tplane:TOP Cplane +D+Z   3D │Gview│WCS │ Planes │ Z: 0.0 ▼ │ ▫ │ ▫ │ ▫ │ Level: 1 ▼ │ Attributes │ ✱ ▼│ ── ▼│ ── ▼│ Groups │ ?

①   ④

▷ Click ① the color button ▫

▷ Click ② the color green

▷ Click ③ the OK button ✓

▷ Click ④ the Level button

▷ Click ⑤ in the Name box; enter CASTING-BODY

▷ Click ⑥ the OK button ✓

Create the casting body as shown above.

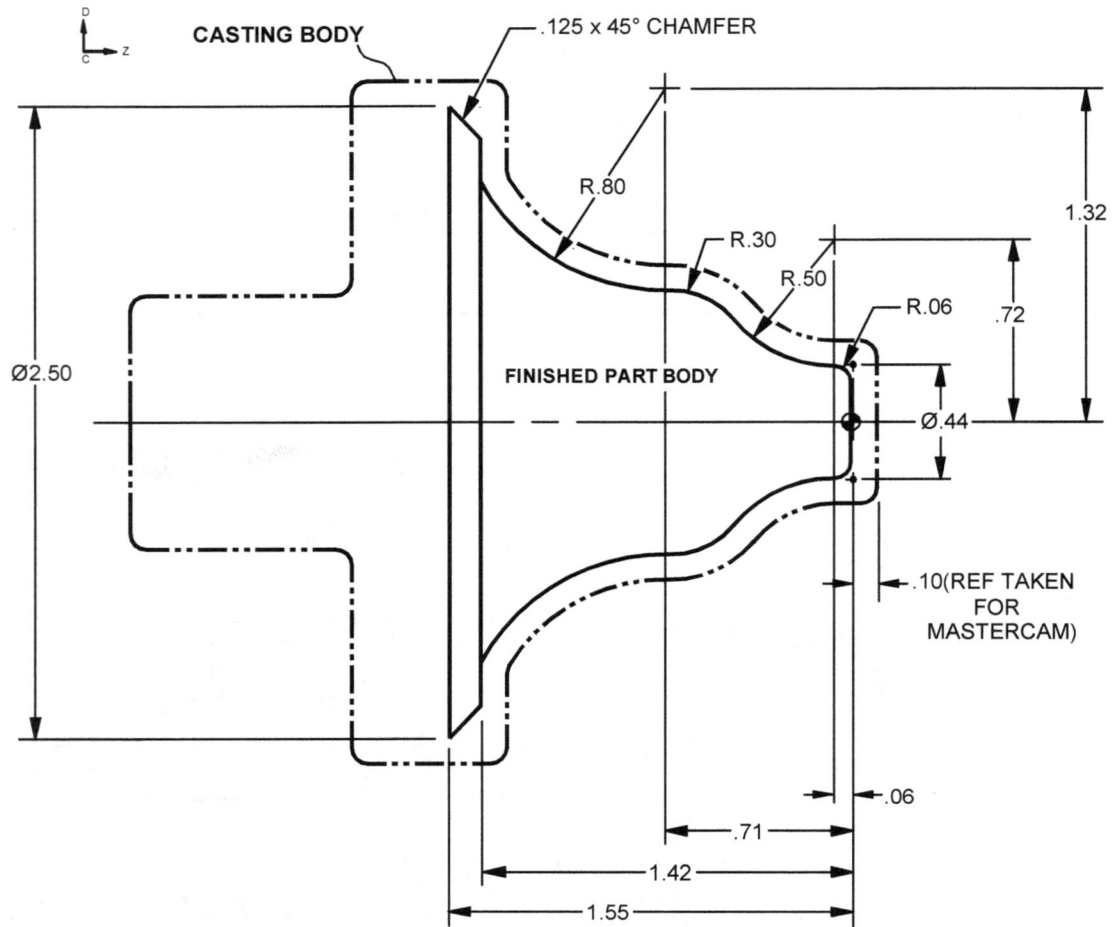

CASTING BODY

.125 x 45° CHAMFER

R.80

R.30

R.50

R.06

FINISHED PART BODY

Ø2.50

Ø.44

1.32

.72

.10(REF TAKEN
FOR
MASTERCAM)

.06

.71

1.42

1.55

2D PROFILE OF FINISHED PART
OPERATOR CREATES IN CAD

CASTING BODY

PICTORIAL

Figure 2p-8

FINISHED PART BODY

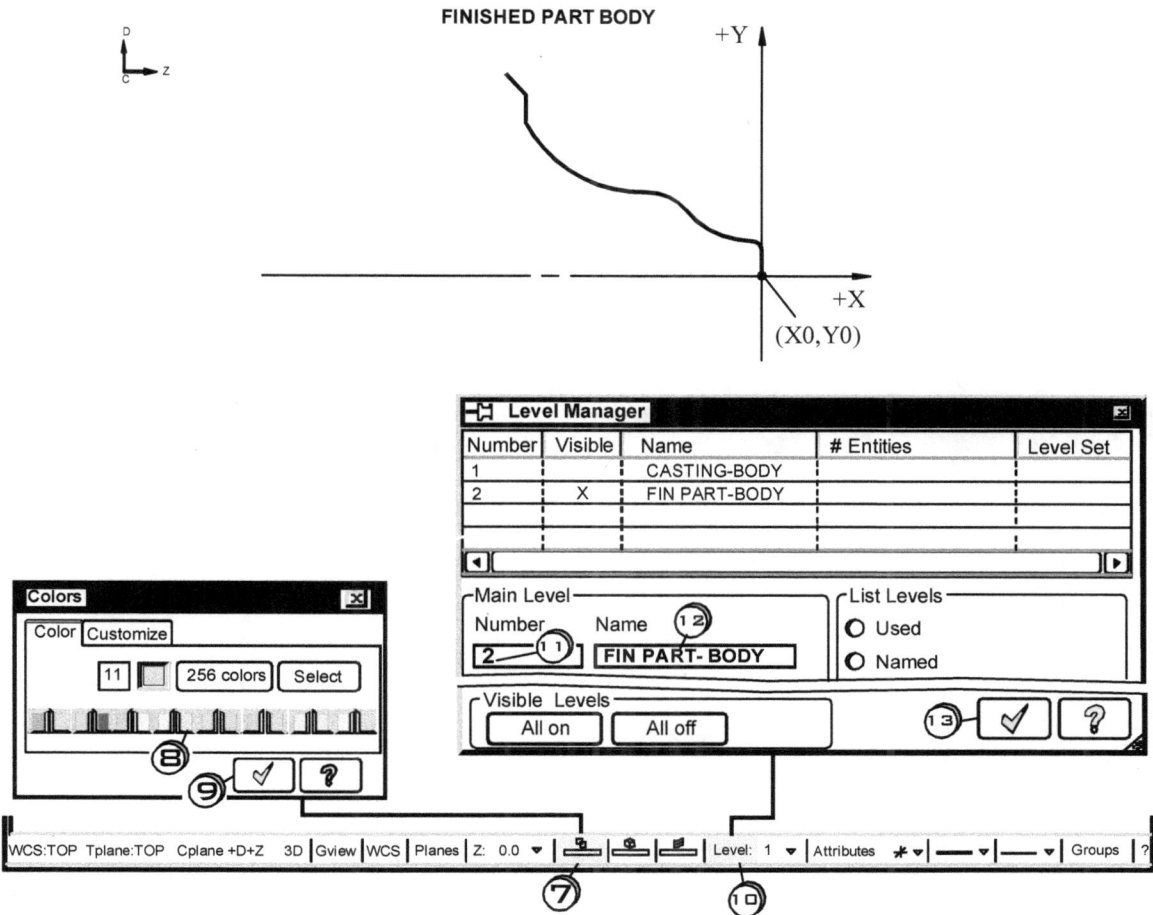

Level Manager

Number	Visible	Name	# Entities	Level Set
1		CASTING-BODY		
2	X	FIN PART-BODY		

Main Level

Number   Name ⑫

2 ⑪   **FIN PART- BODY**

List Levels

○ Used
○ Named

Visible Levels

All on   All off

⑬ ✓   ?

Colors

Color | Customize

11 ☐   256 colors   Select

⑧

⑨ ✓   ?

WCS:TOP Tplane:TOP Cplane +D+Z   3D | Gview | WCS | Planes | Z: 0.0 ▼ | 🔧 ⊕ 📕 | Level: 1 ▼ | Attributes ✳ ▼ | ——— ▼ | ——— ▼ | Groups | ?

⑦   ⑩

➤ Click ⑦ the color button 🔧

➤ Click ⑧ the color red

➤ Click ⑨ the OK button ✓

➤ Click ⑩ the Level button

➤ Click ⑪ in the level Number box; enter 2

➤ Click ⑫ in the Name box; enter FIN PART-BODY

➤ Click ⑬ the OK button ✓

Create the finished part body as shown above.

CHAPTER - 3

BASIC LATHE OPERATIONS

3-1 Chapter Objectives

After completing this chapter you will be able to:

1. Identify the stock to be machined.
2. Specify the chuck boundary.
3. Specify the machining operation and obtain the needed tool.
4. Specify the OD reference point.
5. Specify the operation's machining parameters.
6. Backplot the toolpath.
7. Simulate the operation.
8. Generate the word address part program.

3-2 An Example of Basic Lathe Machining

Get the CAD model file **EX2-1JV** :
 a) From the file generated in exercise 2-1.
                    or
 b) From the CD provided at the back of this text(file is located in the folder   CHAPTER3).

Direct *Mastercam* to create a part program for executing the PROCESS PLAN 3-1 on the part shown in Figure 3-1.

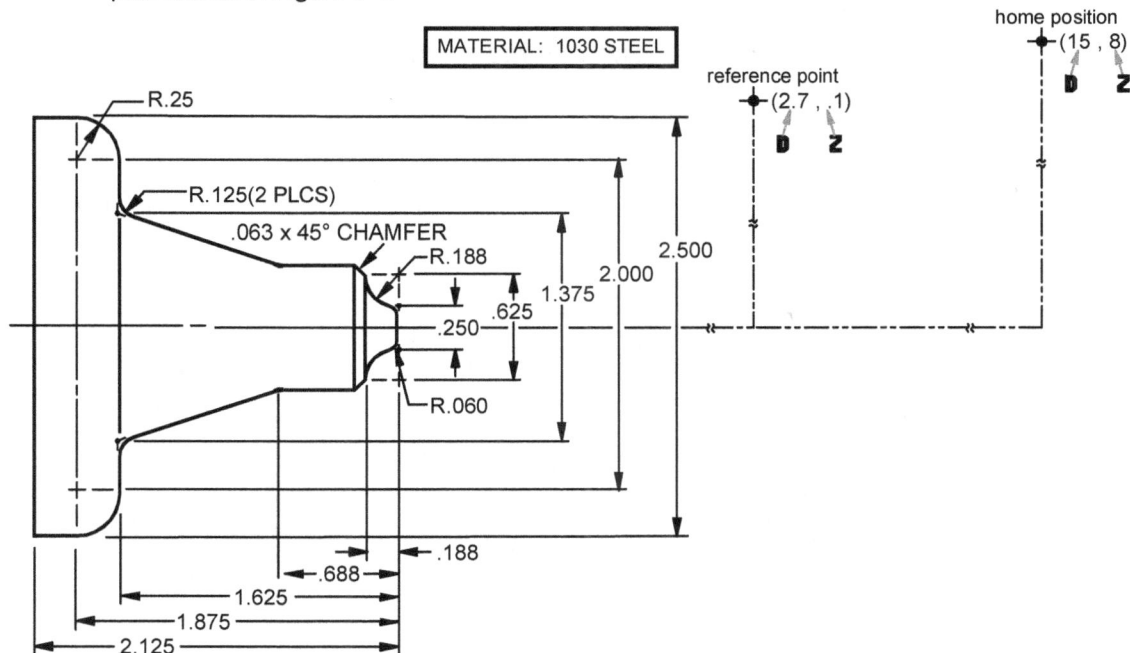

Figure 3-1

PROCESS PLAN 3-1

No.	Operation	Tooling	
1	FACE END TO .10IN	1/32TNR - RH - ROUGH - OD - TURNING TOOL	
2	ROUGH TURN OD; LEAVE .01 IN X AND Z FOR FINISHING	R1/32	
3	FINISH OD CONTOUR	1/64TNR - RH - OD - FINISHING TOOL	R1/64
4	LATHE CUTOFF	.016TNR - RH - OD - CUTOFF TOOL	
		DETAIL A R.016 1.5 A	

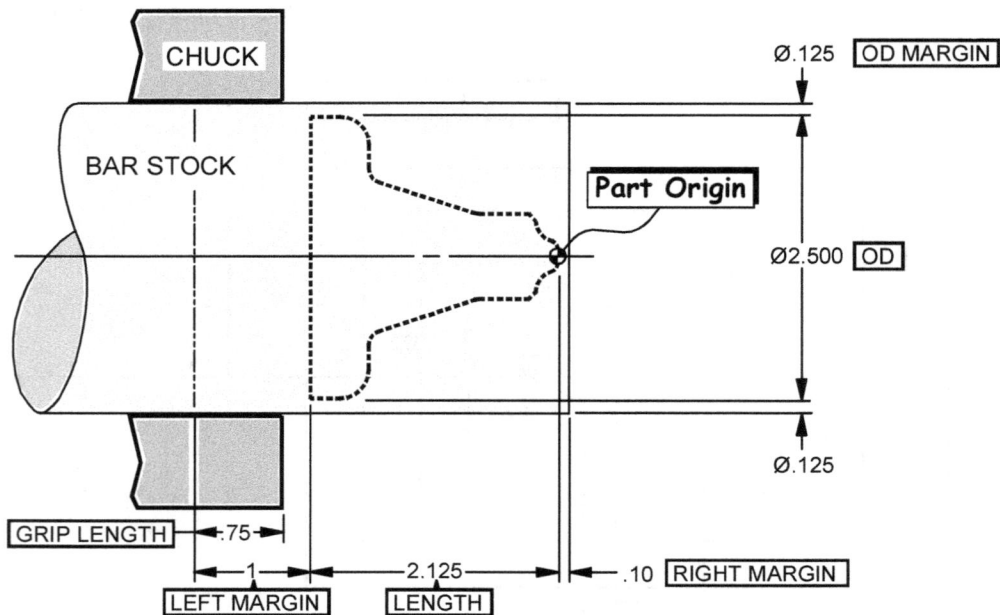

A) GET THE CAD MODEL FILE

◆ OPEN THE FILE EX2-1JV

➤ Click ① the file open icon 📂 in the **File** toolbar

✕ Open					✕
⬅ ➡ ▽ 🗔 ▷ Computer ▷ Windows 7 64-Bit (C:) ▷ ● ● ● ● ● ⚡				Search CHAPTER2 🔍	
Organize ▽ New folder				🔳 ▽ ⑦	

Name	Date modified	Type	Size
📁 Mastercam X8 ②			
📁 my mcamx8 ③ 📁 JVAL-LATHE			
📁 shared mcamx8			
📁 MCX EX2-1JV ④			

File name	EX1-1JV ▽	All Mastercam X8 files(*.MCX-8) ▽
	⑤ Open	Cancel

➤ Click ② 📁 my mcamx7

➤ **Double** Click ③ on the file folder 📁 **JVAL-LATHE**

➤ Click ④ on the file to open   **EX2-1JV**

➤ Click ⑤ the [Open] button

B) IDENTIFY THE STOCK TO BE MACHINED

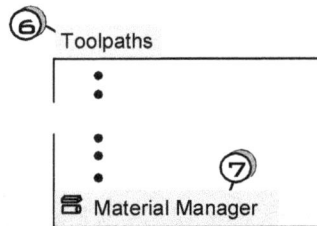

Toolpaths
 •
 •
 •
🖫 Material Manager

➤ Click ⑥ Toolpaths

➤ Click ⑦ 🖫 Material Manager

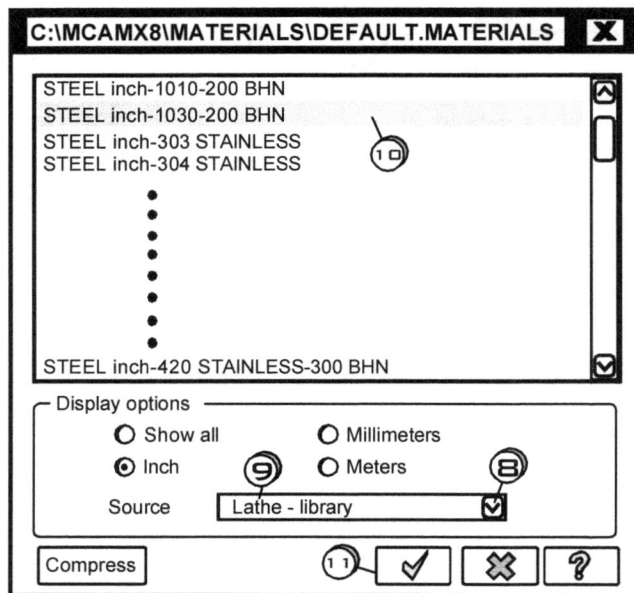

C:\MCAMX8\MATERIALS\DEFAULT.MATERIALS ❎

STEEL inch-1010-200 BHN
STEEL inch-1030-200 BHN
STEEL inch-303 STAINLESS
STEEL inch-304 STAINLESS
 •
 •
 •
 •
 •
 •
STEEL inch-420 STAINLESS-300 BHN

Display options
○ Show all   ○ Millimeters
◉ Inch ⑨   ○ Meters   ⑧
Source   [Lathe - library ▾]

[Compress]   ⑪ [✓] [✖] [?]

➤ Click ⑧ the down arrow ▾

➤ Click ⑨ Lathe - library

➤ Click ⑩ STEEL inch-1030-200 BHN

➤ Click ⑪ the OK button [✓]

C) SPECIFY THE SIZE OF THE STOCK BOUNDARY

Toolpaths ▽ 📌 ✕

- ⊟ Machine Group 1
  - ⊞ Properties - Lathe-Default
    - (12)
      - 📁 Files
      - 🗒 Tool settings
      - ◇ Stock setup
        - (13)
      - ⚠ Safety zone

Machine Group Properties ✕

Files | Tool Settings | Stock Setup

┌─ Stock Plane ─────────────────────────
│ [🖩]   TOP
└────────────────────────────────────

┌─ Stock ──────────────────────────────
│   (14) [Properties]
│   ◉ Left Spindle   ○ Right Spindle   [Delete]
│   (Not Defined)   (Not Defined)
│   ┌─ Chuck Jaws ────────────────────
│   │   [Properties]
│   │ ◉ Left Spindle   ○ Right Spindle   [Delete]
│   │   (Not Defined)   (Not Defined)

┌─ Tailstock Center ──────┐ ┌─ Steady Rest ──────┐
│   [Properties]   │ │   [Properties]   │
│   [Delete]   │ │   [Delete]   │
│   (Not Defined)   │ │   (Not Defined)   │

(26) [✓] [✗] [?]

Machine Component Manager - Stock ✕

Name: Stock:(Left Spindle)

Geometry | Position/Orientation on Machine

Geometry [Cylinder ▼]

Chord tolerance [0.001]   Color [103] [ ] [▦]

┌─ Translucency ──────────────────
│ Solid   Transparent
│ [▮────────────────────]

[Make from 2 points]

(15)
OD: [2.5] [Select]

☐ ID: [0.02] [Select]

Length: [2.125] [Select]
(16)

┌─ Position Along Axis ──────
│ Z: [0.00] [Select]

Axis [-Z ▼]

OD margin
[.125]
(17)

Right margin
[.1]
(18)

Left margin
[1] (19)

☑ Use Margins   [Preview Lathe Boundaries]

(20) [✓] [✗] [?]

➤ Click ⑫ the minus sign ⊟

➤ Click ⑬ ◇ Stock setup

➤ Click ⑭ the [Properties] button

➤ Click ⑮ in the OD box; enter | **2.5** |

➤ Click ⑯ in the Length box; enter | **2.125** |

➤ Click ⑰ in the OD margin box; enter | **.125** |

➤ Click ⑱ in the Right margin box; enter | **.1** |

➤ Click ⑲ in the Left margin box; enter | **1** |

➤ Click ⑳ the OK button [✓]

D) SPECIFY THE CHUCK BOUNDARY

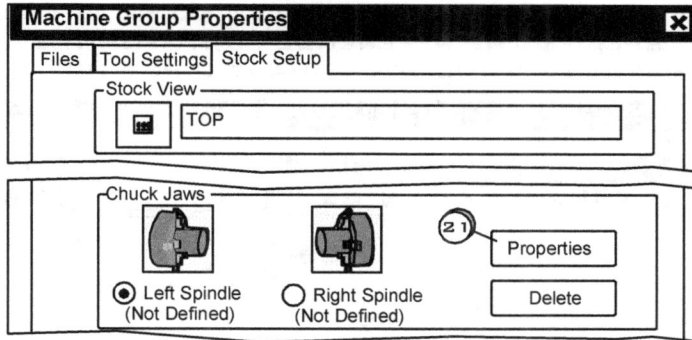

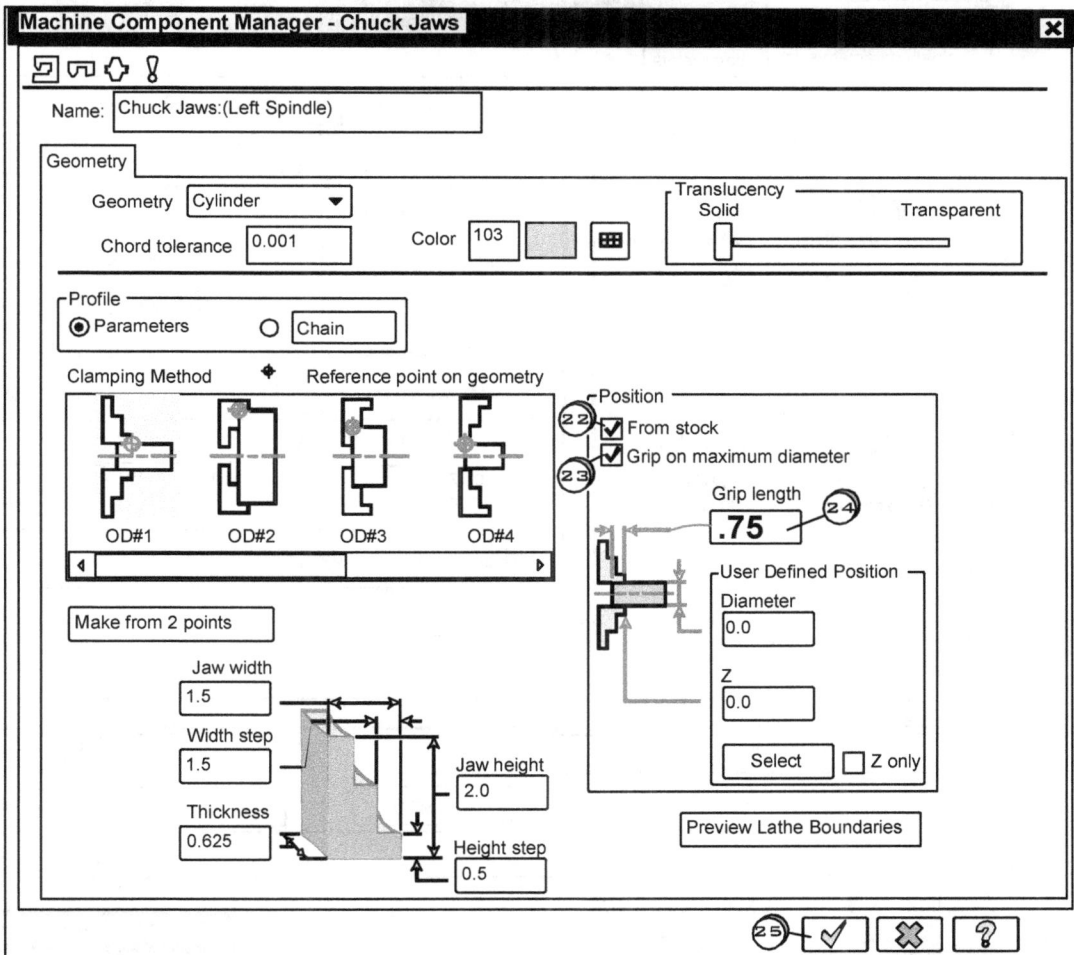

Machine Group Properties

Files | Tool Settings | Stock Setup

Stock View
TOP

Chuck Jaws

(21) Properties

Delete

◉ Left Spindle (Not Defined)   ○ Right Spindle (Not Defined)

Machine Component Manager - Chuck Jaws

Name: Chuck Jaws:(Left Spindle)

Geometry

Geometry | Cylinder ▼

Chord tolerance | 0.001

Color | 103

Translucency
Solid ———————————— Transparent

Profile
◉ Parameters   ○ Chain

Clamping Method   ✦   Reference point on geometry

OD#1   OD#2   OD#3   OD#4

Position
(22) ☑ From stock
(23) ☑ Grip on maximum diameter

Grip length (24)
.75

User Defined Position
Diameter
0.0

Z
0.0

Select   ☐ Z only

Preview Lathe Boundaries

Make from 2 points

Jaw width
1.5

Width step
1.5

Jaw height
2.0

Thickness
0.625

Height step
0.5

(25) ✓ | ✗ | ?

▶ Click (21) the [Properties] button

▶ Click (22) the check *on* ☑ From stock

▶ Click (23) the check *on* ☑ Grip on maximum diameter

▶ Click (24) enter Grip length **.75**

▶ Click (25) the OK button ✓

▶ Click (26) the OK button ✓

C) OPERATION#1- FACE THE END

- ◆ OBTAIN THE NEEDED R1/32 OD ROUGH RIGHT TOOL
- ◆ SPECIFY THE REFERENCE POINT FOR THE TOOL
- ◆ SPECIFY HOME(TOOL CHANGE) POSITION FOR THE TOOL

⯈ Click (27) TOOLPATHS] ⯈ Click (28) ▥ Face

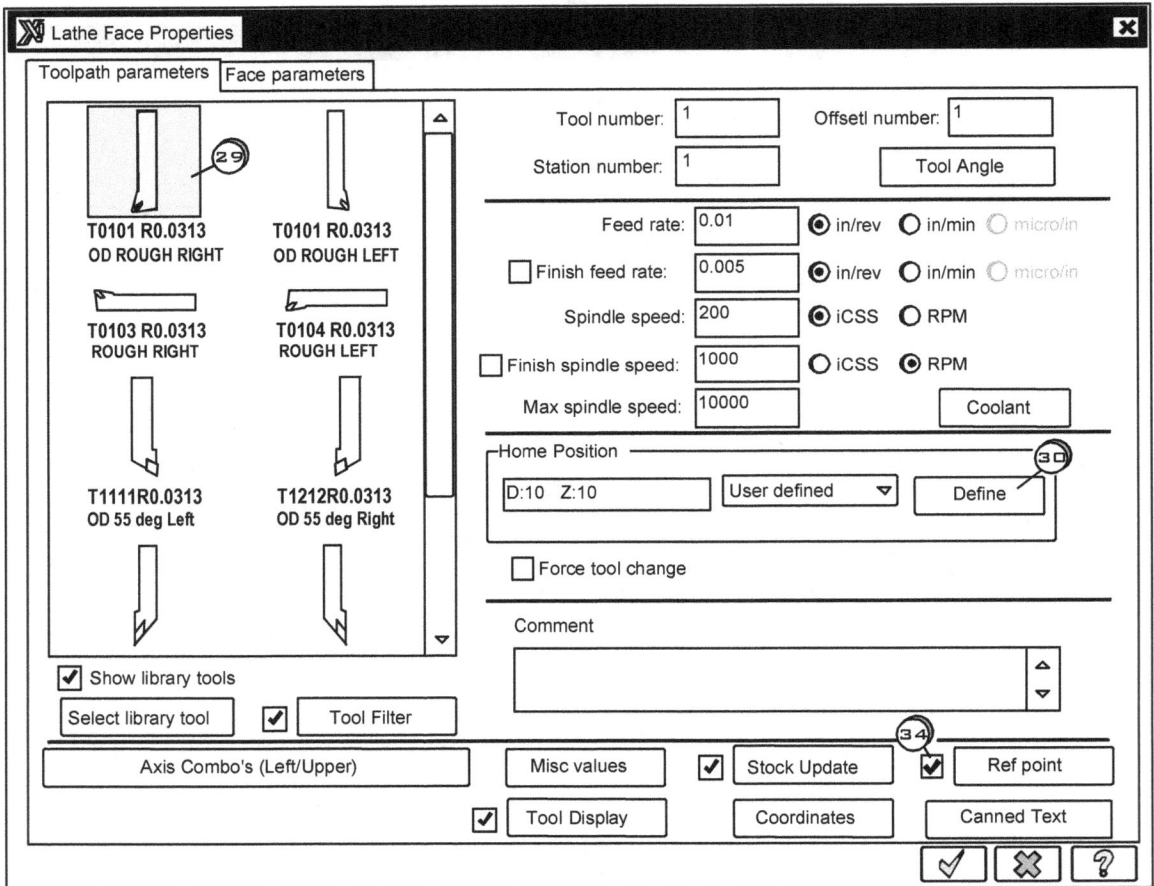

⯈ Click (29) the OD ROUGH RIGHT tool
⯈ Click (30) the [Define] button
⯈ Click (31) in the D box; enter [15]
⯈ Click (32) in the Z box; enter [8]
⯈ Click (33) the OK button [✓]

▶ Click ㉞ the check on ☑ | Ref point |

▶ Click ㉟ the check on ☑ Approach

▶ Click ㊱ in the D box; enter | 2.7 |

▶ Click ㊲ in the Z box; enter | .1 |

▶ Click ㊳ the copy button | → |

▶ Click ㊴ the OK button | ✓ |

♦ ENTER THE FACE MACHINING PARAMETERS

Lathe Face Properties   ☒

| Toolpath parameters | Face parameters | ④ⓞ |

Tool Compensation
Compensation type:
[Computer ▽]

Compensation direction:
[Left ▽]

Roll cutter around corners
[All ▽]

○ Select Points

⦿ Use stock

Finish Z:
[0.0]

Entry amount:
[0.1]

☐ Rough stepover:
[.1] ④①

☑ Finish stepover: Maximum number of finish passes:
[.01] ④②   [1]

Overcut amount:
[.015] ④③

Retract amount:
[0.1]

Stock to leave:
[0.1]
☐ Cut away from center line

☐ Corner
☐ Lead In/Out ..
☐ Filter

④④ [✓] [✗] [?]

➤ Click ④ⓞ the ⌐Face parameters tab

➤ Click ④① in the Rough stepover: box; enter [.1]

➤ Click ④② in the Finish stepover: box; enter [.01]

➤ Click ④③ in the Overcut amount: box; enter [.015]

➤ Click ④④ the OK button [✓]

♦ BACKPLOT THE TOOLPATH TO VERIFY THAT THE FACE END
  OPERATION IS FREE FROM ANY ERRORS

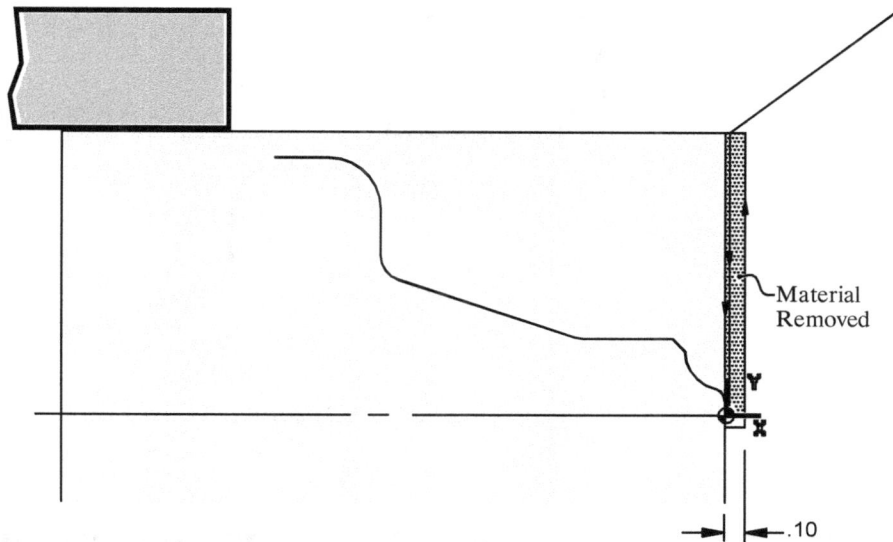

Material
Removed

.10

Toolpaths

- Machine Group 1
  - Properties - Lathe Default
  - Toolpath Group 1
    - 1 - Lathe Face-[WCS:T
      - Parameters
      - T0101-General Tur
      - Geometry
      - Toolpath-4.9K-E2X1JV
      - Update stock

Backplot

▶ Click (45) the Backplot button ≋

▶ Click (46) the OK button ✓

D) OPERATION#2- ROUGH TURN THE OD, LEAVE .01 IN X AND Z FOR FINISHING

◆ CHAIN THE REQUIRED OD GEOMETRY

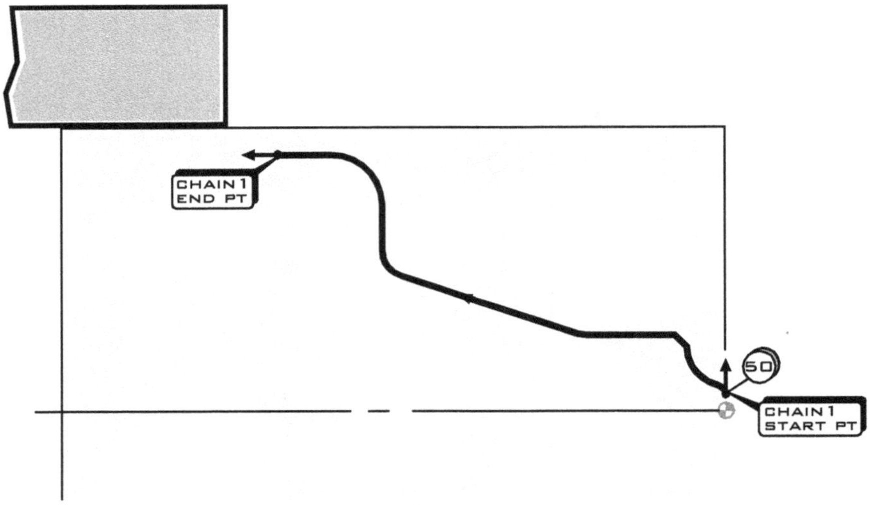

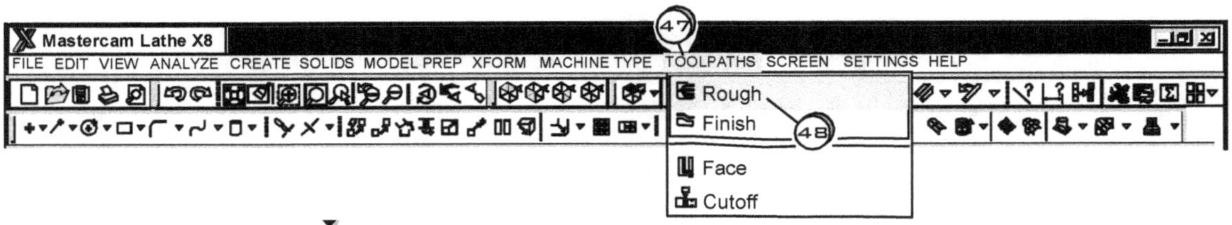

➤ Click (47) TOOLPATHS ➤ Click (48) Rough

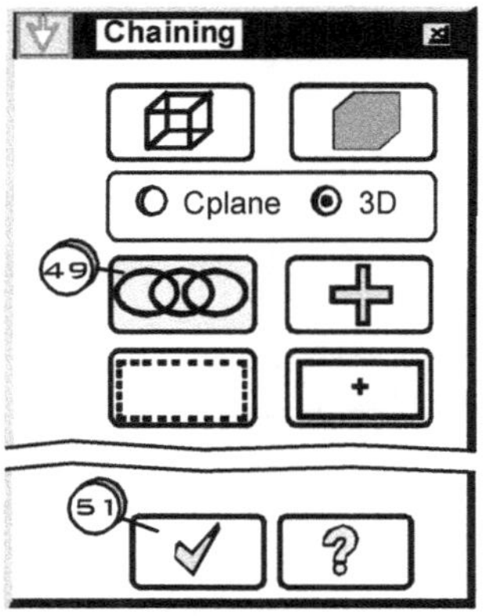

➤ Click (49) the chain button

Select the entry point or chain the inner boundary

➤ Click (50) the entry point

➤ Click (51) the OK button

◆ ACCEPT THE DEFAULT T0101 R0.0313 OD ROUGH RIGHT TOOL

◆ ENTER THE ROUGH OD MACHINING PARAMETERS

Click (52) the Rough parameters tab

Click (53) in the Stock to leave in X: box; enter .01

Click (54) in the Stock to leave in Z: box; enter .01

Click (55) the Compensation type down button ▽

Click (56) Wear

Click (57) check on ✔ for Lead In/Out

Click (58) the [Lead In/Out ..] button

Click (59) the Lead In tab

Click (60) check on ✔ for Use entry vector

Click (61) in the Length box; enter 0
lead in will be from the reference point (2.7,.1)

Click (62) the Lead Out tab

Click (63) check on ✔ for Extend/shorten

Click (64) the Extend radio button ◉

Click (65) in the Ammount: box; enter .13

Click (66) check on ✔ for Use exit vector
lead out will be a polar line at a 45° angle from the
horizontal of length 0.1

Click (67) the OK button ✓

Click (68) the OK button ✓

♦ BACKPLOT THE TOOLPATH TO VERIFY THAT THE ROUGH OD
  OPERATION IS FREE FROM ANY ERRORS

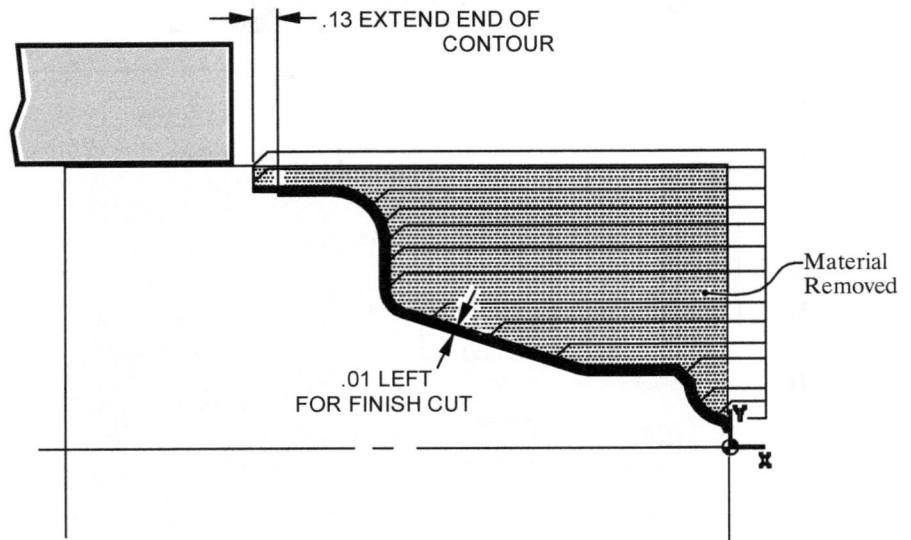

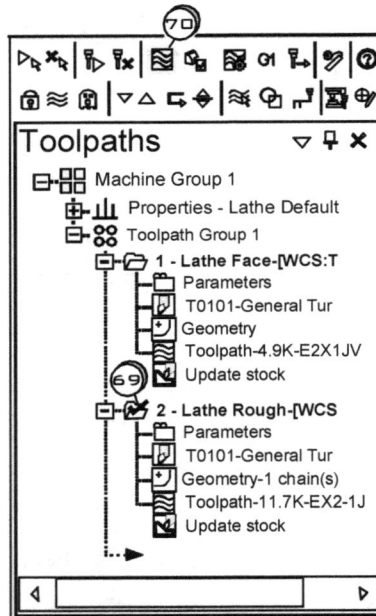

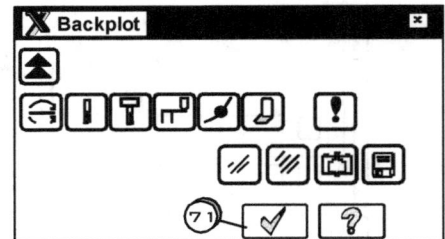

.13 EXTEND END OF
        CONTOUR

Material
Removed

.01 LEFT
FOR FINISH CUT

Y

X

Toolpaths   ▽ ⚲ ✕

Machine Group 1
  Properties - Lathe Default
  Toolpath Group 1
    1 - Lathe Face-[WCS:T
      Parameters
      T0101-General Tur
      Geometry
      Toolpath-4.9K-E2X1JV
      Update stock
    2 - Lathe Rough-[WCS
      Parameters
      T0101-General Tur
      Geometry-1 chain(s)
      Toolpath-11.7K-EX2-1J
      Update stock

X Backplot   ✕

⮞ Click ⓺⑨ the check *on* 📝 2 - Lathe Rough-[WCS

⮞ Click ⑦⓪ the Backplot button ≋

⮞ Click ⑦① the OK button ☑

D) OPERATION#3- FINISH THE OD CONTOUR

- ◆ OBTAIN THE NEEDED R1/64 OD FINISH RIGHT TOOL
- ◆ SPECIFY HOME(TOOL CHANGE) POSITION FOR THE TOOL

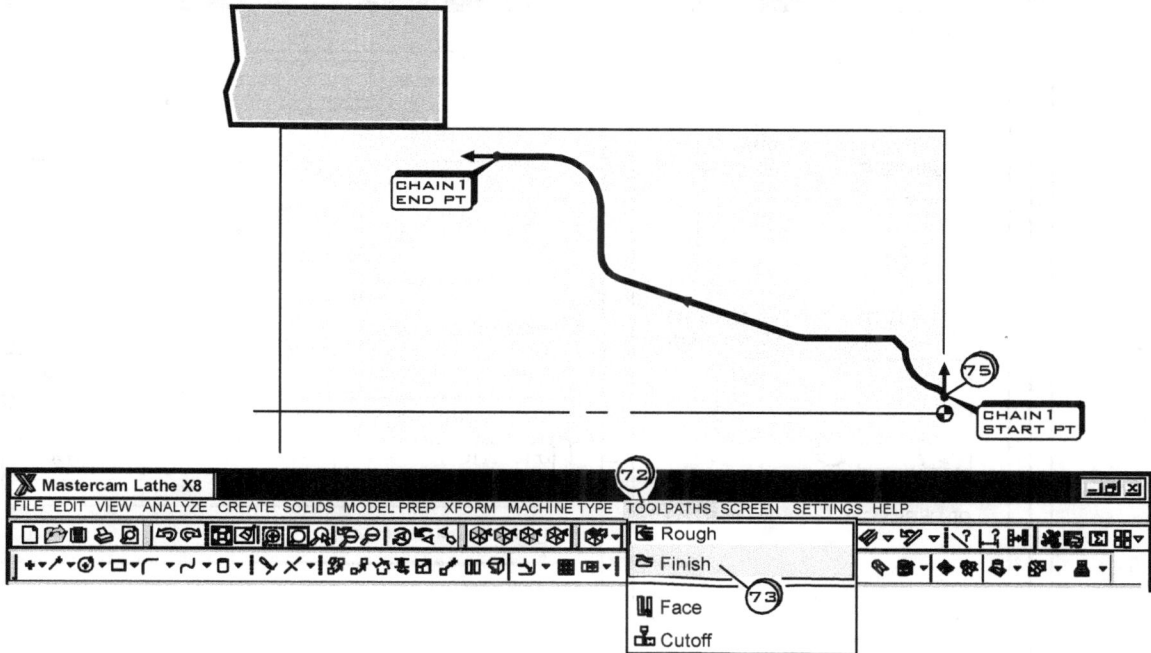

⟫ Click ⑦② TOOLPATHS ⟫ Click ⑦③ ☞ Finish

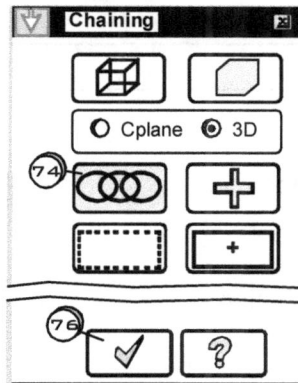

⟫ Click ⑦④ the chain button 〖⦿⦿⦿〗

| Select the entry point or chain the inner boundary |

⟫ Click ⑦⑤ the entry point

⟫ Click ⑦⑥ the OK button 〖✓〗

Lathe Finish Properties

Toolpath parameters | Finish parameters

T0103 R0.0313 ROUGH RIGHT
T0104 R0.0313 ROUGH LEFT
T1111R0.0313 OD 55 deg Left
T1212R0.0313 OD 55 deg Right
T2121R0.0156 OD FINISH RIGHT
T2222R0.0156 OD FINISH LEFT

Tool number: 1   Offsetl number: 1
Station number: 1   Tool Angle

Feed rate: 0.01   ● in/rev ○ in/min ○ micro/in
☐ Finish feed rate: 0.005   ● in/rev ○ in/min ○ micro/in
Spindle speed: 200   ● iCSS ○ RPM
☐ Finish spindle speed: 1000   ○ iCSS ● RPM
Max spindle speed: 10000   Coolant

Home Position
D:10 Z:10   User defined ▽   Define

☐ Force tool change

Comment

☑ Show library tools
Select library tool ☑ Tool Filter

Axis Combo's (Left/Upper)   Misc values ☑ Stock Update ☑ Ref point
☑ Tool Display   Coordinates   Canned Text

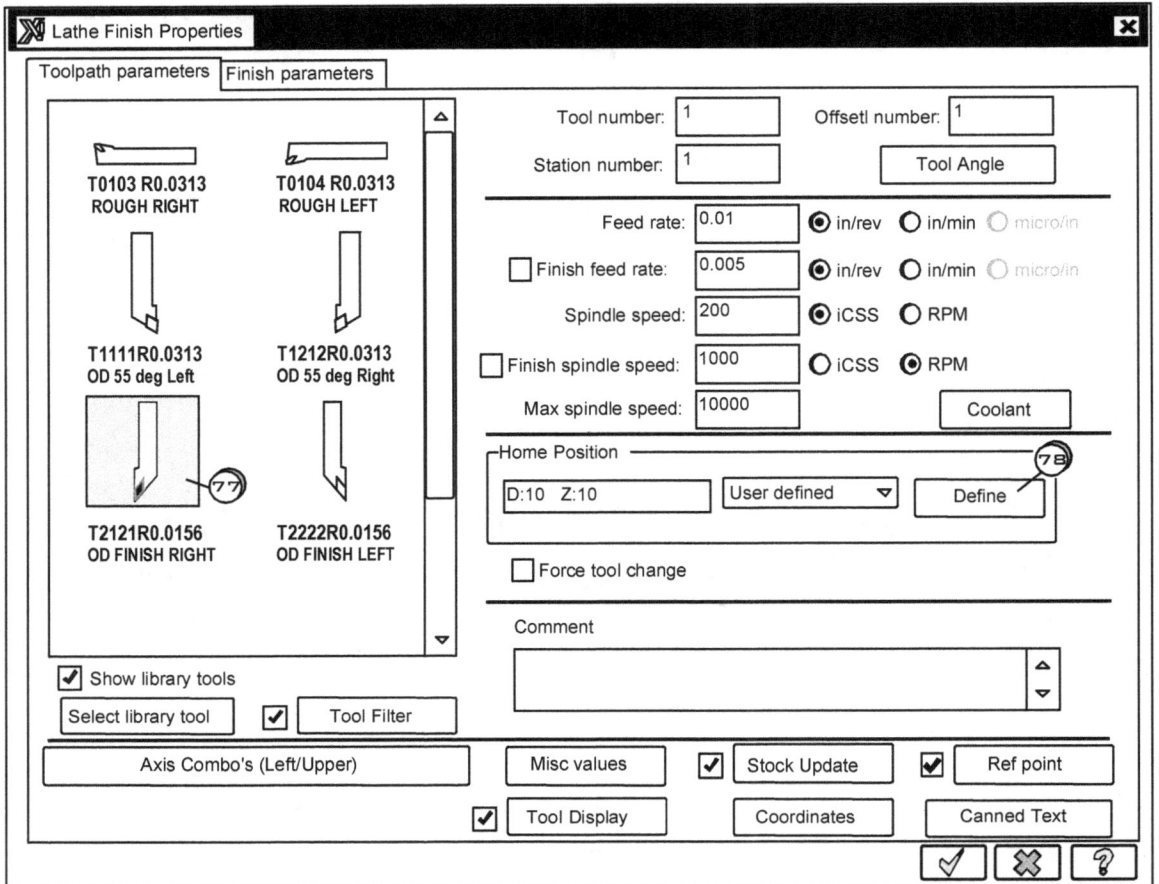

➤ Click 77 the OD FINISH RIGHT tool

➤ Click 78 the Define button

➤ Click 79 in the D box; enter 15

➤ Click 80 in the Z box; enter 8

➤ Click 81 the OK button

Home Position - User Defined
D: 15 — 79   Select
Z: 8 — 80   From Machine
81

◆ ENTER THE FINISH OD MACHINING PARAMETERS

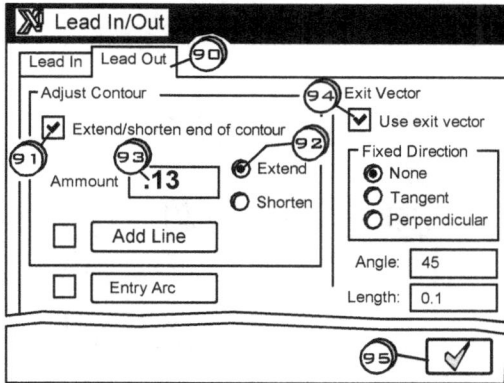

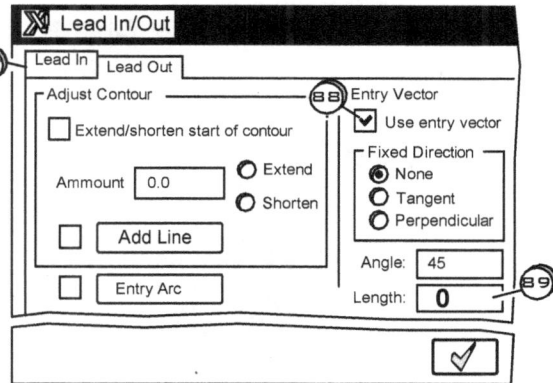

Lathe Finish Properties

Toolpath parameters | **Finish parameters** (82)

Tool Compensation
Compensation type:
(84) Wear ▽ (83)

☑ Optimize cutter comp in control

Finish stepover:
.1

Number of finish passes
1

Compensation direction:
Right ▽

Roll cutter around corners
All ▽

Stock to leave in X:
0.0

☐ Corner Break

☐ Down Cutting

Stock to leave in Z:
0.0

(85) ☑ Lead In/Out .. (86)

(96) ✓ | ✗ | ?

Lead In/Out

Lead In | **Lead Out** (90)

Adjust Contour ─── (94) Exit Vector
☑ Use exit vector

☑ Extend/shorten end of contour
(91) (93)
Ammount .13

Fixed Direction
◉ None
○ Tangent
○ Perpendicular

○ Extend (92)
○ Shorten

☐ Add Line

Angle: 45

☐ Entry Arc

Length: 0.1

(95) ✓

Lead In/Out

(87) **Lead In** | Lead Out

Adjust Contour ─── (88) Entry Vector
☑ Use entry vector

☐ Extend/shorten start of contour

Ammount 0.0

○ Extend
○ Shorten

Fixed Direction
◉ None
○ Tangent
○ Perpendicular

☐ Add Line

Angle: 45

☐ Entry Arc

Length: 0 (89)

✓

➤ Click (82) the ⌐Finish parameters tab

➤ Click (83) the Compensation type down button ▽

➤ Click (84) Wear

➤ Click (85) check on ☑ for Lead In/Out

➤ Click (86) the ⌐Lead In/Out ..⌐ button

➤ Click (87) the ⌐Lead In tab

➤ Click (88) check on ☑ for Use entry vector

➤ Click (89) in the Length box; enter 0
   lead in will be from the reference point (2.7,.1)

➤ Click (90) the ⌐Lead Out tab

➤ Click (91) check on ☑ for Extend/shorten

➤ Click (92) the Extend radio button ◉

➤ Click (93) in the Ammount: box; enter .13

➤ Click (94) check on ☑ for Use exit vector
   lead out will be a polar line at a 45° angle from the horizontal of length 0.1

➤ Click (95) the OK button ✓

➤ Click (96) the OK button ✓

♦ BACKPLOT THE TOOLPATH TO VERIFY THAT THE FINISH OD
  OPERATION IS FREE FROM ANY ERRORS

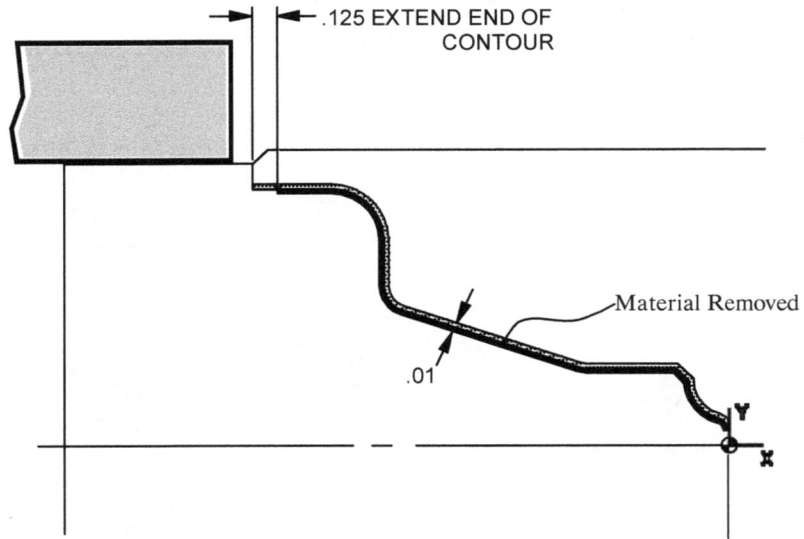

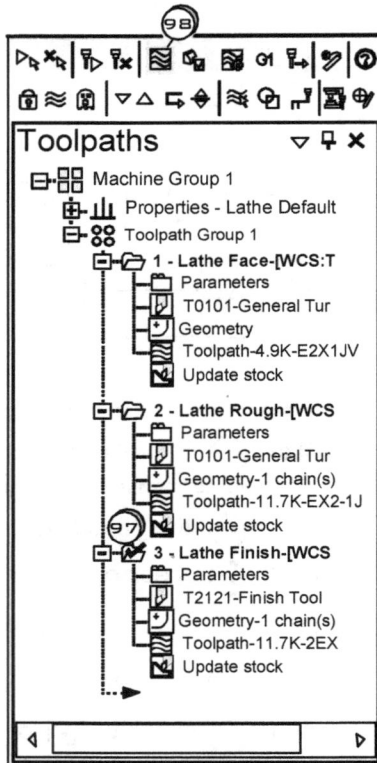

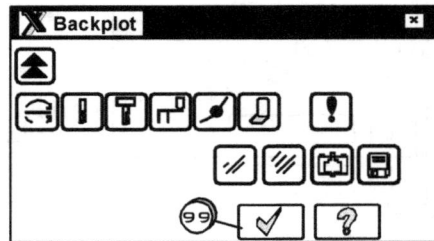

.125 EXTEND END OF
CONTOUR

Material Removed

.01

Toolpaths

Machine Group 1
  Properties - Lathe Default
  Toolpath Group 1
    1 - Lathe Face-[WCS:T
      Parameters
      T0101-General Tur
      Geometry
      Toolpath-4.9K-E2X1JV
      Update stock

    2 - Lathe Rough-[WCS
      Parameters
      T0101-General Tur
      Geometry-1 chain(s)
      Toolpath-11.7K-EX2-1J
      Update stock

    3 - Lathe Finish-[WCS
      Parameters
      T2121-Finish Tool
      Geometry-1 chain(s)
      Toolpath-11.7K-2EX
      Update stock

Backplot

➤ Click ㉗ the check *on* ⟋ 3 - Lathe Finish-[WCS

➤ Click ㉘ the Backplot button ≋

➤ Click ㉙ the OK button ☑

E) OPERATION#4- LATHE CUTOFF

- ◆ OBTAIN THE NEEDED R1/64 TNR, RIGHT CUTOFF TOOL
- ◆ SPECIFY HOME(TOOL CHANGE) POSITION FOR THE TOOL

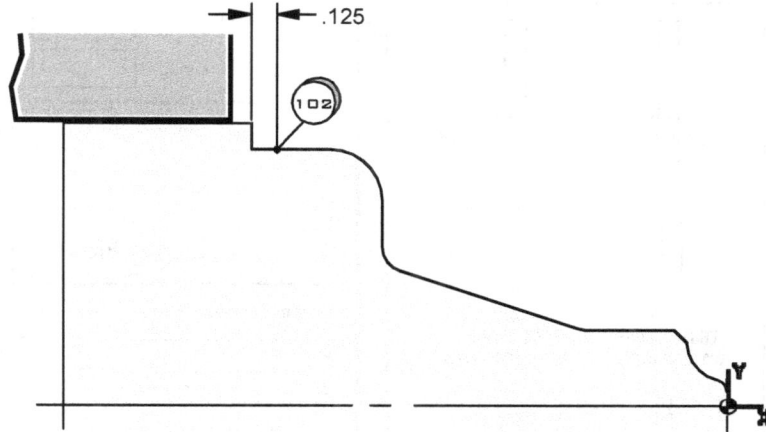

━▶ Click (100)   TOOLPATHS

━▶ Click (101)   ⬛ Cutoff

| Select cutoff boundary point |

━▶ Click (102)

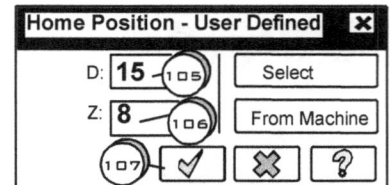

Lathe Cutoff Properties dialog

➤ Click ⟨103⟩ the W0.125OD CUTOFF RIGHT tool

➤ Click ⟨104⟩ the [Define] button

➤ Click ⟨105⟩ in the D box; enter [15]

➤ Click ⟨106⟩ in the Z box; enter [8]

➤ Click ⟨107⟩ the OK button [✓]

Home Position - User Defined

D: 15 ⟨105⟩   Select
Z: 8 ⟨106⟩   From Machine
⟨107⟩ ✓ ✗ ?

♦ ENTER THE LATHE CUTOFF MACHINING PARAMETERS

➤ Click (108) the [Cutoff parameters] tab

➤ Click (109) the Compensation type down button ▽

➤ Click (110) Wear

➤ Click (111) the Cut to Back radius radio button ◉

➤ Click (112) the OK button ✓

♦ BACKPLOT THE TOOLPATH TO VERIFY THAT THE CUTOFF
  OPERATION IS FREE FROM ANY ERRORS

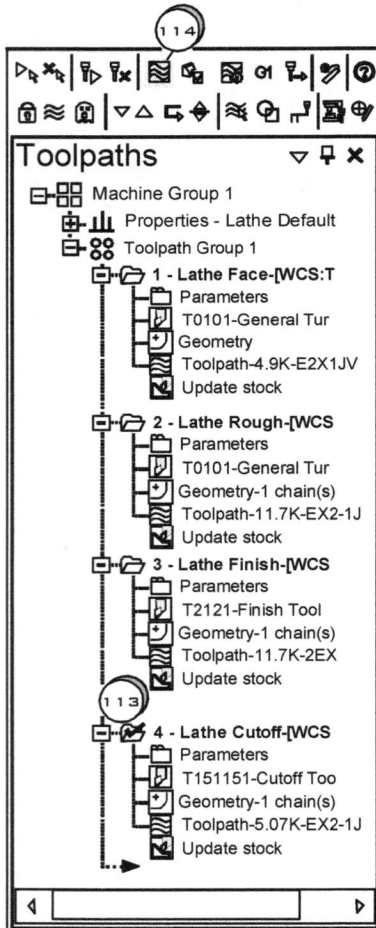

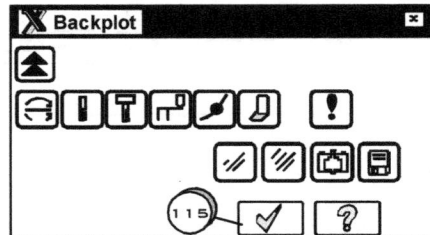

Material Removed

Toolpaths

Machine Group 1
  Properties - Lathe Default
  Toolpath Group 1
    1 - Lathe Face-[WCS:T
      Parameters
      T0101-General Tur
      Geometry
      Toolpath-4.9K-E2X1JV
      Update stock
    2 - Lathe Rough-[WCS
      Parameters
      T0101-General Tur
      Geometry-1 chain(s)
      Toolpath-11.7K-EX2-1J
      Update stock
    3 - Lathe Finish-[WCS
      Parameters
      T2121-Finish Tool
      Geometry-1 chain(s)
      Toolpath-11.7K-2EX
      Update stock
    4 - Lathe Cutoff-[WCS
      Parameters
      T151151-Cutoff Too
      Geometry-1 chain(s)
      Toolpath-5.07K-EX2-1J
      Update stock

Backplot

➤ Click (113) the check on 4 - Lathe Cotoff-[WCS

➤ Click (114) the Backplot button

➤ Click (115) the OK button

F) Verify the toolpaths for all the machining operations

Toolpaths

Machine Group 1
  Properties - Lathe Default
  Toolpath Group 1
    1 - Lathe Face-[WCS:T
      Parameters
      T0101-General Tur
      Geometry
      Toolpath-4.9K-E2X1JV
      Update stock
    2 - Lathe Rough-[WCS
      Parameters
      T0101-General Tur
      Geometry-1 chain(s)
      Toolpath-11.7K-EX2-1J
      Update stock
    3 - Lathe Finish-[WCS
      Parameters
      T2121-Finish Tool
      Geometry-1 chain(s)
      Toolpath-11.7K-2EX
      Update stock
    4 - Lathe Cutoff-[WCS
      Parameters
      T151151-Cutoff Too
      Geometry-1 chain(s)
      Toolpath-5.07K-EX2-1J
      Update stock

Click (116) select all operations

Click (117) the Verify button

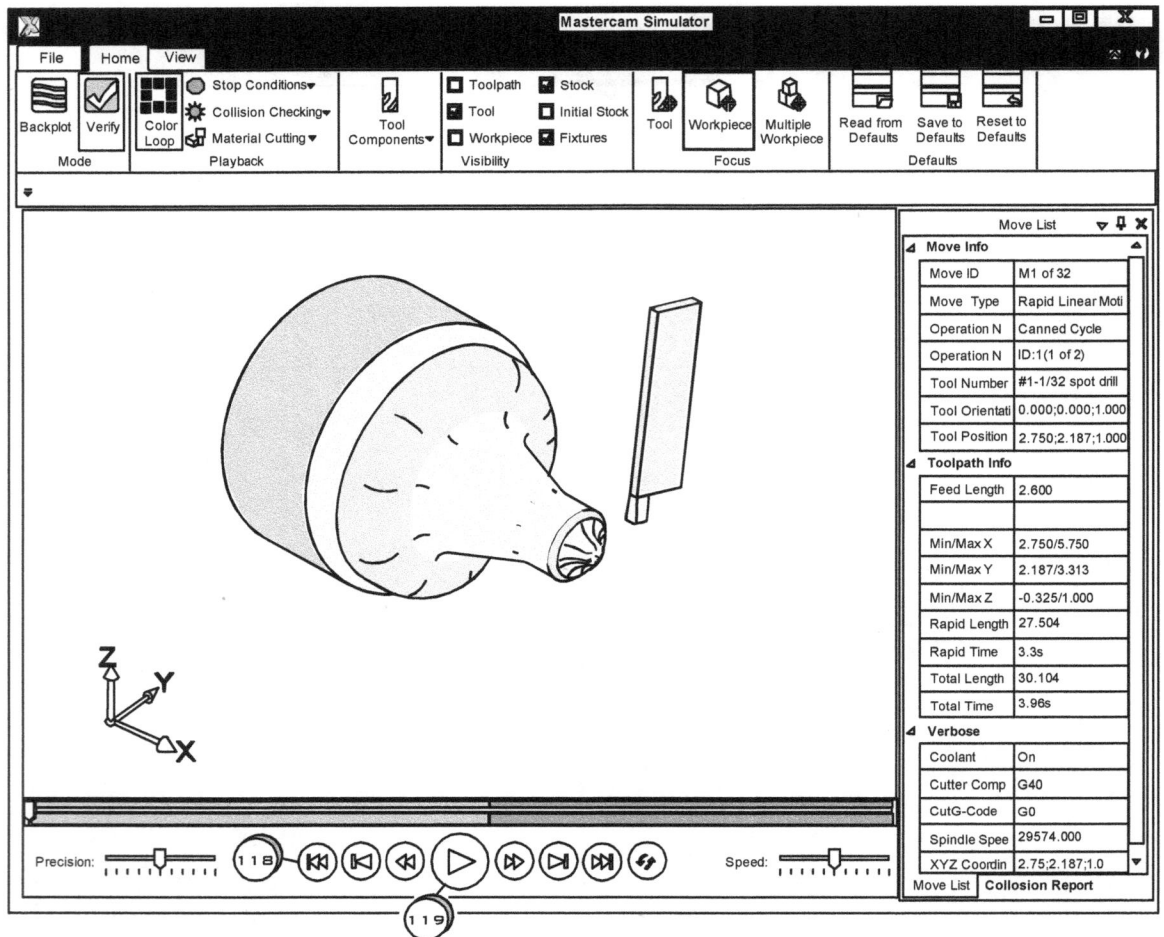

➤ Click ⓘⓘⓑ the Start (◄◄) button

➤ Click ⓘⓘⓨ the Play (▷) button

G) GENERATE THE PART PROGRAM FOR ALL THE MACHINING OPERATIONS

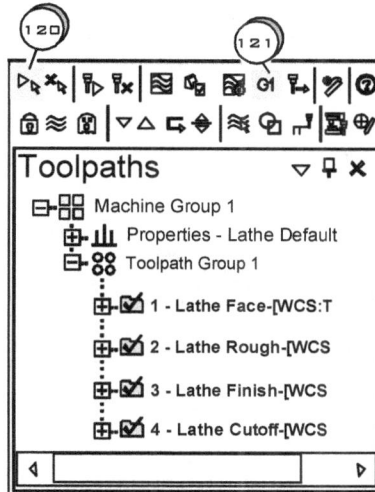

Toolpaths

- Machine Group 1
  - Properties - Lathe Default
  - Toolpath Group 1
    - 1 - Lathe Face-[WCS:T
    - 2 - Lathe Rough-[WCS
    - 3 - Lathe Finish-[WCS
    - 4 - Lathe Cutoff-[WCS

➤ Click ⑫⓪ the Select all operations button

➤ Click ⑫① the Post selected operations button G1

Post Processing

Active post   Select Post

GENERIC FANUC 3X MILL.PST

☐ Output MCX file descriptor   Properties

⑫② ☑ NC file
   ○ Overwrite   ☑ Edit
   ● Ask   NC extension
                    .NC

   ☐ Send to machine   Communications

⑫③ ☑ NCI file
   ○ Overwrite   ☐ Edit
   ● Ask   ☑ Output Tplanes
                       relative to WCS

⑫④ ✓   ✗   ?

➤ Click ⑫② the check on for ☑ NC file

➤ Click ⑫③ the check on for ☑ NCI file

➤ Click ⑫④ the OK button ✓

Click (125) the File name box; enter **EX2-1JV**

Click (126) the [Save] button

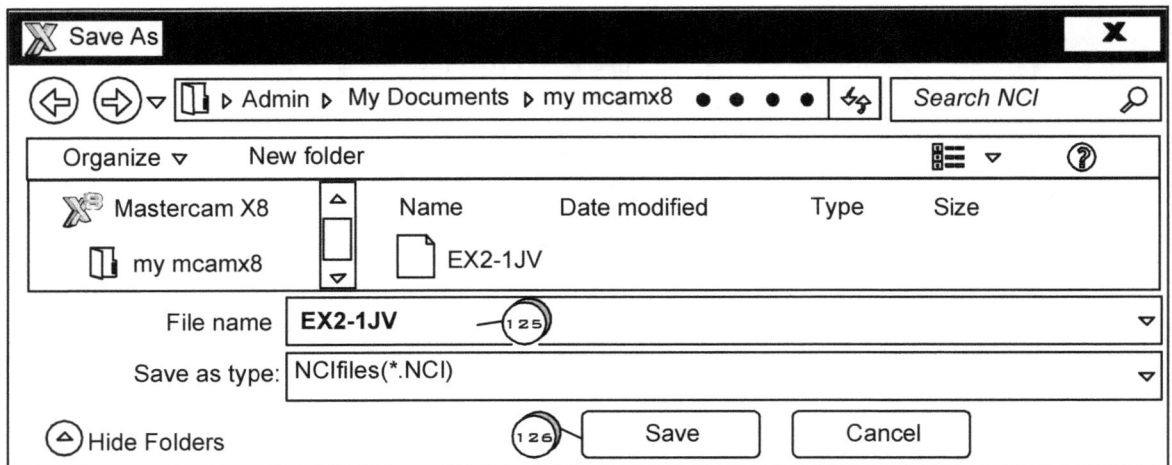

Click (127) the [Save] button

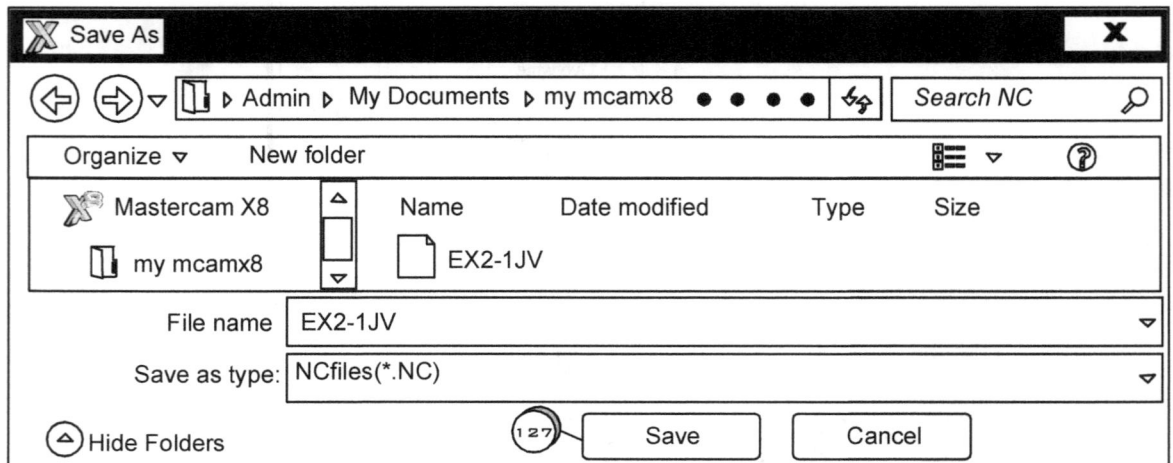

Mastercam will then *generate* the word address part program, open the Code Expert and display it. The operator can edit the program and/or send it to machine tool by selecting ⬆ Send file

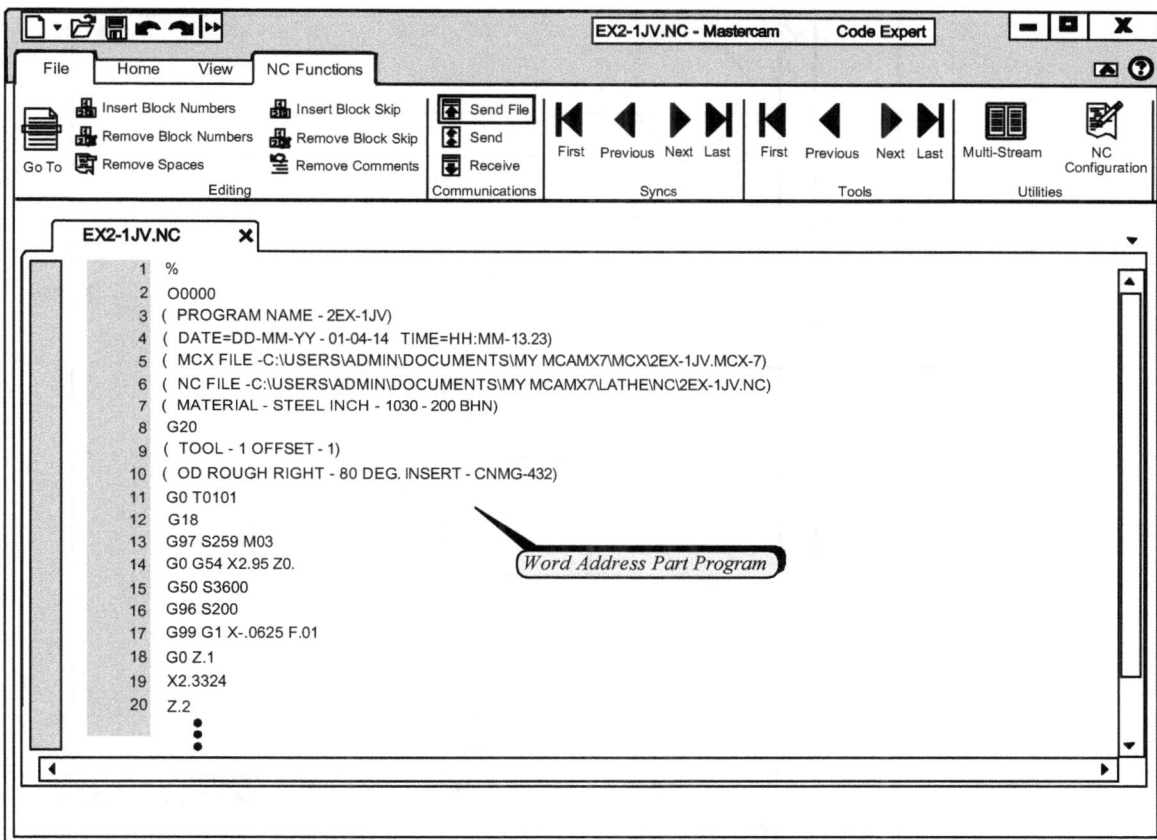

EXERCISES

3-1) Get the CAD model file EX2-2 created in Chapter2 . Using PROCESS PLAN 3P1
generate the part program to produce the part shown in Figure 3p-1.

Material: 1030 Steel

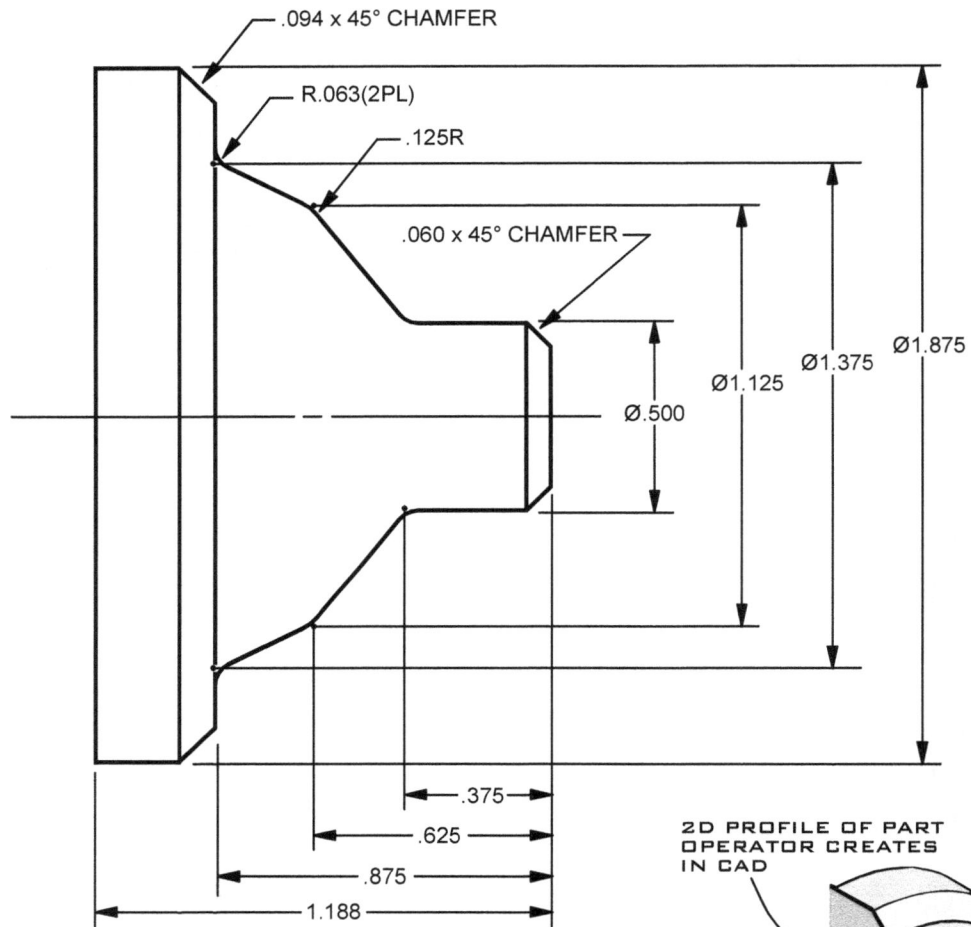

.094 x 45° CHAMFER

R.063(2PL)

.125R

.060 x 45° CHAMFER

Ø.500

Ø1.125

Ø1.375

Ø1.875

.375

.625

.875

1.188

2D PROFILE OF PART
OPERATOR CREATES
IN CAD

Figure 3p-1

CHUCK

BAR STOCK

Part Origin

Ø.0625   OD MARGIN

Ø1.875   OD

Ø.0625

GRIP LENGTH   .75

1   1.188   .10   RIGHT MARGIN

LEFT MARGIN   LENGTH

PROCESS PLAN 3P-1

No.	Operation	Tooling
1	FACE END Material Removed .10	1/32 TNR, RH OD TURNING TOOL
2	ROUGH TURN OD; LEAVE .01 IN X AND Z FOR FINISHING ☑ Lead In/Out Lead In | Lead Out ┌ Adjust Contour ☑ Extend/shorten end of contour Ammount **.125** ◉ Extend ○ Shorten ☐ Add Line .125 EXTEND END OF CONTOUR CHAIN 1 END PT ② Material Removed .01 LEFT FOR FINISH CUT ① CHAIN 1 START PT	R1/32

PROCESS PLAN 3P-1(*continued*)

No.	Operation	Tooling
3	**Finish OD Contour** .125 EXTEND END OF CONTOUR Material Removed CHAIN 1 END PT ② .01 CHAIN 1 START PT ①	1/64 TNR, RH OD FINISHING TOOL R1/64
4	**Lathe Cutoff** Material Removed	.016 TNR, RH OD CUTOFF TOOL 1.5 A DETAIL A R.016

3-2) Get the CAD model file EX2-3 created in Chapter2. Using PROCESS PLAN 3P-2 generate the part program to produce the part shown in Figure 3p-2.

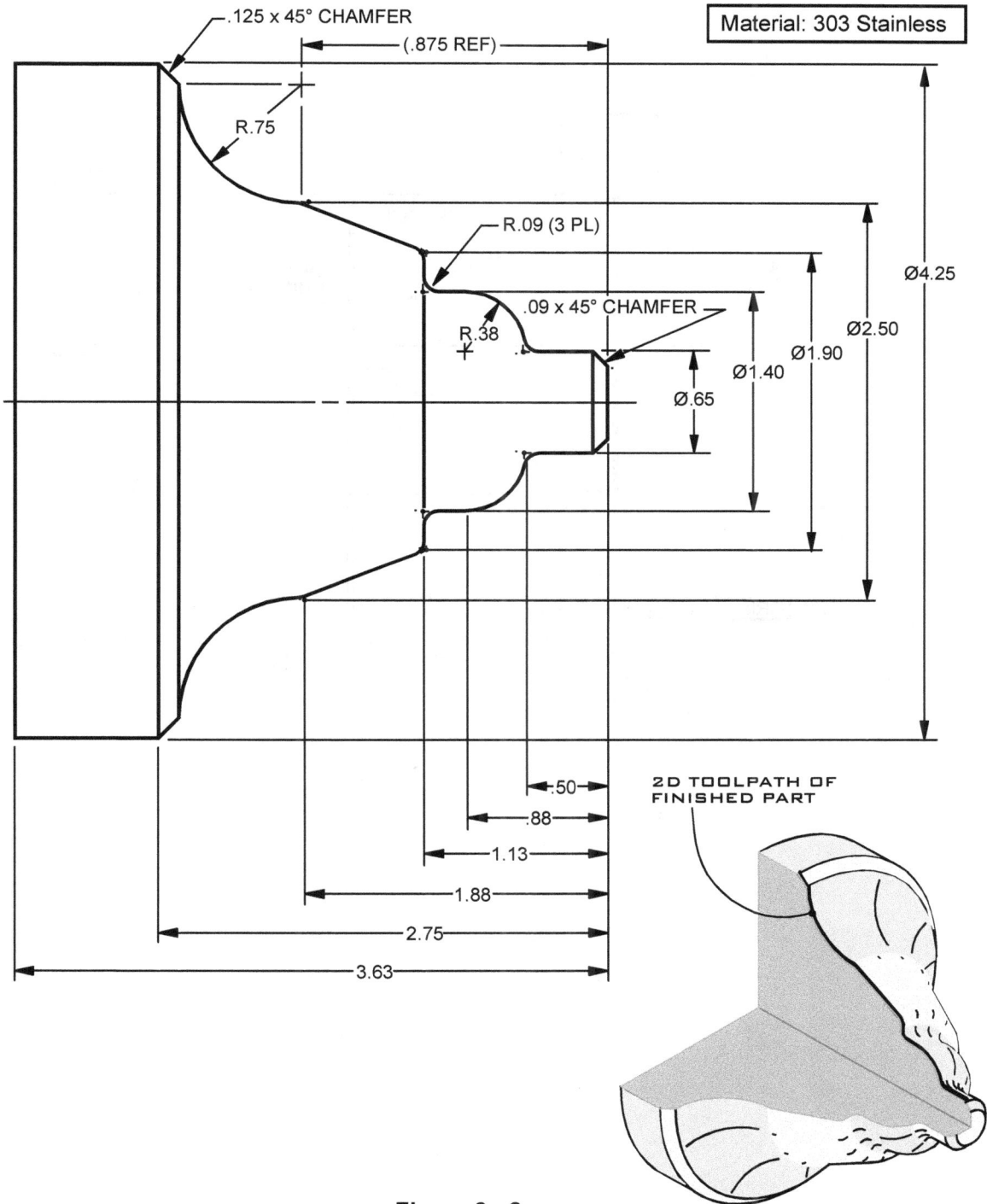

.125 x 45° CHAMFER
(.875 REF)
Material: 303 Stainless
R.75
R.09 (3 PL)
.09 x 45° CHAMFER
R.38
Ø4.25
Ø2.50
Ø1.90
Ø1.40
Ø.65
Ø1.40
.50
.88
1.13
1.88
2.75
3.63
2D TOOLPATH OF FINISHED PART

Figure 3p-2

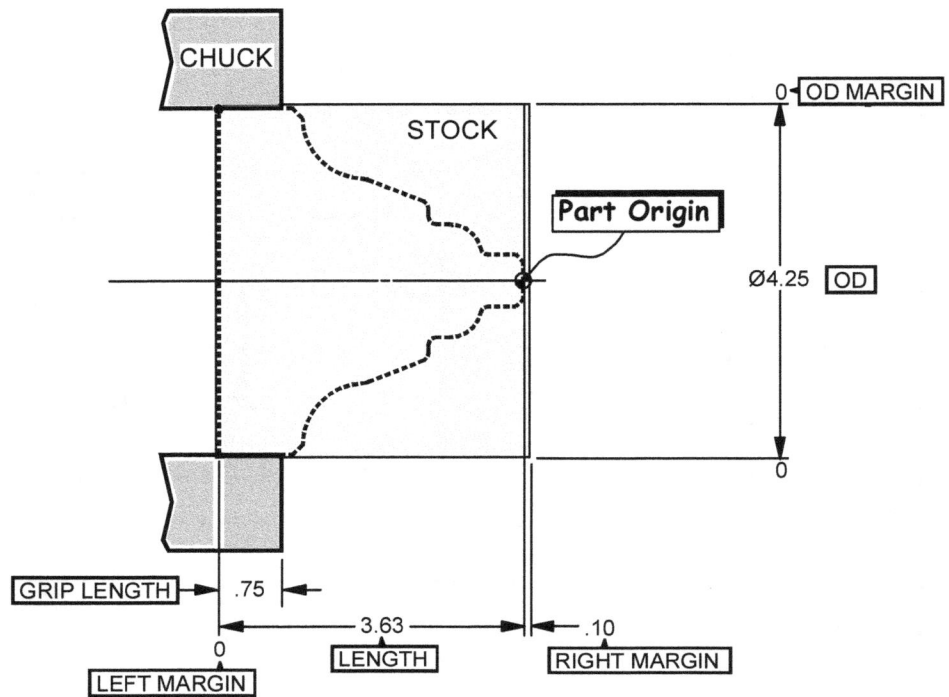

PROCESS PLAN 3P-2

No.	Operation	Tooling
1	FACE END	1/32 TNR, RH OD TURNING TOOL
3	ROUGH TURN OD; LEAVE .01 IN X AND Z FOR FINISHING.	1/32 TNR, RH OD TURNING TOOL

STOCK
Material Removed
.10

R1/32

☑ Lead In/Out

Lead In | Lead Out
Adjust Contour
☑ Extend/shorten end of contour
Ammount .1   ● Extend   ○ Shorten
☑ Add Line

New Contour Line ✕
Length .1
Angle 90
Define
✓ ✖ ?

.10 ADD LINE
.10 EXTEND END OF CONTOUR
Material Removed
CHAIN 1 END PT ②
.01
① CHAIN 1 START PT

R1/32

PROCESS PLAN 3P-2(*continued*)

No.	Operation	Tooling
3	Finish OD Contour	1/64 TNR, RH OD FINISHING TOOL

3-3) Get the CAD model file EX2-4 created in Chapter 2. Using PROCESS PLAN 3P-3 generate the part program to produce the part shown in Figure 3p-3.

Material: 303 Stainless

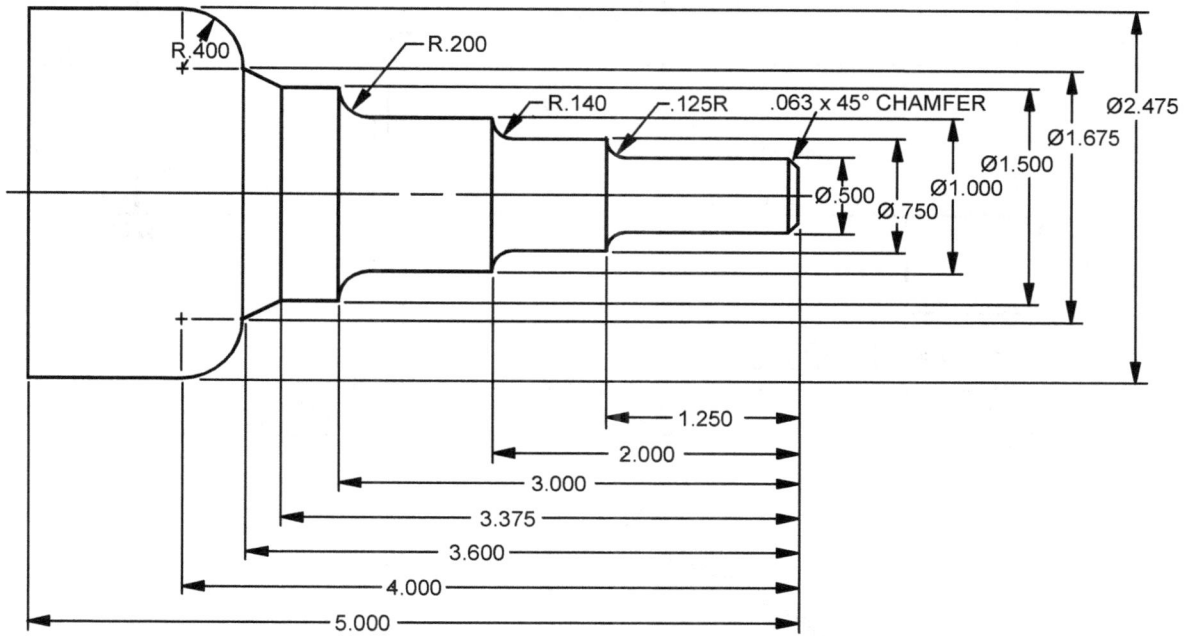

R.400
R.200
R.140
.125R
.063 x 45° CHAMFER
Ø2.475
Ø1.675
Ø1.500
Ø1.000
Ø.500
Ø.750

1.250
2.000
3.000
3.375
3.600
4.000
5.000

2D PROFILE OF PART
OPERATOR CREATES
IN CAD

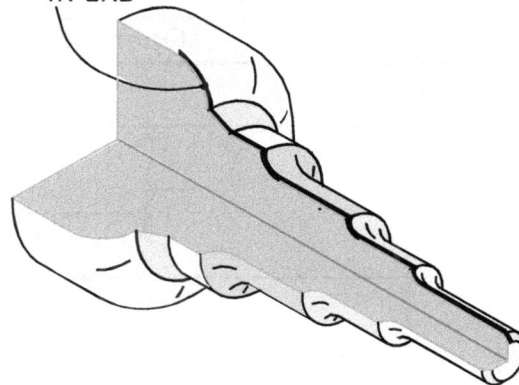

Figure 3p-3

CHUCK

Ø.0 OD MARGIN

STOCK

Part Origin

Ø2.48 OD

Ø.0

.875 GRIP LENGTH

5
LENGTH

.10
RIGHT MARGIN

0
LEFT MARGIN

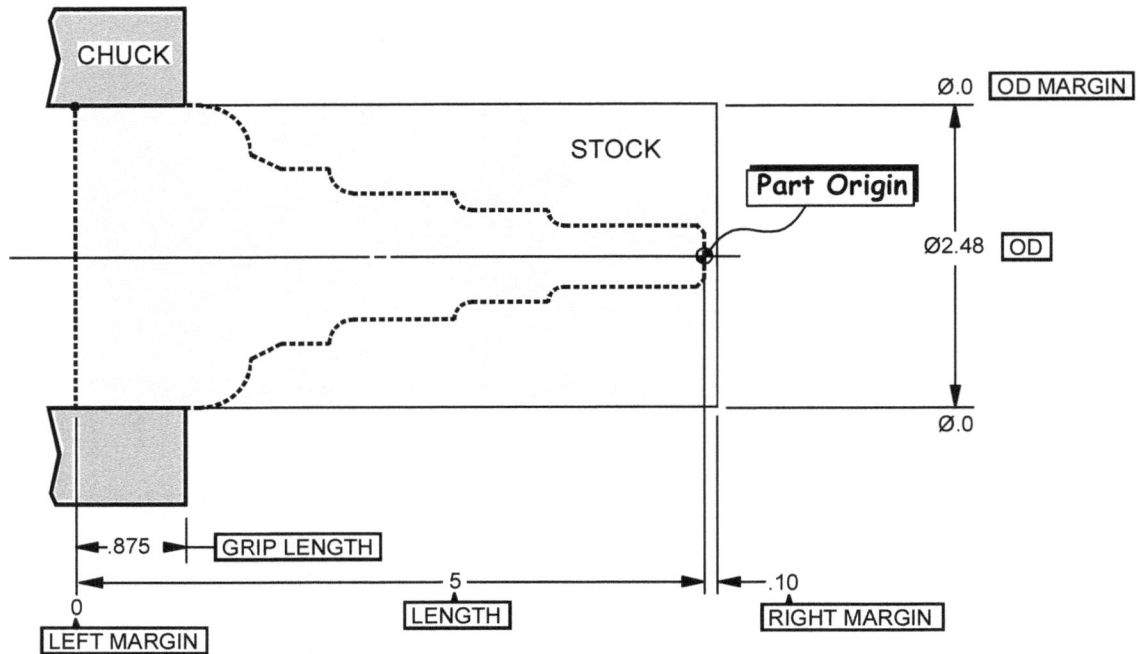

PROCESS PLAN 3P-3

No.	Operation	Tooling
1	FACE END Material Removed STOCK .10	1/32 TNR, RH OD TURNING TOOL R1/32

PROCESS PLAN 3P-3(*continued*)

No.	Operation	Tooling
2	ROUGH TURN OD AND SEMIFINISH OD CONTOUR; LEAVE .002 FOR FINISH CUT.	1/32 TNR, RH OD TURNING TOOL

☑ Lead In/Out

| Lead In | Lead Out |

┌─ Adjust Contour ──────────
☑ Extend/shorten end of contour
Ammount [0.0]   ● Extend
                    ○ Shorten
☑ [Add Line]

New Contour Line ☒

Length [.1]
Angle   [90]
        [Define]
[✓] [✗] [?]

Stock to leave in X: [0.01]
Stock to leave in Z: [0.00]

R1/32

☑ Semi Finish

Semi Finish Parameters ☒

Number of passes [1]

Stepover [.005]

Stock to leave in X: [.002]

Stock to leave in Z: [0]

[✓] [✗] [?]

.10 ADD LINE

② CHAIN 1 END PT

Material Removed

INCLUDE .10 LINE ELEMENT IN CHAIN

.002 LEFT AFTER SEMI-FINISH CUT

① CHAIN 1 START PT

PROCESS PLAN 3P-3(*continued*)

No.	Operation	Tooling
3	Finish OD Contour 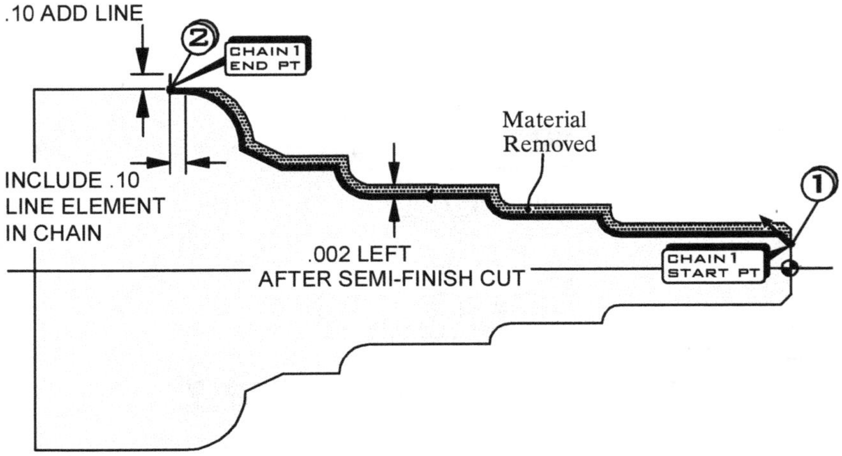	1/64 TNR, RH OD FINISHING TOOL 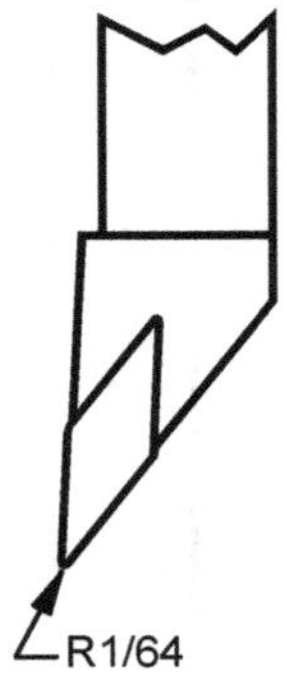

Diagram labels:

- .10 ADD LINE
- ② CHAIN 1 END PT
- INCLUDE .10 LINE ELEMENT IN CHAIN
- .002 LEFT AFTER SEMI-FINISH CUT
- Material Removed
- ① CHAIN 1 START PT
- R1/64

3-4) Get the CAD model file EX2-5 created in Chapter 2 . Using PROCESS PLAN 3P-4 generate the part program to produce the part shown in Figure 3p-4.

Material: 420 Stainless

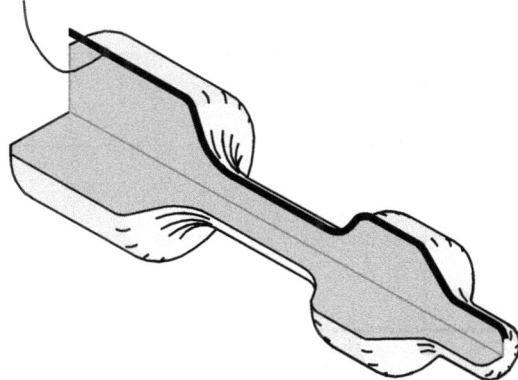

Figure 3p-4

CHUCK

BAR STOCK

Part Origin

OD MARGIN
Ø.125

Ø1.00 OD

Ø.125

.75
GRIP LENGTH

1.50
LEFT MARGIN

3.20
LENGTH

.10 RIGHT MARGIN

PROCESS PLAN 3P-4

No.	Operation	Tooling
1	FACE END Material Removed .10	1/32 TNR, RH OD TURNING TOOL R1/32

PROCESS PLAN 3P-4(*continued*)

No.	Operation	Tooling
2	ROUGH TURN OD; LEAVE .01 IN X AND Z FOR FINISHING.	1/64 TNR, 17.5° V OD TURNING TOOL

Operation detail (No. 2):

Cutting Method
- ○ One-way
- ◉ Zig-zag

☑ Lead In/Out

Stock to leave in X: **0.01**

Stock to leave in Z: **0.00**

Lead In | Lead Out

Adjust Contour
- ☐ Extend/shorten end of contour

Ammount 0.0
- ◉ Extend
- ○ Shorten

[Add Line]

Entry Vector
- ☑ Use exit vector

Fixed Direction
- ◉ None
- ○ Tangent
- ○ Perpendicular

Angle: -135

Length: 0.1   Resolution(deg) 45

① [Inetlliset...]

Lead In | **Lead Out**

Adjust Contour
- ☑ Extend/shorten end of contour

Ammount **.13**
- ◉ Extend
- ○ Shorten

[Add Line]

Exit Vector
- ☑ Use exit vector

Fixed Direction
- ◉ None
- ○ Tangent
- ○ Perpendicular

Angle: 45.0

Length: 0.1   Resolution(deg) 45

② [Inetlliset...]

[Plunge parameters]

Material Removed

CHAIN 1 END PT ④

.13 EXTEND END OF CONTOUR

.01 LEFT FOR FINISH

CHAIN 1 START PT ③

Tooling detail:

A   DETAIL A

Ø1/4

R1/64

35°DIAMOND

PROCESS PLAN 3P-4(*continued*)

No.	Operation	Tooling
3	Finish OD Contour ☑ Lead In/Out Lead In │ Lead Out **Adjust Contour** ☑ Extend/shorten end of contour Ammount [.13] ● Extend ○ Shorten [Add Line] **Exit Vector** ☑ Use exit vector **Fixed Direction** ● None ○ Tangent ○ Perpendicular Angle: 45.0 Length: 0.1 Resolution(deg) [45] ③ [Inetlliset...] [Plunge parameters] CHAIN 1 END PT ② .13 EXTEND END OF CONTOUR Material Removed .01 ① CHAIN 1 START PT	1/64 TNR, 17.5° V OD TURNING TOOL

PROCESS PLAN 3P-4(*continued*)

No.	Operation	Tooling
4	**LATHE CUTOFF**	.016 TNR, RH OD CUTOFF TOOL

Cutoff: Select boundary point:

➤ Click ① the boundary point

Corner Geometry
- ○ None
- ○ Radius `0.003`
- ◉ Chamfer [Parameters..] →
- ☑ [Clearance Cut..]

Cutoff Chamfer ✕

Width/Height
- ○ Width `.0625`
- ◉ Height **.0625**
- [Select chamfer line..]

Angle	**45**
Top radius	`0.003`
Bottom radius	`0.003`

[✓] [✗] [?]

Clearance Cut ✕

Entry ammount	`.1`	☐ From stock
X increment	`.01`	
Z increment	`.01`	

☐ [Peck]

[✓] [✗] [?]

Clearance Cut

Material Removed

① .101

.01

Cutoff Chamfer

Material Removed

Tooling:

1.5

A —

DETAIL A

— R.016
— .125

3-5) Get the CAD model file EX2-6 created in Chapter 2. Using PROCESS PLAN 3P-5 generate the part program to produce the part shown in Figure 3p-5.

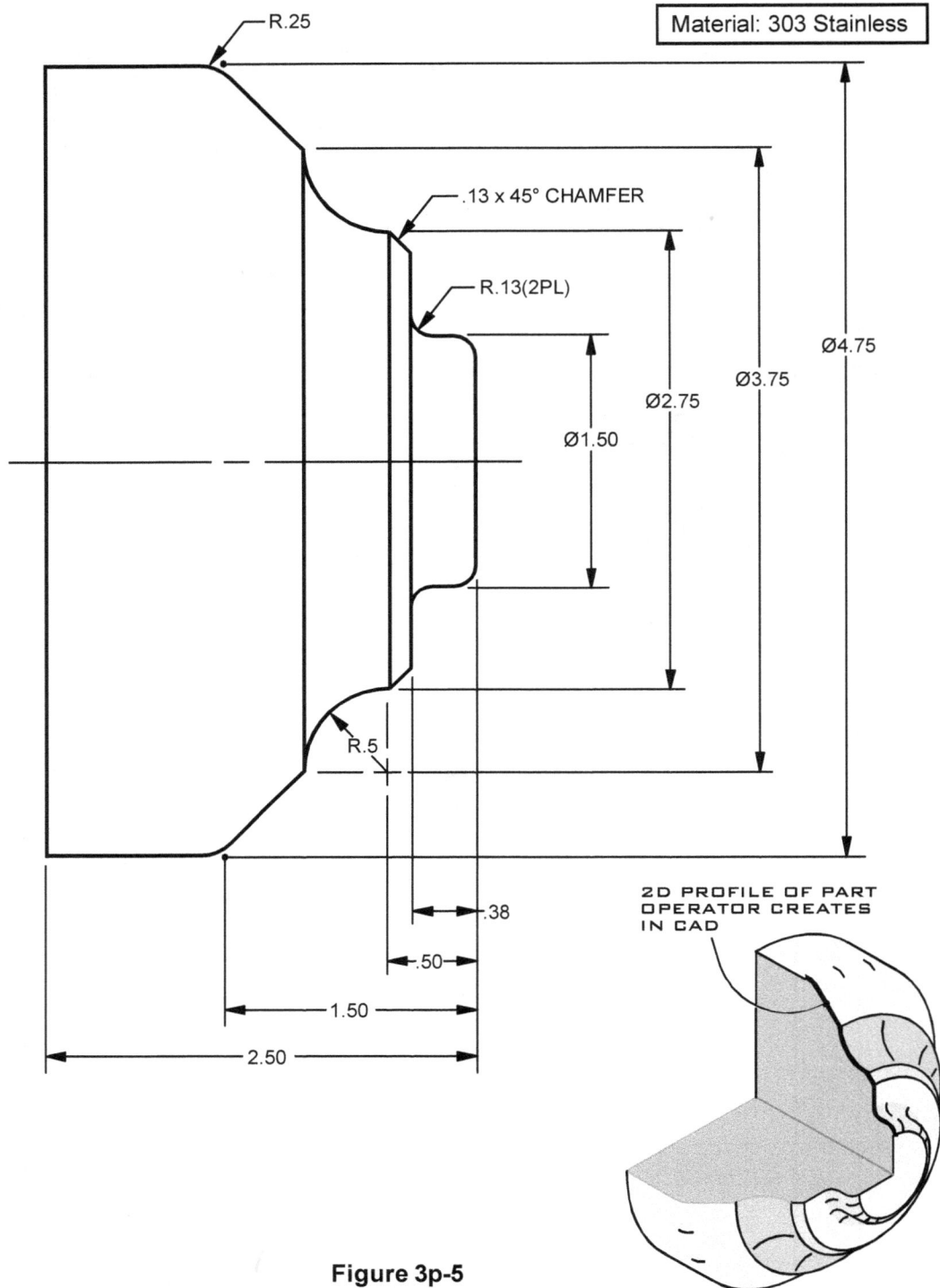

Material: 303 Stainless

R.25

.13 x 45° CHAMFER

R.13(2PL)

Ø4.75

Ø3.75

Ø2.75

Ø1.50

R.5

.38

.50

1.50

2.50

2D PROFILE OF PART OPERATOR CREATES IN CAD

Figure 3p-5

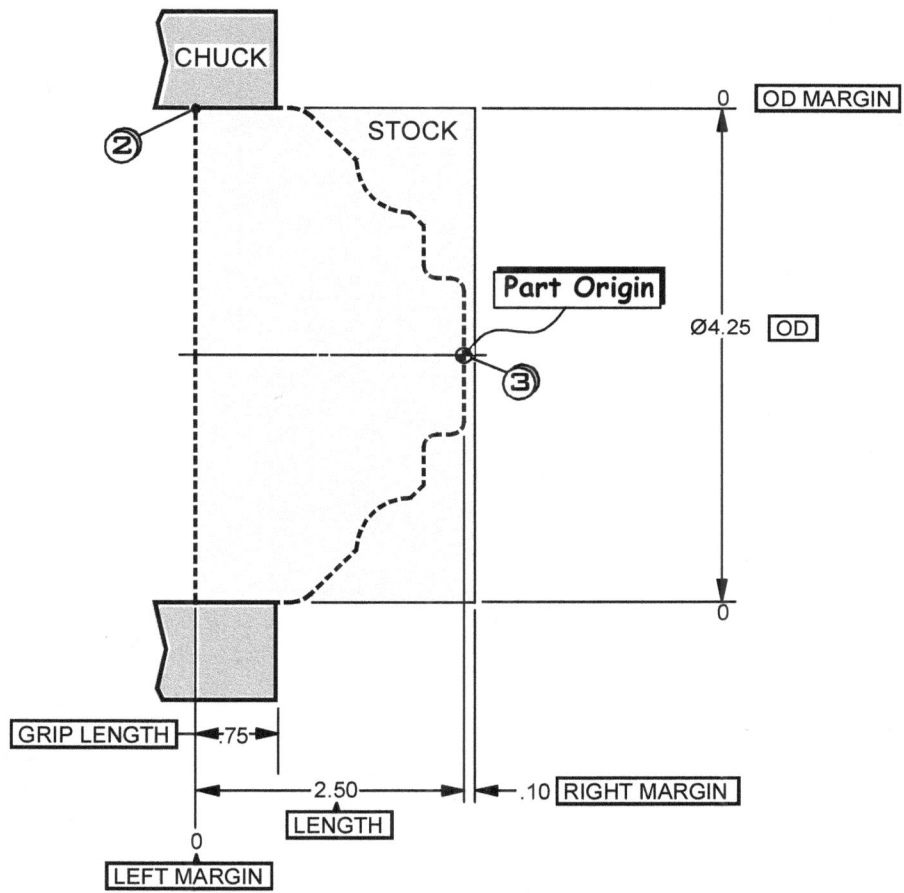

PROCESS PLAN 3P-5

No.	Operation	Tooling
1	FACE END STOCK — Material Removed ←— .10	1/32 TNR, RH OD TURNING TOOL R1/32
2	ROUGH TURN FACE; LEAVE .01 IN X AND Z FOR FINISHING. Rough Direction/Angle Angle.. 0.0 ☑ Lead In/Out Lead In | Lead Out Adjust Contour ☐ Extend/shorten end of contour Ammount 0.0 ◉ Extend ○ Shorten ☐ Add Line Exit Vector ☑ Use exit vector Fixed Direction ◉ None ○ Tangent ○ Perpendicular Angle: 45.0 Length: 0.1 CHAIN 1 START PT ① Material Removed .01 LEFT FOR FINISH CUT ① CHAIN 1 END PT	1/32 TNR, RH OD TURNING TOOL R1/32

PROCESS PLAN 3P-5(*continued*)

No.	Operation	Tooling
3	Finish OD Contour	1/64 TNR, RH OD FINISHING TOOL

3-6) Get the CAD model file EX2-7 created in Chapter 2 . Using PROCESS PLAN 3P-7 generate the part program to produce the part shown in Figure3-p6.

Material: Cast Iron Austensitic-160BHN

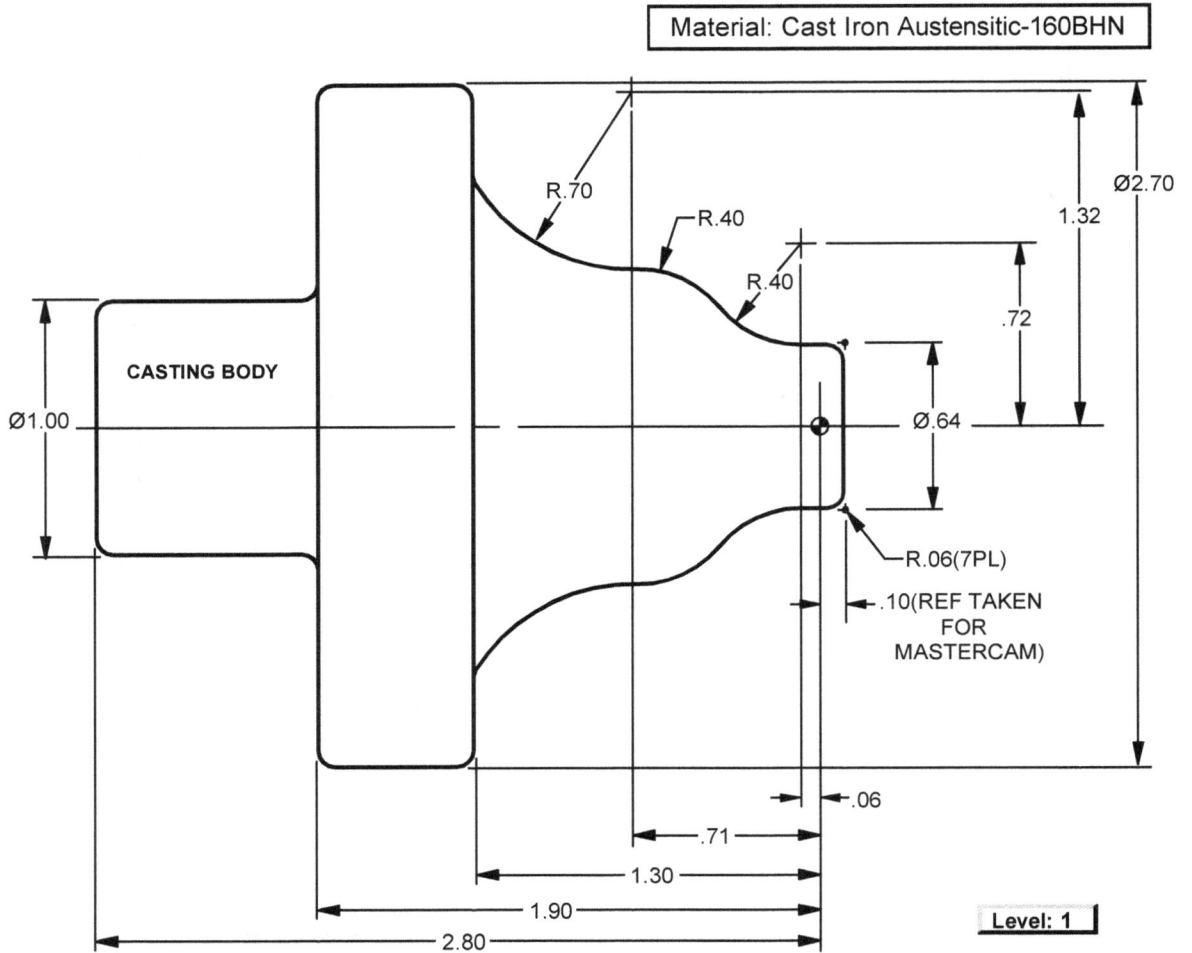

R.70

R.40

R.40

CASTING BODY

Ø1.00

Ø2.70

1.32

.72

Ø.64

R.06(7PL)

.10(REF TAKEN FOR MASTERCAM)

.06

.71

1.30

1.90

2.80

Level: 1

CASTING BODY

2D PROFILE OF CASTING OPERATOR CREATES IN CAD

Figure 3-p6

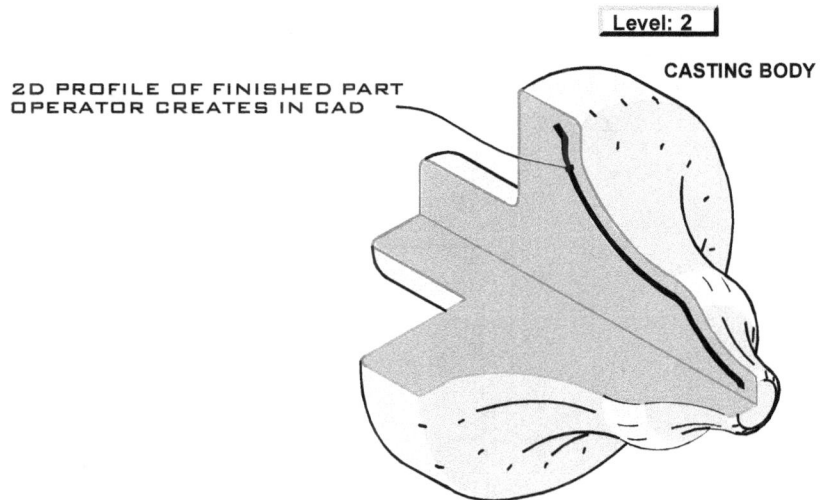

CASTING OUTLINE

.125 x 45° CHAMFER

R.80

R.30

R.50

R.06

1.32

.72

FINISHED PART BODY

Ø2.50

Ø.44

.10(REF TAKEN
FOR
MASTERCAM)

.06

.71

1.42

1.55

Level: 2

CASTING BODY

2D PROFILE OF FINISHED PART
OPERATOR CREATES IN CAD

Figure 3p-7

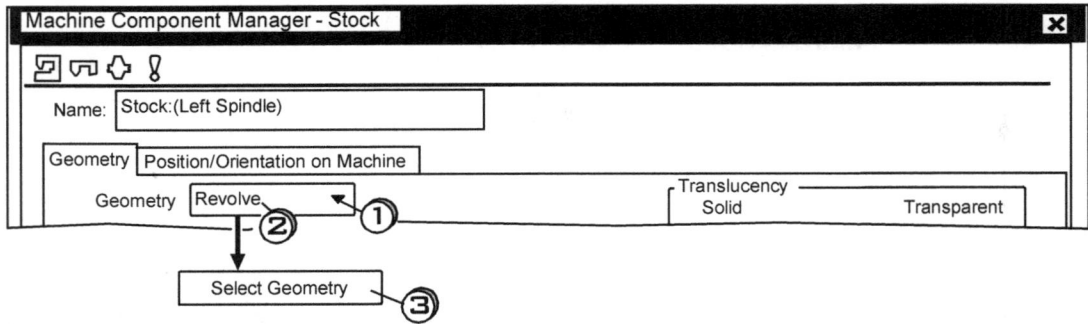

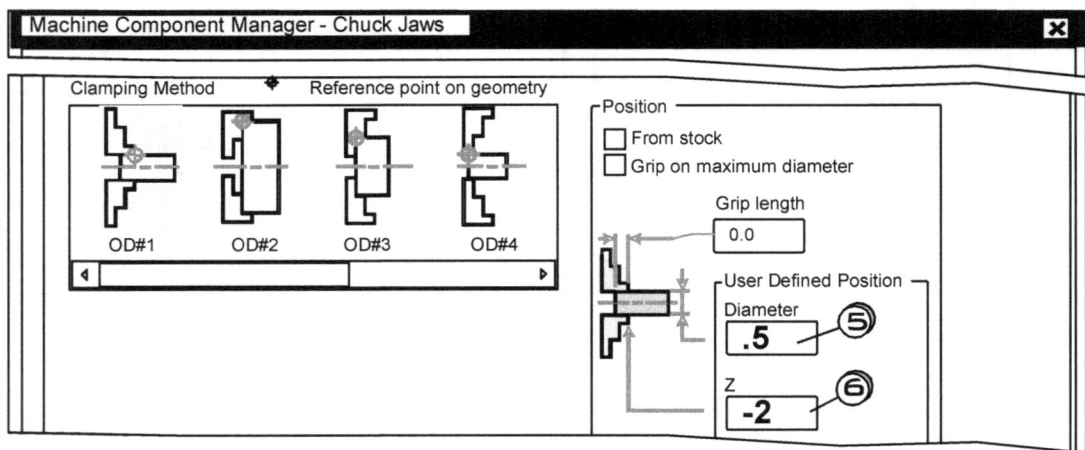

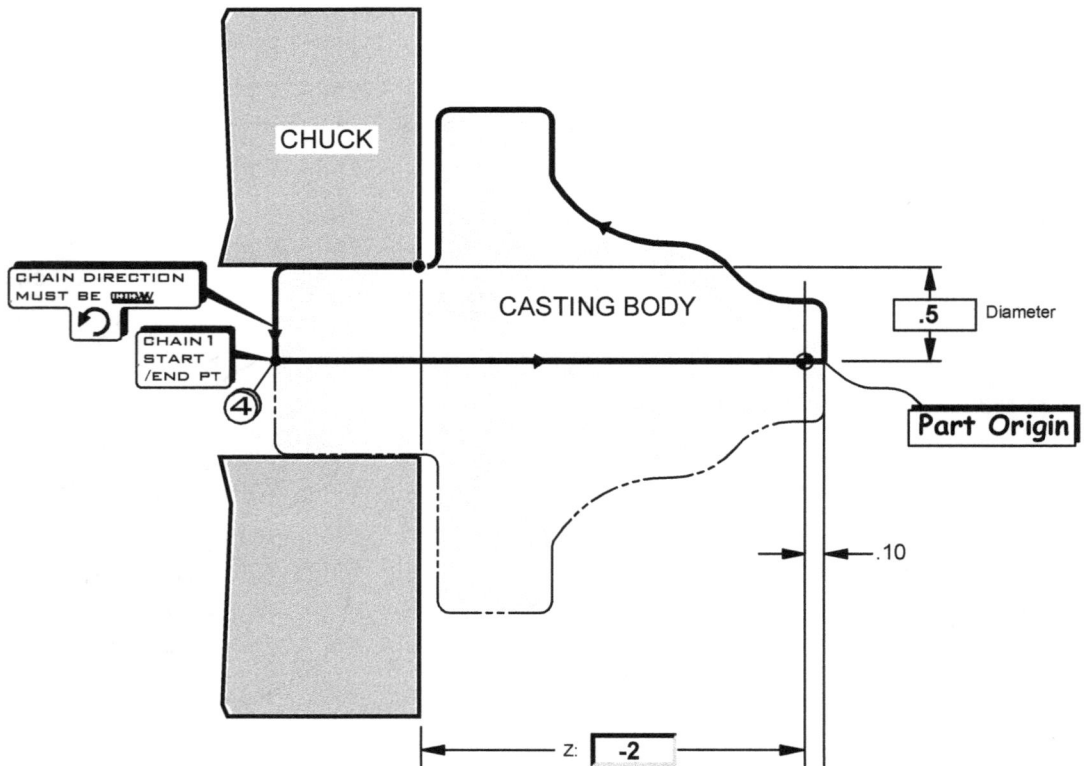

PROCESS PLAN 3P-6(*continued*)

No.	Operation	Tooling
1	**FACE END** FINISHED PART OUTLINE CASTING BODY Material Removed .10	1/32 TNR, RH OD TURNING TOOL R1/32
2	**ROUGH TURN OD AND SEMI-FINISH OD CONTOUR; LEAVE .003 FOR FINISH CUT.** Stock to leave in X: **0.01** Stock to leave in Z: **0.00** ☑ Lead In/Out **Lead In** — Lead Out Adjust Contour ☐ Extend/shorten end of contour Ammount 0.0 ⦿ Extend ○ Shorten ☐ Add Line Exit Vector ☑ Use entry vector Fixed Direction ○ None ○ Tangent ⦿ Perpendicular Angle: 180 Length: 0.1 Lead In — **Lead Out** Adjust Contour ☐ Extend/shorten end of contour Ammount 0.0 ⦿ Extend ○ Shorten ☐ Add Line Exit Vector ☑ Use exit vector Fixed Direction ○ None ○ Tangent ⦿ Perpendicular Angle: 180 Length: 0.1	1/64 TNR, RH OD FINISHING TOOL R1/64

PROCESS PLAN 3P-6(*continued*)

No.	Operation	Tooling

☑ Semi Finish → **Semi Finish Parameters** ✖

Number of passes

1

Stepover

.005

Stock to leave in X:

.003

Stock to leave in Z:

0

✓ ✖ ?

Stock Recognition

Use stock for outer boundary ▼

Adjust Stock →

End to Adjust

☐ ○
✚ ◉

🔄 📋 ◀ ◀ ◀

☐ Auto hide

✓ ✖ ?

Rough / Semi-Finish Option-1

◀

CHAIN 1 END PT

② Material Removed

.003 LEFT
AFTER SEMI-FINISH CUT

① CHAIN 1 START PT

PROCESS PLAN 3P-6(*continued*)

No.	Operation	Tooling
	Rough / Semi-Finish Option-2	
	CHAIN 1 END PT ② Material Removed ① CHAIN 1 START PT .003 LEFT AFTER SEMI-FINISH CUT	
	Rough / Semi-Finish Option-3	
	CHAIN 1 END PT ② Material Removed ① CHAIN 1 START PT .003 LEFT AFTER SEMI-FINISH CUT	

PROCESS PLAN 3P-6(*continued*)

No.	Operation	Tooling
3	**Finish OD Contour** ☑ Extend contour to stock [Adjust Contour Ends] ☑ Add lead-out line to contour End to Adjust ☐ Auto hide **Finish Option-1** CHAIN 1 END PT ② Material Removed .003 ① CHAIN 1 START PT	1/64 TNR, RH OD FINISHING TOOL R1/64

PROCESS PLAN 3P-6(*continued*)

No.	Operation	Tooling

Finish Option-2

CHAIN1 END PT ②

Material Removed

.003

① CHAIN1 START PT

Finish Option-3

CHAIN1 END PT ②

Material Removed

.003

① CHAIN1 START PT

CHAPTER - 4

GROOVING AND THREADING OPERATIONS

4-1 Chapter Objectives

After completing this chapter you will be able to:

1. Evoke the groove command
2. Specify the grooving options
3. Specify the grooving tool.
4. Specify the rough groove parameters.
5. Evoke the thread command.
6. Obtain the threading tool and insert.
7. Specify the thread shape parameters.
8. Specify the thread cut parameters.

4-2 An Example of Grooving and Threading

EXAMPLE 4-1

Open the file EX4-1 in the folder ⬜CHAPTER 4

Using PROCESS PLAN 4-1 direct Mastercam to execute grooving and threading operations on the part shown in Figure 4-1.

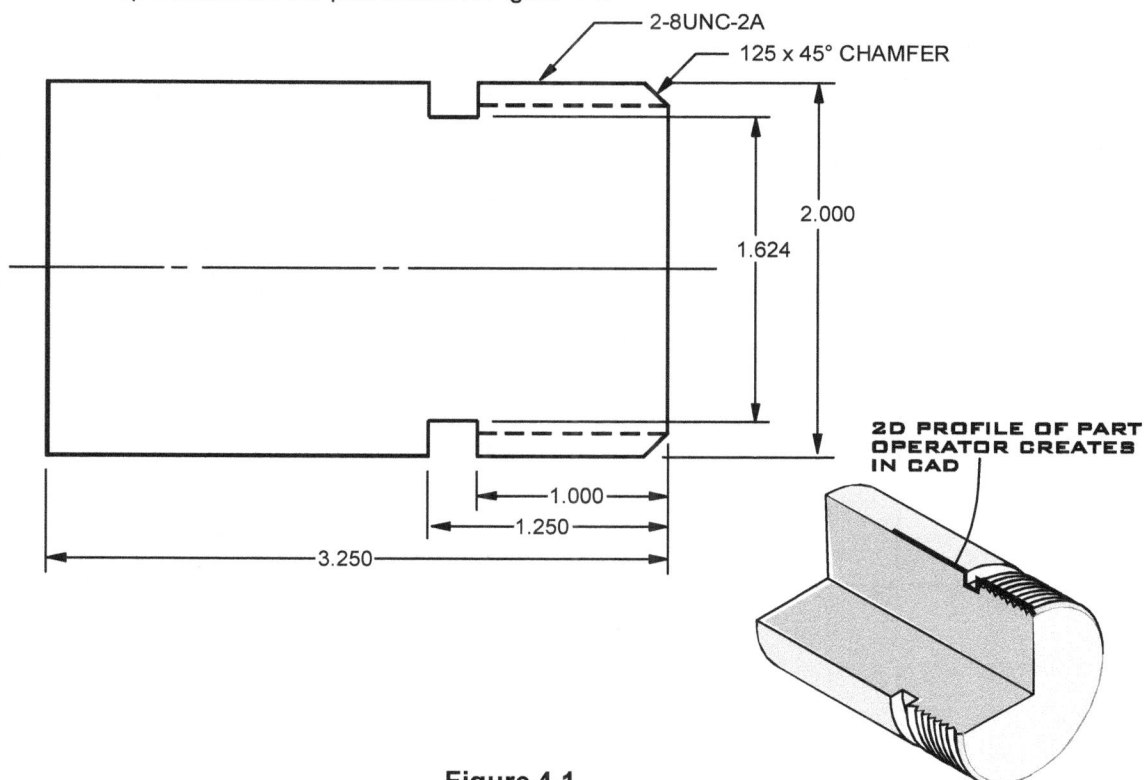

Figure 4-1

PROCESS PLAN 4-1

No.	Operation	Tooling
1	ROUGH OD GROOVE IN A SINGLE PASS	R.01, W .25 OD GROOVING RIGHT
2	CUT 2-8UNC OD THREAD	60°V, RH, OD THREADING TOOL

DETAIL A

.195

.117

1

R.018

LT-16ER-8UN

A) OPERATION#1- ROUGH GROOVE THE OD

◆ OBTAIN THE NEEDED R1/32 OD ROUGH RIGHT TOOL

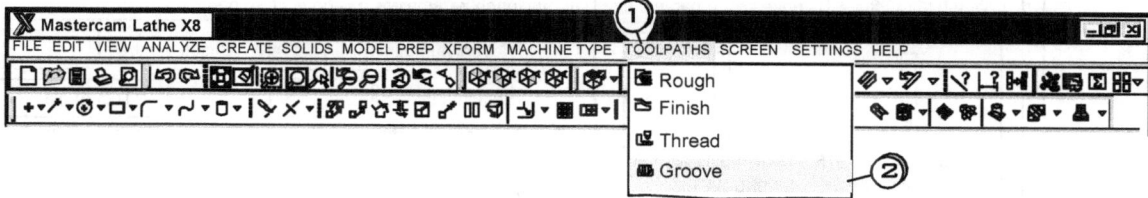

Click ① TOOLPATHS] Click ② 🔳 Groove

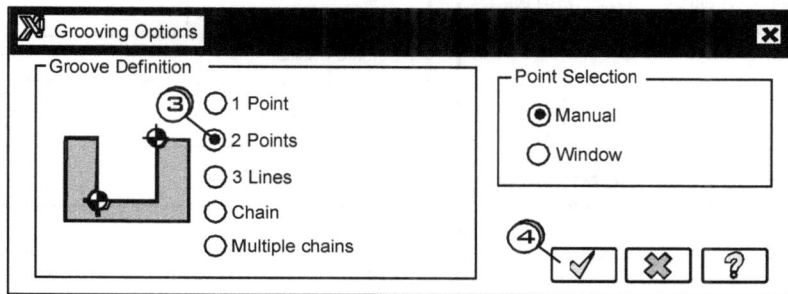

Click ③ the radio button ◉ 2 Points] Click ④ the OK button ✓

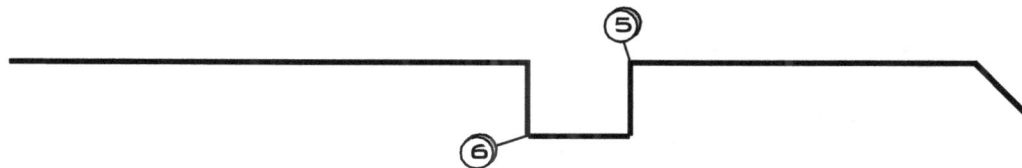

Two point groove: select first point

Click ⑤

Select second point

Click ⑥

Select first point or press <Enter> when done

Press | Enter |

Lathe Groove Properties   ✕

| Toolpath parameters | Groove shape parameters | Groove rough parameters | Groove finish parameters |

T4343 R0.01 W0.375
OD GROOVE CEN

T4444 R0.01 W0.125
OD GROOVE RIG

⑦

T4545 R0.01 W0.25
OD GROOVE RIGHT

T4646 R0.01 W0.375
OD GROOVE RIG

T4747 R0.01 W0.125
OD GROOVE LEF

T4848 R0.01 W0.25
OD GROOVE LEF

Tool number: 1   Offsetl number: 1

Station number: 1   [Tool Angle]

Feed rate: 0.01   ● in/rev   ○ in/min   ○ micro/in

☐ Finish feed rate: 0.005   ● in/rev   ○ in/min   ○ micro/in

Spindle speed: 200   ● iCSS   ○ RPM

☐ Finish spindle speed: 1000   ○ iCSS   ● RPM

Max spindle speed: 10000   [Coolant]

┌─ Home Position ─────────────────────────┐
│ D:10   Z:10 | User defined ▽ | [Define] │
└───┘

☐ Force tool change

Comment
[   ] ▲ ▼

☑ Show library tools

[Select library tool]   ☑ [Tool Filter]

[Axis Combo's (Left/Upper)]   [Misc values]   ☑ [Stock Update]   ☑ [Ref point]

☑ [Tool Display]   [Coordinates]   [Canned Text]

[✓] [✖] [？]

➤ Click ⑦ the **T4545 R0.01 W0.25** tool
                OD GROOVE RIGHT

◆ ENTER THE ROUGH AND FINISH GROOVING MACHINING PARAMETERS

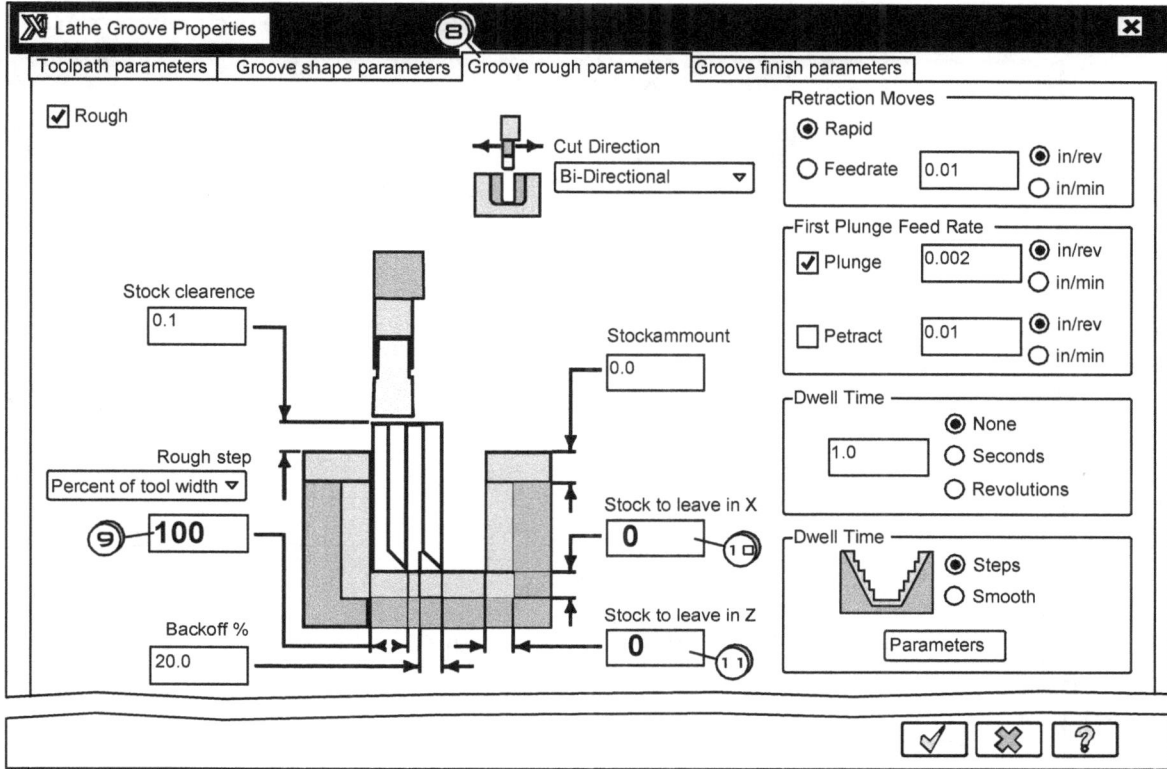

➤ Click ⑧ ⌐Groove rough parameters⌐ tab

➤ Click ⑨ in the [Percent of tool width ▼] box; enter **100**

➤ Click ⑩ in the Stock to leave in X box; enter **0**

➤ Click ⑪ in the Stock to leave in Z box; enter **0**

➤ Click ⑫ ⌐Groove finish parameters⌐ tab

➤ Click ⑬ check *off* ☐ Finish

➤ Click ⑭ the OK button ☑

B) OPERATION#2- CUT 2-8UNC OD THREAD

◆ OBTAIN THE NEEDED 8 UNC OD THREADING TOOL

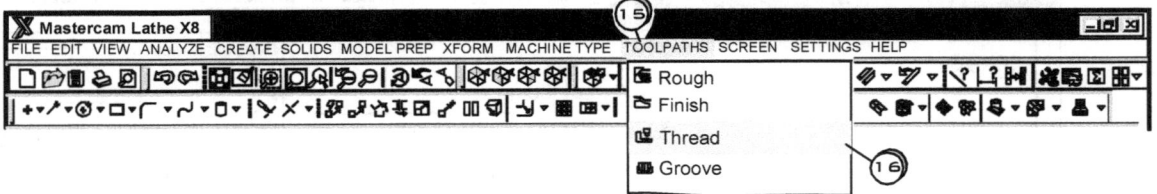

> Click ⑮ [Toolpaths]] > Click ⑯ 💾 Thread

> Click ⑰ the T9797 R0.0144 tool
> OD THREAD RIGHT

> *Right* Click; Click ⑱ Edit tool

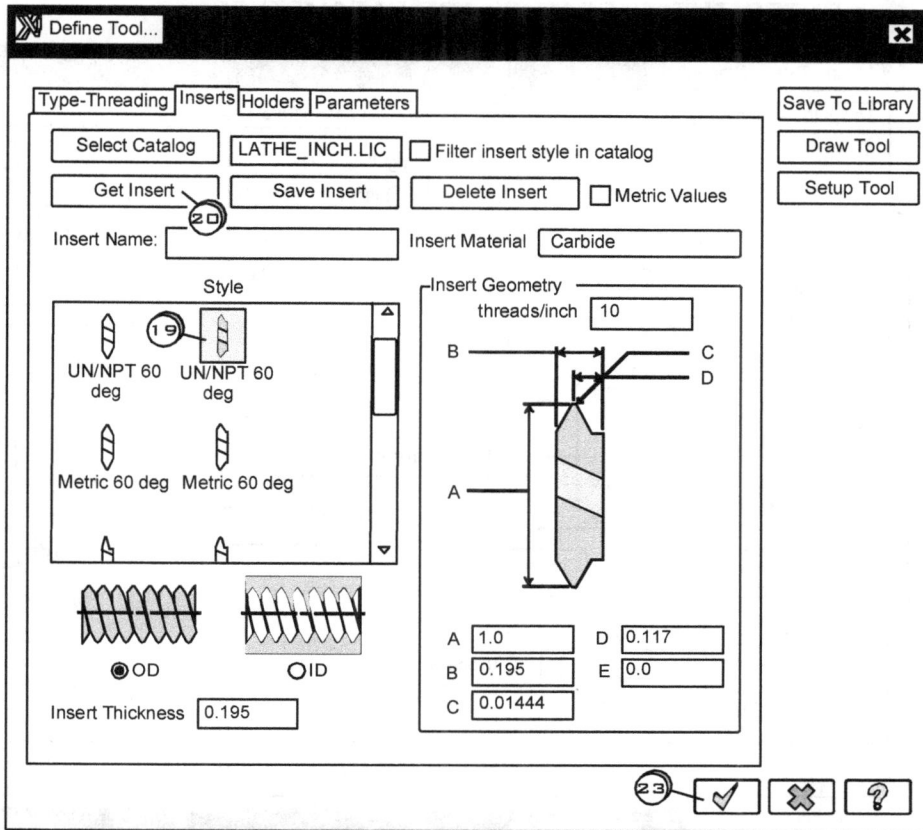

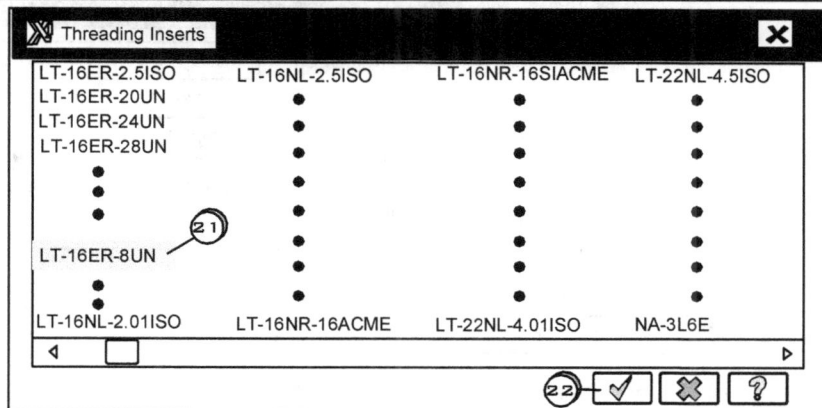

Click ⑲ the UN/NPT 60 deg insert

Click ⑳ Get Insert

Click ㉑ the LT-16ER-8UN insert

Click ㉒ the OK button ✓

Click ㉓ the OK button ✓

◆ ENTER THE THREAD SHAPE PARAMETERS

Lathe Thread Properties

| Toolpath parameters | Thread shape parameters | Thread cut parameters | ㉚ |

Lead
8.0
○ threads/inch
○ inches/thread

Included angle
60.0

Thread angle
30.0

Major Diameter   2.0
Minor Diameter   1.8512

Thread depth   0.0744

End Position
㉙ -1

Start Position
㉘ 0

Thread orientation   OD ▽   ☐ Negative X

Taper angle   0.0

Threads Form
Select from table   ㉔
Compute from formula
Draw thread

Adjusted Diameters
2.
1.8512

Major/Minor Diameters
◉ Large end of taper
○ Small end of taper

Adjusted Diameters
☑ Select From Table
Major allowance   0.0
Minor allowance   0.0
Allowance tolerance   0.0

Thread Table

Thread form   Unified - UNR, UNRC, UNRF, UNREF ▽

Common diameter/lead combinations up to 4 inches

Basic Major	Lead	Major Diameter	Minor Diameter	Comment	
2.0000 ㉖	8.0000	2.0000	1.8512	2, 8 ㉕	

㉗

▶ Click ㉔   Select from table

▶ Click ㉕ the down arrow ▽

▶ Click ㉖   2.0000   8.0000   2.0000   1.8512   2, 8   threads

▶ Click ㉗ the OK button

▶ Click ㉘ in the Start Position box; enter   0

▶ Click ㉙ in the End Position box; enter   -1

▶ Click ㉚ the   Thread cut parameters   tab

◆ ENTER THE THREAD CUT PARAMETERS

⤳ Click ③① in the Overcut box; enter   .06

⤳ Click ③② the OK button ✓

EXERCISES

4-1) Generate the 2D profile of the part shown in Figure 4p-1. Use PROCESS PLAN 4P-1 as an aid in generating the part program to produce the part.

Material: 1030 Steel

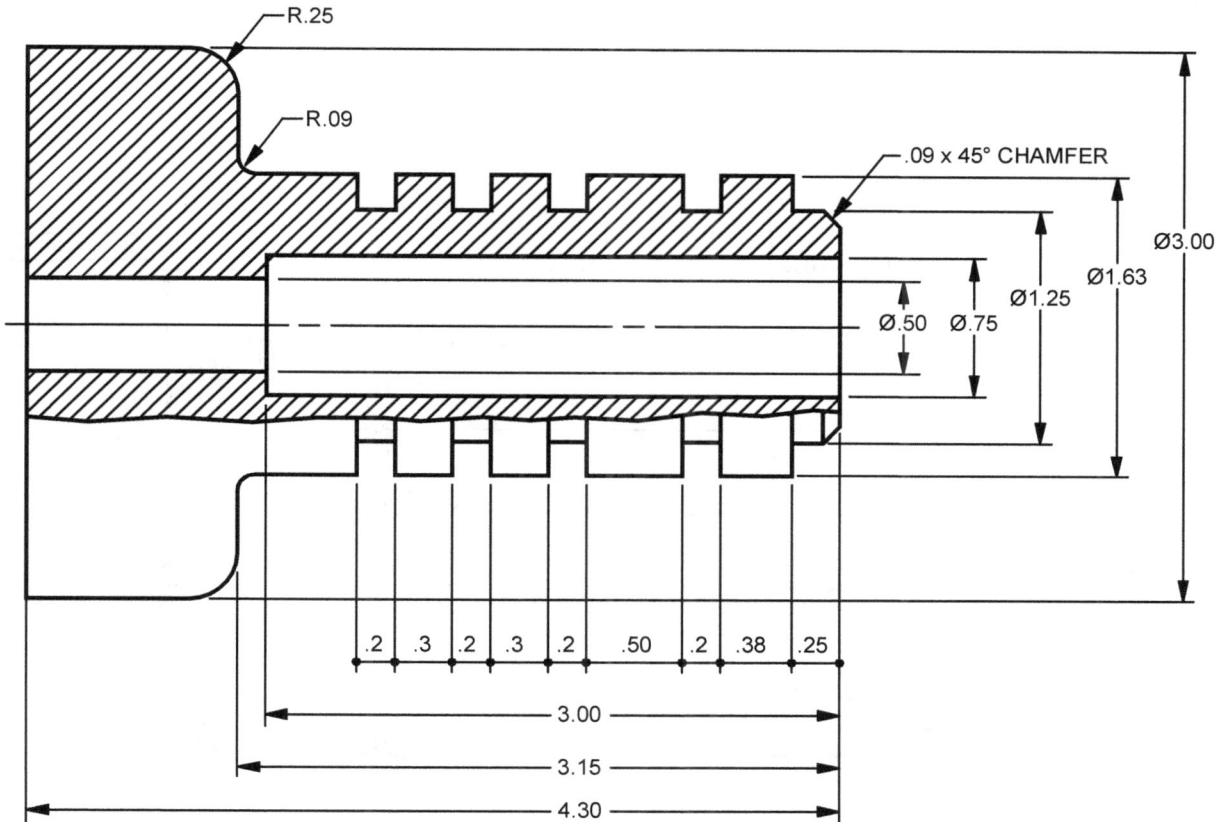

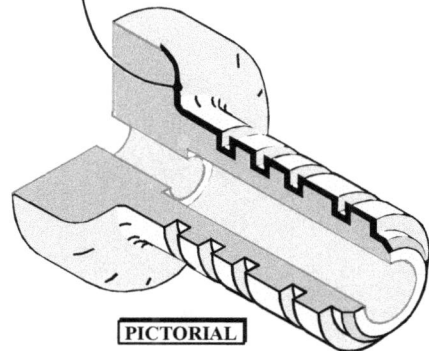

2D PROFILE OF PART
OPERATOR CREATES
IN CAD

PICTORIAL

Figure 4p-1

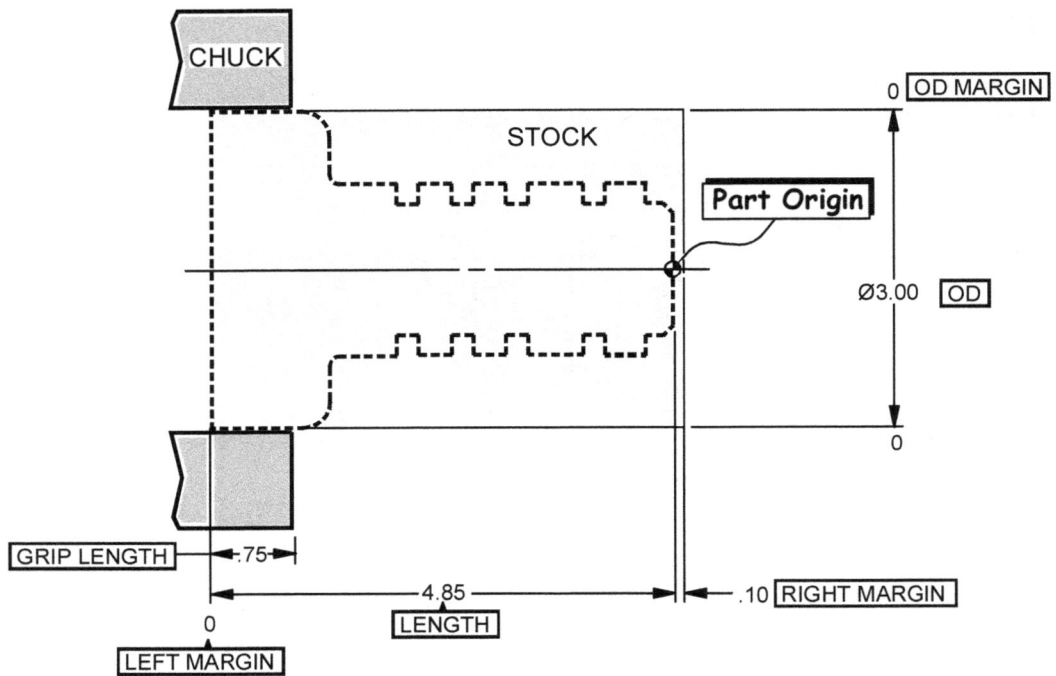

PROCESS PLAN 4P-1

No.	Operation	Tooling
1	FACE END STOCK Material Removed .10	1/32 TNR, RH OD TURNING TOOL
2	ROUGH TURN OD; LEAVE .01 IN X AND Z FOR FINISHING Plunge Cutting No Plunge Cutting allowed CHAIN1 END PT ② Material Removed CHAIN1 START PT .01 LEFT FOR FINISH CUT ①	R1/32
3	FINISH OD CONTOUR Plunge Cutting No Plunge Cutting allowed CHAIN1 END PT ② Material Removed CHAIN1 START PT .01 ①	1/64 TNR, RH OD FINISHING TOOL R1/64

PROCESS PLAN 4P-1(*continued*)

No.	Operation	Tooling
4	ROUGH AND FINISH OD GROOVE ; LEAVE .01 IN X AND Z FOR FINISHING.	R.01,W .125 OD GROOVING RIGHT

Grooving Options

Groove Definition: ● 1 Point, ○ 2 Points, ○ 3 Lines, ○ Chain, ○ Multiple chains

Point Selection: ● Manual, ○ Window

☑ Lead In

Entry Vector — ☑ Use Entry Vector
Fixed Direction: ○ None, ● Tangent, ○ Perpendicular

Material Removed ④ ③ ② ①

DETAIL A
.093, .5, .125, R.01, .125

GC-4125

PROCESS PLAN 4P-1(*continued*)

No.	Operation	Tooling
5	**LATHE DRILL, THRU** Depth **-4.0** ● Absolute ○ Incremental Drill Cycle Parameters Cycle Peck drill ▼ Ist peck **.125** Subsequent peck **.125** CHECK ☑ Drill tip compensation Breakthrough ammount **.125** Material Removed	**.5 DIA DRILL** ◄─Ø2─► 4.38 9 Ø.5
6	**LATHE DRILL X 3 DEEP** NO CHECK ☐ Drill tip compensation Material Removed	**.75 DIA END MILL** ◄─Ø2─► 4.38 8 Ø.75

4-2) Generate the 2D profiles of the part shown in Figure 4p-2. Use PROCESS PLAN 4P-2 as an aid in generating the part program to produce the part.

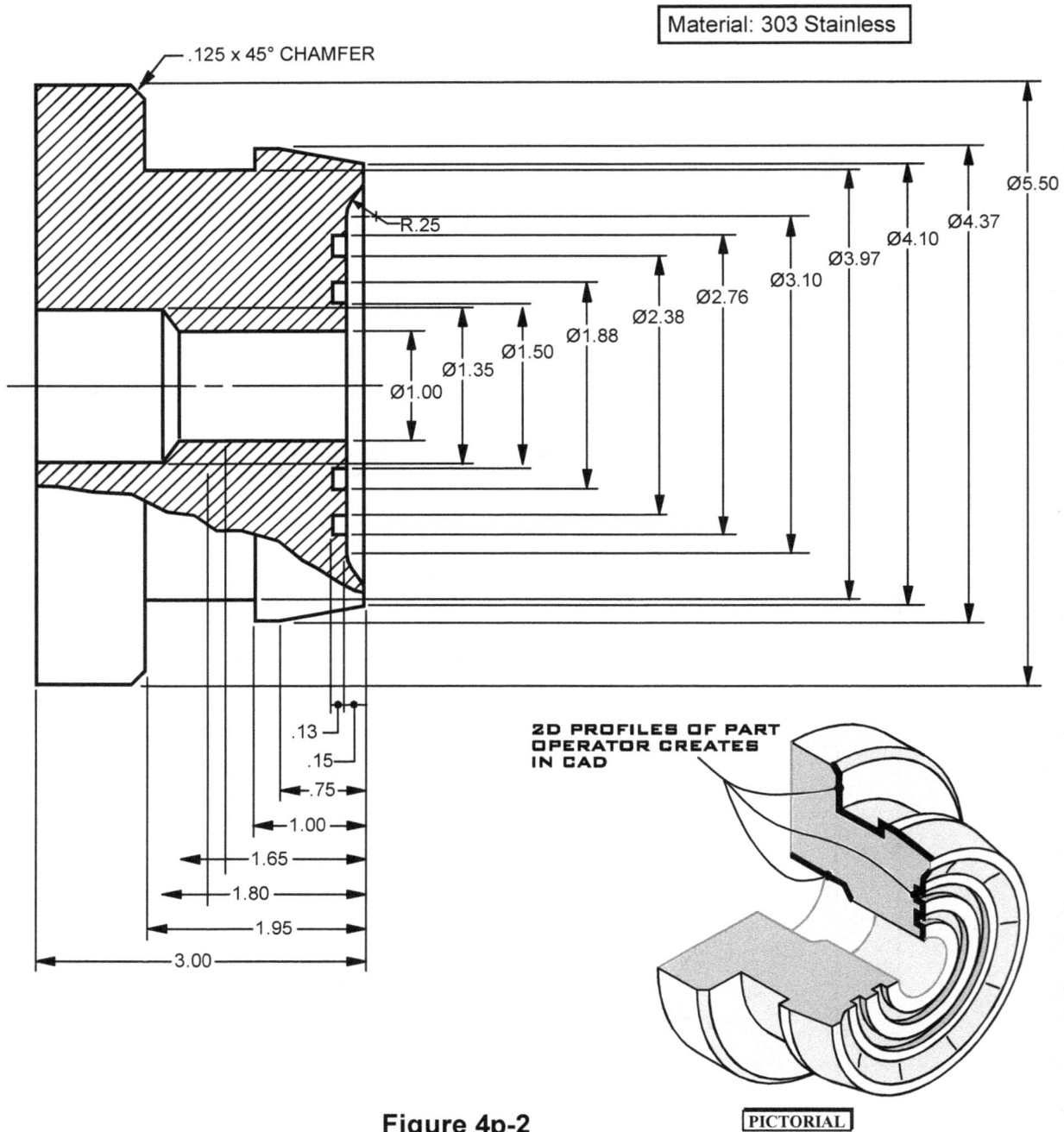

Material: 303 Stainless

.125 x 45° CHAMFER

Ø5.50
Ø4.37
Ø4.10
Ø3.97
Ø3.10
Ø2.76
Ø2.38
Ø1.88
Ø1.50
Ø1.35
Ø1.00
R.25

.13
.15
.75
1.00
1.65
1.80
1.95
3.00

2D PROFILES OF PART OPERATOR CREATES IN CAD

Figure 4p-2

PICTORIAL

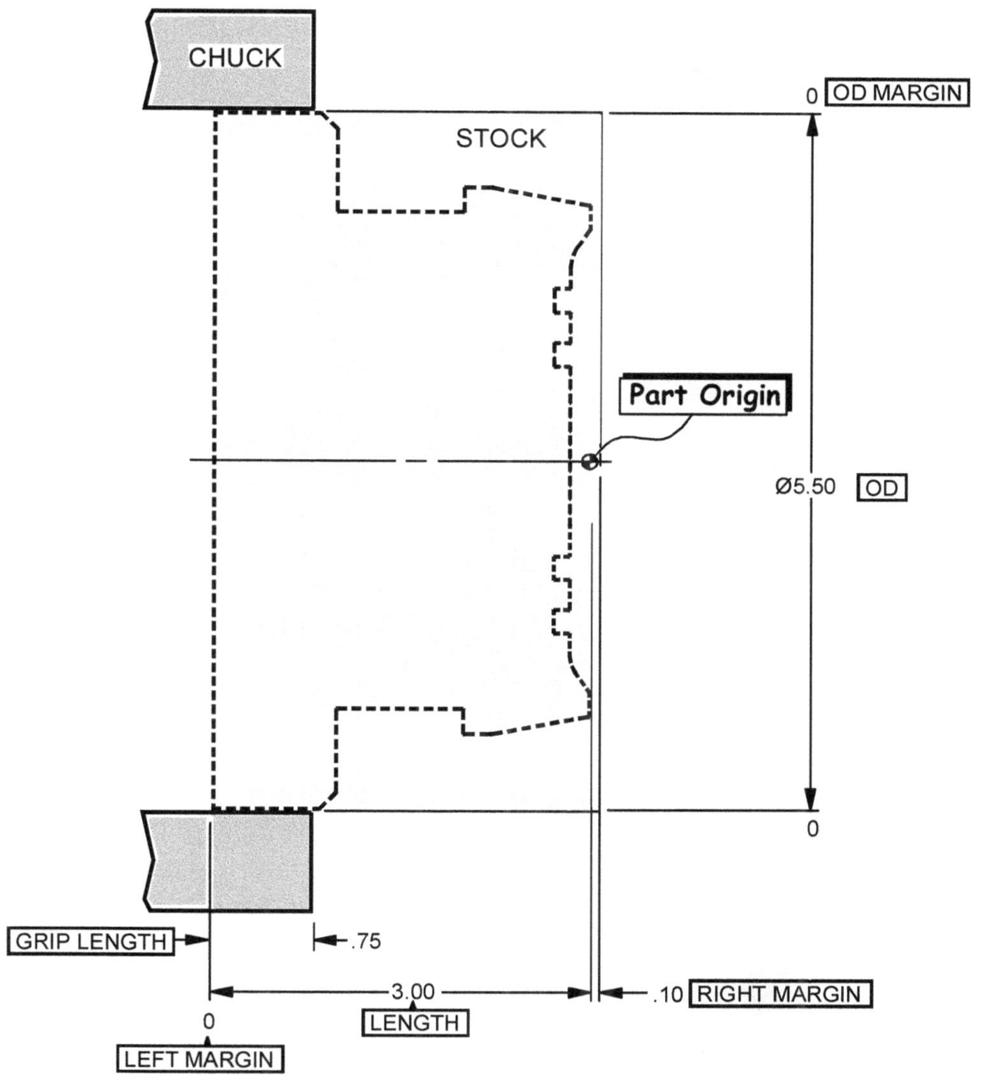

CHUCK

STOCK

0 | OD MARGIN |

Part Origin

Ø5.50 | OD |

0

GRIP LENGTH | → | .75

| 3.00 |
| LENGTH |

.10 | RIGHT MARGIN |

0

| LEFT MARGIN |

PROCESS PLAN 4P-2

No.	Operation	Tooling
1	Face End	1/32 TNR, RH OD TURNING TOOL

STOCK

Material Removed

R1/32

.10

PROCESS PLAN 4P-2(*continued*)

No.	Operation	Tooling
2	**ROUGH TURN OD; LEAVE .01 IN X AND Z FOR FINISHING** ☑ Lead In/Out Lead In ／ **Lead Out** — Adjust Contour — ☐ Extend/shorten end of contour Ammount `0.0` ◉ Extend ○ Shorten ☐ Add Line — Exit Vector — ☑ Use exit vector —Fixed Direction— ○ None ◉ Tangent ○ Perpendicular Angle: `45` Length: `0.1` CHAIN 1 END PT ② Material Removed ① CHAIN 1 START PT .01 LEFT FOR FINISH CUT — Plunge Cutting — No Plunge Cutting allowed ◉	**1/32 TNR, RH OD TURNING TOOL** R1/32
3	**FINISH OD CONTOUR** — Plunge Cutting — No Plunge Cutting allowed ◉ CHAIN 1 END PT ② Material Removed ① CHAIN 1 START PT .01	**1/64 TNR, RH OD FINISHING TOOL** R1/64

PROCESS PLAN 4P-2(*continued*)

No.	Operation	Tooling
4	ROUGH AND FINISH OD GROOVE; LEAVE .01 IN X AND Z FOR FINISHING	R.01, W .25 OD GROOVING RIGHT

Operation 4 details:

☑ Lead In/Out

Lead In | Lead Out

─ Adjust Contour ─
☐ Extend/shorten end of contour

Ammount [0.0]   ◉ Extend
                    ◯ Shorten

☐ [Add Line]

─ Exit Vector ─
☑ Use exit vector

┌Fixed Direction┐
◯ None
◉ Tangent
◯ Perpendicular

Angle: [45]

Length: [0.1]

─ Groove Definition ─
◯ 1 Point
◯ 2 Points
◯ 3 Lines
◉ Chain
◯ Multiple chains

CHAIN 1 END PT ②

Material Removed

① CHAIN 1 START PT

Tooling (Operation 4):

A—

DETAIL A
.144
.5
.250
.250
R.01

GCP-4250

| 5 | LATHE DRILL X .18 DEEP | .25 DIA CENTER DRILL |

Operation 5 details:

Depth   [-.18]

◉ Absolute ◯ Incremental

Material Removed

NO CHECK
☐ Drill tip compensation

Ø.25

PROCESS PLAN 4P-2(*continued*)

No.	Operation	Tooling
6	**LATHE DRILL THRU** Depth **-3** ⦿ Absolute ◯ Incremental Drill Cycle Parameters Cycle Peck drill ▼ CHECK ☑ Drill tip compensation Breakthrough ammount **.125** Material Removed	1 DIA DRILL 3.75 7.5 Ø1.00
7	**ROUGH AND FINISH OD GROOVE; LEAVE .01 IN X AND Z FOR FINISHING** Groove Definition ◯ 1 Point ◯ 2 Points ◯ 3 Lines ⦿ Chain ◯ Multiple chains CHAIN 1 END PT ② CHAIN 1 START PT ① Material Removed ☑ Lead In Entry Vector ☑ Use Entry Vector Fixed Direction ◯ None ⦿ Tangent ◯ Perpendicular .10	R.01, W .125 ID GROOVING RIGHT HAND .562 A Ø.625 .15 8 A10-NEL2 DETAIL A .144 .5 .250 .250 R.01 GCP-4250

PROCESS PLAN 4P-2(*continued*)

No.	Operation	Tooling
8	Rough and Finish Face groove; leave .01 in X and Z for finishing	R.01, W.125 FACE GROOVING RIGHT

☑ Lead In/Out

Lead In | Lead Out

┌─ Adjust Contour ─┐
☐ Extend/shorten end of contour

Ammount [0.0]   ● Extend
                    ○ Shorten

☐ [Add Line]

┌─ Exit Vector ─┐
☑ Use exit vector

Fixed Direction
○ None
● Tangent
○ Perpendicular

Angle: [45]

Length: [0.1]

┌─ Groove Definition ─┐
○ 1 Point
○ 2 Points
○ 3 Lines
● Chain
○ Multiple chains

② CHAIN1 END PT

Material Removed

CHAIN1 START PT ①

.75
.5
6
.093

KGDSL-164D

DETAIL A
.093
.5
.125
.125
R.01

GC-4125

4-3) Generate the 2D profiles of the part shown in Figure 4p-3. Use PROCESS PLAN 4P-3 as an aid in generating the part program to produce the part.

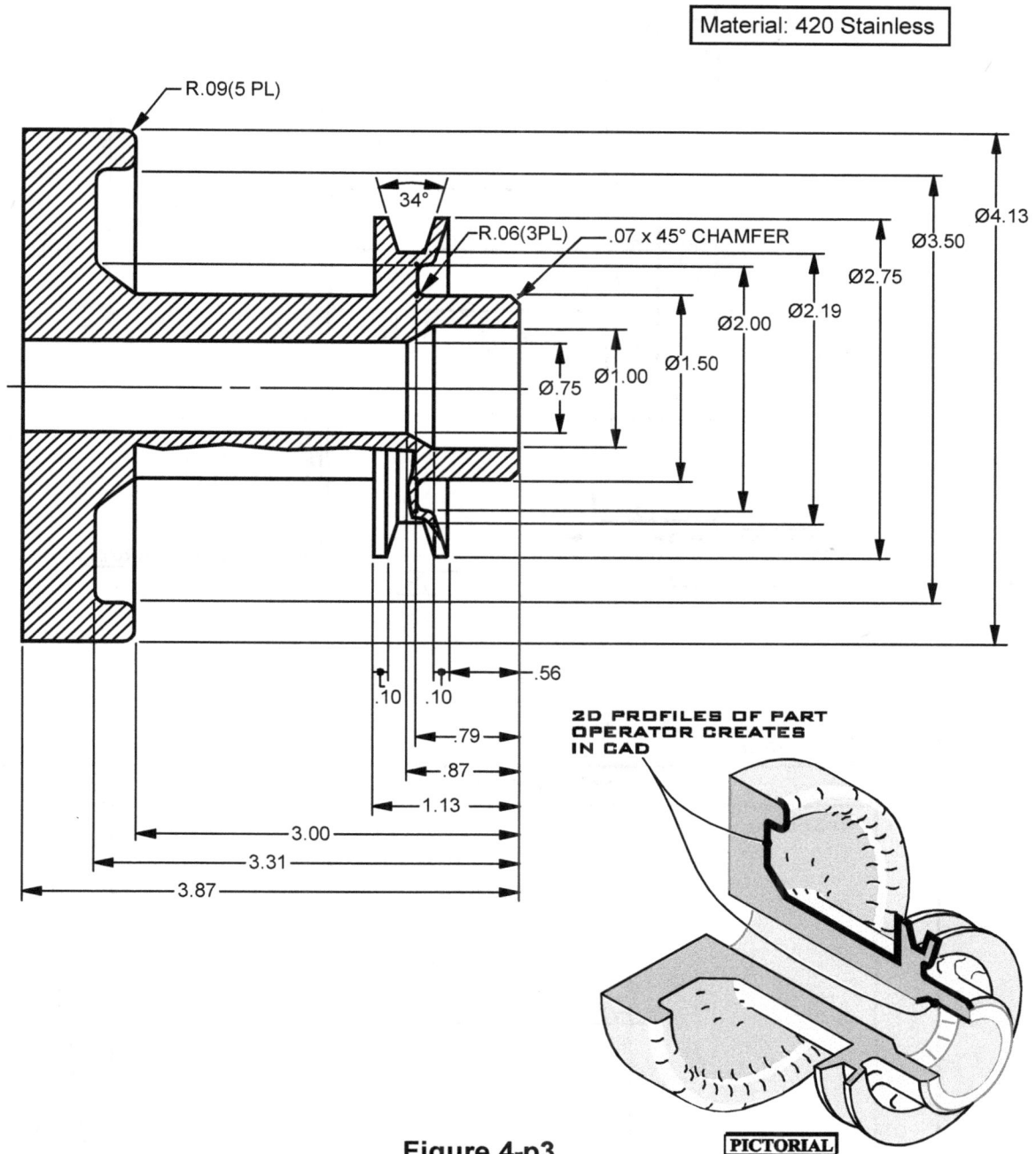

Material: 420 Stainless

R.09(5 PL)

34°

R.06(3PL) .07 x 45° CHAMFER

Ø4.13

Ø3.50

Ø2.75

Ø2.19

Ø2.00

Ø1.50

Ø1.00

Ø.75

.56

.10 .10

.79

.87

1.13

3.00

3.31

3.87

2D PROFILES OF PART OPERATOR CREATES IN CAD

PICTORIAL

Figure 4-p3

CHUCK

0 OD MARGIN

STOCK

Part Origin

Ø4.13 OD

0

GRIP LENGTH ← .7

LEFT MARGIN 0   ← 3.87 →   .10 RIGHT MARGIN

LENGTH

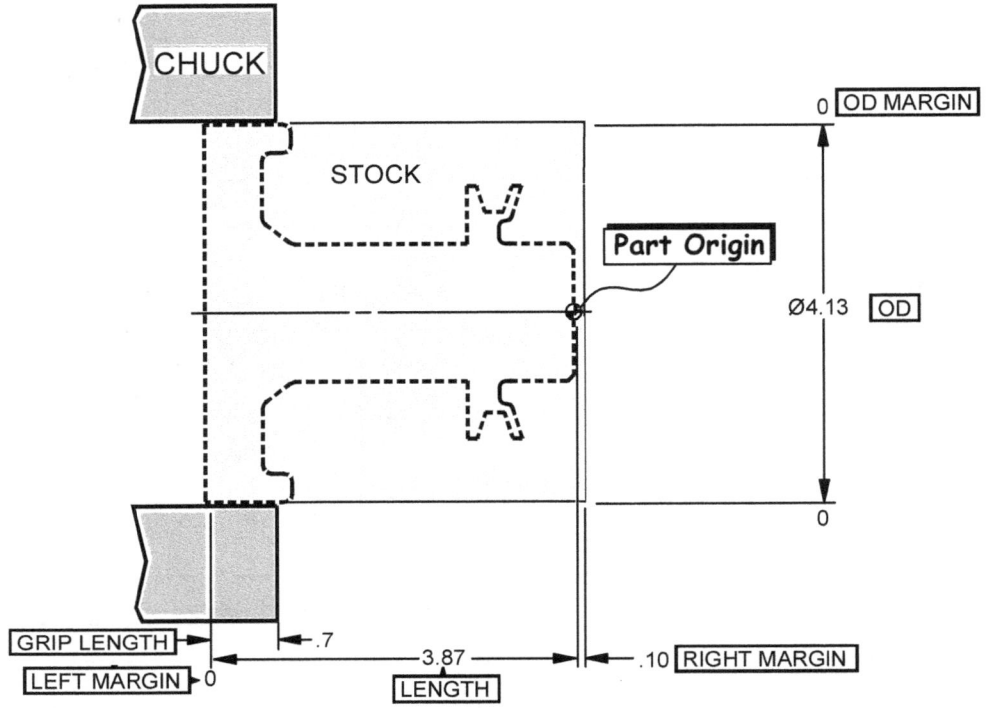

PROCESS PLAN4P-3

No.	Operation	Tooling
1	FACE END Material Removed / STOCK / .10	1/32 TNR, RH OD TURNING TOOL R1/32

PROCESS PLAN 4P-3(*continued*)

No.	Operation	Tooling
2	ROUGH TURN OD, LEAVE .01 IN X AND Z FOR SEMI-FINISH. SEMI-FINISH CONTOUR IN TWO PASSES USING A STEPOVER OF .005	1/64 TNR, RH OD FINISHING TOOL

☑ Lead In/Out

Lead In | Lead Out

Adjust Contour
☐ Extend/shorten end of contour

Ammount 0.0   ⦿ Extend   ○ Shorten

☐ Add Line

Exit Vector
☑ Use exit vector

Fixed Direction
⦿ None
○ Tangent
○ Perpendicular

Angle: 45
Length: 0.1

Plunge Cutting
No Plunge Cutting allowed
⦿

R1/64

☑ Semi Finish →

Semi Finish Parameters ☒

Number of passes
2

Stepover
.005

Stock to leave in X:
0

Stock to leave in Z:
0

✓ ✗ ?

CHAIN1 END PT
②

Material Removed

CHAIN1 START PT
①

PROCESS PLAN 4P-3(*continued*)

No.	Operation	Tooling
3	ROUGH AND FINISH OD GROOVE ; LEAVE .01 IN X AND Z FOR FINISHING.	R.01, W .25 OD GROOVING RIGHT

Grooving Options

Groove Definition
○ 1 Point
● 2 Points
○ 3 Lines
○ Chain
○ Multiple chains

Point Selection
● Manual
○ Window

☑ Lead In

Entry Vector
☑ Use Entry Vector
Fixed Direction
○ None
● Tangent
○ Perpendicular

Material Removed

.75   .15

DETAIL A
.144
.625
.250
.250   R.01

GFG-250

PROCESS PLAN 4P-3(*continued*)

No.	Operation	Tooling
4	Rough and Finish OD Groove ; leave .01 in X and Z for finishing.	R.01, W .25 OD GROOVING RIGHT DETAIL A GC-4125

PROCESS PLAN 4P-3(*continued*)

No.	Operation	Tooling
5	Rough and Finish OD Groove ; leave .008 in X and Z for finishing.	R.01, W .25 OD Grooving Right

Groove Definition

O 1 Point
O 2 Points
O 3 Lines
◉ Chain
O Multiple chains

Groove Walls

O Steps
◉ Smooth

☑ Lead In

Entry Vector
☑ Use Entry Vector

Fixed Direction
O None
◉ Tangent
O Perpendicular

Material Removed

② ①

CHAIN 1 END PT

CHAIN 1 START PT

1

5

.10

A

DETAIL A

.093

.5

.125

.125

R.01

GC-4125

PROCESS PLAN 4P-3(*continued*)

No.	Operation	Tooling
6	ROUGH AND FINISH OD GROOVE ; LEAVE .01 IN X AND Z FOR FINISHING.	R.02, DOUBLE END OD GROOVING RIGHT

Groove Definition
- ○ 1 Point
- ○ 2 Points
- ○ 3 Lines
- ◉ Chain
- ○ Multiple chains

Groove Walls
- ◉ Steps
- ○ Smooth

☑ Lead In

Entry Vector
- ☑ Use Entry Vector

Fixed Direction
- ○ None
- ○ Tangent
- ◉ Perpendicular

CHAIN 1 START PT ①

Material Removed

CHAIN 1 END PT ②

VGSOR-24-8250E

DETAIL A

.255

.25

1.00

R.063

NRD-4062R

PROCESS PLAN 4P-3(*continued*)

No.	Operation	Tooling
7	LATHE DRILL X .18 DEEP Depth [-.18] ⦿ Absolute ○ Incremental Material Removed [NO CHECK] ☐ Drill tip compensation	.25 DIA CENTER DRILL ⌀.25
8	LATHE DRILL THRU Depth [-3.87] ⦿ Absolute ○ Incremental ┌ Drill Cycle Parameters ┐ Cycle [Peck drill ▼] [CHECK] ☑ Drill tip compensation Breakthrough ammount [.125] Material Removed	3/4 DIA DRILL 4.5 8.75 ⌀.75

PROCESS PLAN 4P-3(*continued*)

No.	Operation	Tooling
9	ROUGH BORE ID, LEAVE .01 FOR FINISH CUT	1/64 TNR, ID ROUGH BORING TOOL ,.5 DIA
10	FINISH BORE ID	1/64 TNR, ID FINISH BORING TOOL ,.5 DIA

Operation 9 detail

☑ Lead In/Out

Lead In | Lead Out

Adjust Contour
☐ Extend/shorten end of contour
Ammount [0.0]   ◉ Extend   ◯ Shorten
☐ Add Line

Exit Vector
☑ Use exit vector
Fixed Direction
◯ None
◉ Tangent
◯ Perpendicular
Angle: [45]
Length: [0.1]

CHAIN1 START PT ①
CHAIN1 END PT ②
.01 LEFT FOR FINISH CUT
Material Removed

R1/64 Ø.5

Operation 10 detail

☑ Lead In/Out

Lead In | Lead Out

Adjust Contour
☐ Extend/shorten end of contour
Ammount [0.0]   ◉ Extend   ◯ Shorten
☐ Add Line

Exit Vector
☑ Use exit vector
Fixed Direction
◯ None
◉ Tangent
◯ Perpendicular
Angle: [45]
Length: [0.1]

CHAIN1 START PT ①
CHAIN1 END PT ②
.01 LEFT FOR FINISH CUT
Material Removed

R1/64 Ø.5

4-4) Generate the 2D profiles of the part shown in Figure 4p-4. Use PROCESS PLAN 4P-4 as an aid in generating the part program to produce the part.

Material: 7075 Aluminum

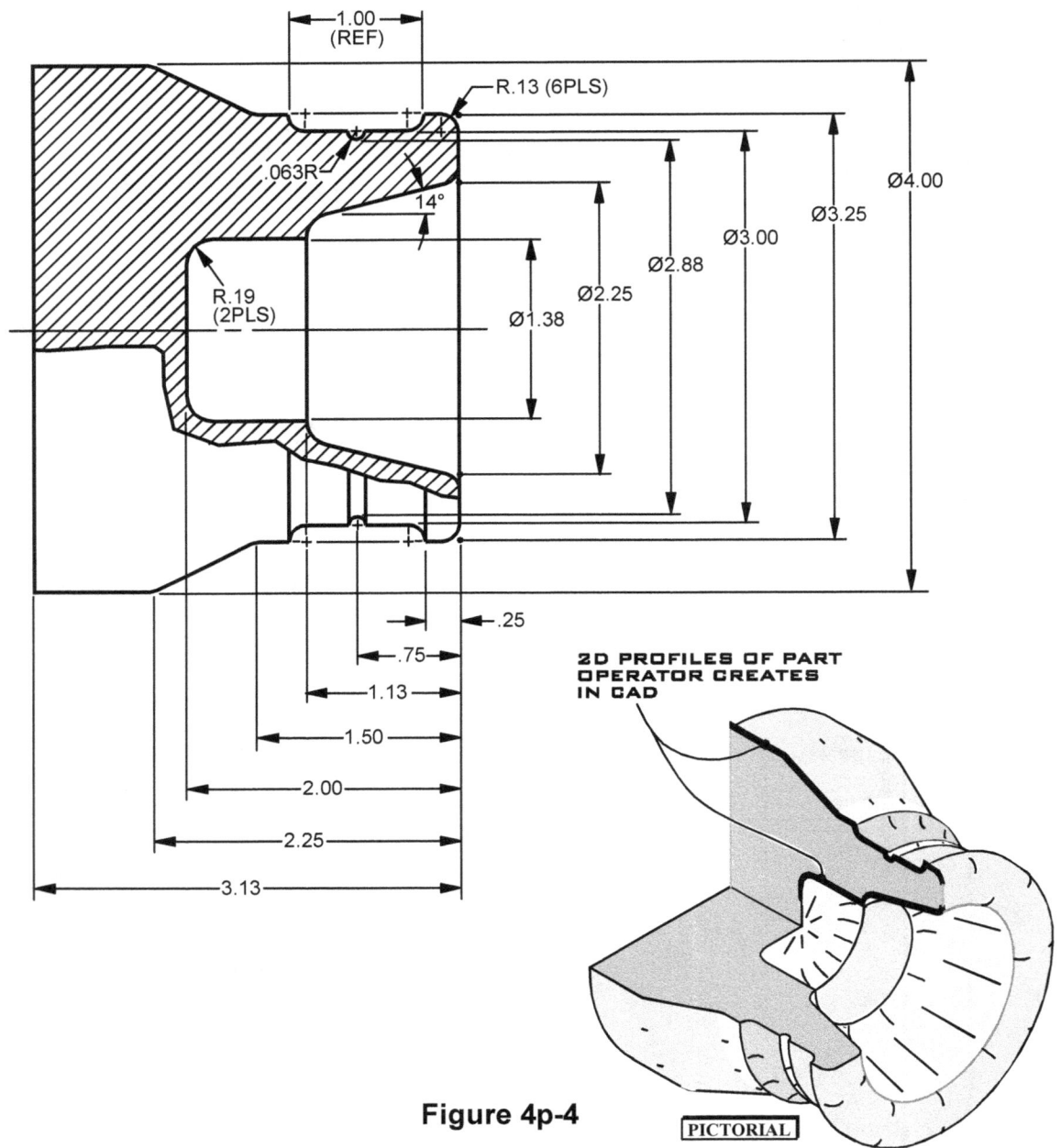

1.00
(REF)

R.13 (6PLS)

.063R

14°

R.19
(2PLS)

Ø4.00

Ø3.25

Ø3.00

Ø2.88

Ø2.25

Ø1.38

.25

.75

1.13

1.50

2.00

2.25

3.13

2D PROFILES OF PART
OPERATOR CREATES
IN CAD

Figure 4p-4

PICTORIAL

CHUCK

0 OD MARGIN

Part Origin

STOCK

Ø4.00 OD

0

GRIP LENGTH .75

LEFT MARGIN 0

3.13

LENGTH

.10 RIGHT MARGIN

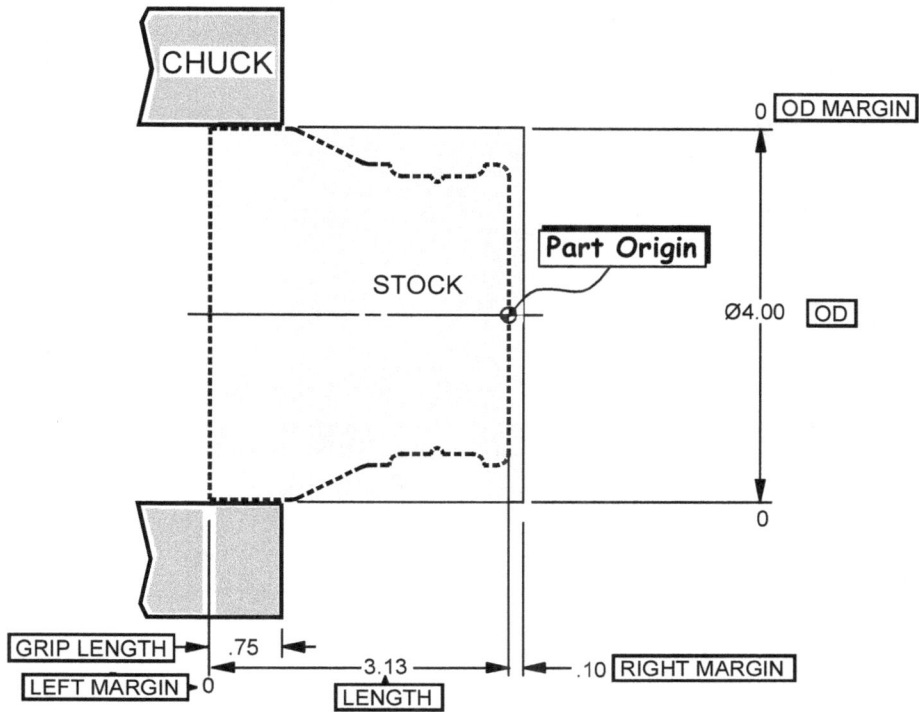

PROCESS PLAN 4P-4

No.	Operation	Tooling
1	FACE END Material Removed STOCK .10	1/32 TNR, RH OD TURNING TOOL R1/32

PROCESS PLAN 4P-4(*continued*)

No.	Operation	Tooling
2	ROUGH TURN OD, LEAVE .01 IN X AND Z FOR SEMI-FINISH. SEMI-FINISH CONTOUR IN TWO PASSES USING A STEPOVER OF .005	1/64 TNR, RH OD FINISHING TOOL

☑ Lead In/Out

Lead In | Lead Out

Adjust Contour
☑ Extend/shorten end of contour
Ammount **.05**   ◉ Extend   ○ Shorten
☐ Add Line

Exit Vector
☑ Use exit vector
Fixed Direction
○ None
○ Tangent
◉ Perpendicular
Angle: 45
Length: 0.1

Plunge parameters

☑ Semi Finish → **Semi Finish Parameters** ☒

Number of passes
2

Stepover
.005

Stock to leave in X:
0

Stock to leave in Z:
0

② CHAIN 1 END PT

① CHAIN 1 START PT

Material Removed

Material Missed by Tool

R1/64

PROCESS PLAN 4P-4(*continued*)

No.	Operation	Tooling
3	ROUGH AND FINISH OD GROOVE ; LEAVE .008 IN X AND Z FOR FINISHING.	R.063, W .093, OD GROOVING RIGHT

Groove Definition
○ 1 Point
○ 2 Points
○ 3 Lines
◉ Chain
○ Multiple chains

Groove Walls
○ Steps
◉ Smooth

☑ Lead In

Entry Vector
☑ Use Entry Vector
Fixed Direction
○ None
◉ Tangent
○ Perpendicular

Material Removed

CHAIN 1 END PT ②

CHAIN 1 START PT ①

.10 add line to chain

.093

.75

A

1

DETAIL A

R.063

.5

.093

GR-4125

PROCESS PLAN 4P-4(*continued*)

No.	Operation	Tooling
4	LATHE DRILL X .18 DEEP Depth [-.18] ⦿ Absolute ◯ Incremental [NO CHECK] ☐ Drill tip compensation Material Removed	.25 DIA CENTER DRILL Ø.25
5	LATHE DRILL X 2 DEEP Depth [-2] ⦿ Absolute ◯ Incremental Drill Cycle Parameters Cycle [Peck drill ▼] [CHECK] ☑ Drill tip compensation Breakthrough ammount [.125] Material Removed	1.25 DIA DRILL ◄─Ø2.25─► 5 8.75 Ø1.25►

PROCESS PLAN 4P-4(*continued*)

No.	Operation	Tooling
6	**LATHE DRILL X 2 DEEP**	**1 DIA END MILL**

Depth	**-2**

◉ Absolute   ○ Incremental

Drill Cycle Parameters

Cycle

Drill/Counterbore ▼

[NO CHECK]

☐ Drill tip compensation

Material Removed

Ø2.25

2.5

1.5

7

3

2.5

Ø1

| 7 | **ROUGH BORE ID, LEAVE .01 FOR FINISH CUT** | **1/64 TNR, ID ROUGH BORING TOOL ,.5 DIA** |

☑ Lead In/Out

| Lead In | Lead Out |

Adjust Contour

☐ Extend/shorten end of contour

Ammount   0.0   ◉ Extend
                   ○ Shorten

☐ Add Line

Exit Vector

☑ Use exit vector

Fixed Direction
◉ None
○ Tangent
○ Perpendicular

Angle: 45

Length: 0.1

Plunge Cutting

No Plunge Cutting allowed
◉

CHAIN 1 START PT
①

.01 LEFT FOR FOR FINISH CUT

②
CHAIN 1 END PT

Material Removed

R1/64

Ø.5

PROCESS PLAN 4P-4(*continued*)

No.	Operation	Tooling
8	Finish Bore ID Contour	1/64 TNR, ID FINISH BORING TOOL ,.5 DIA

4-5) Generate the 2D profiles of the part shown in Figure 4p-5. Use PROCESS PLAN 4P-5 as an aid in generating the part program to produce the part.

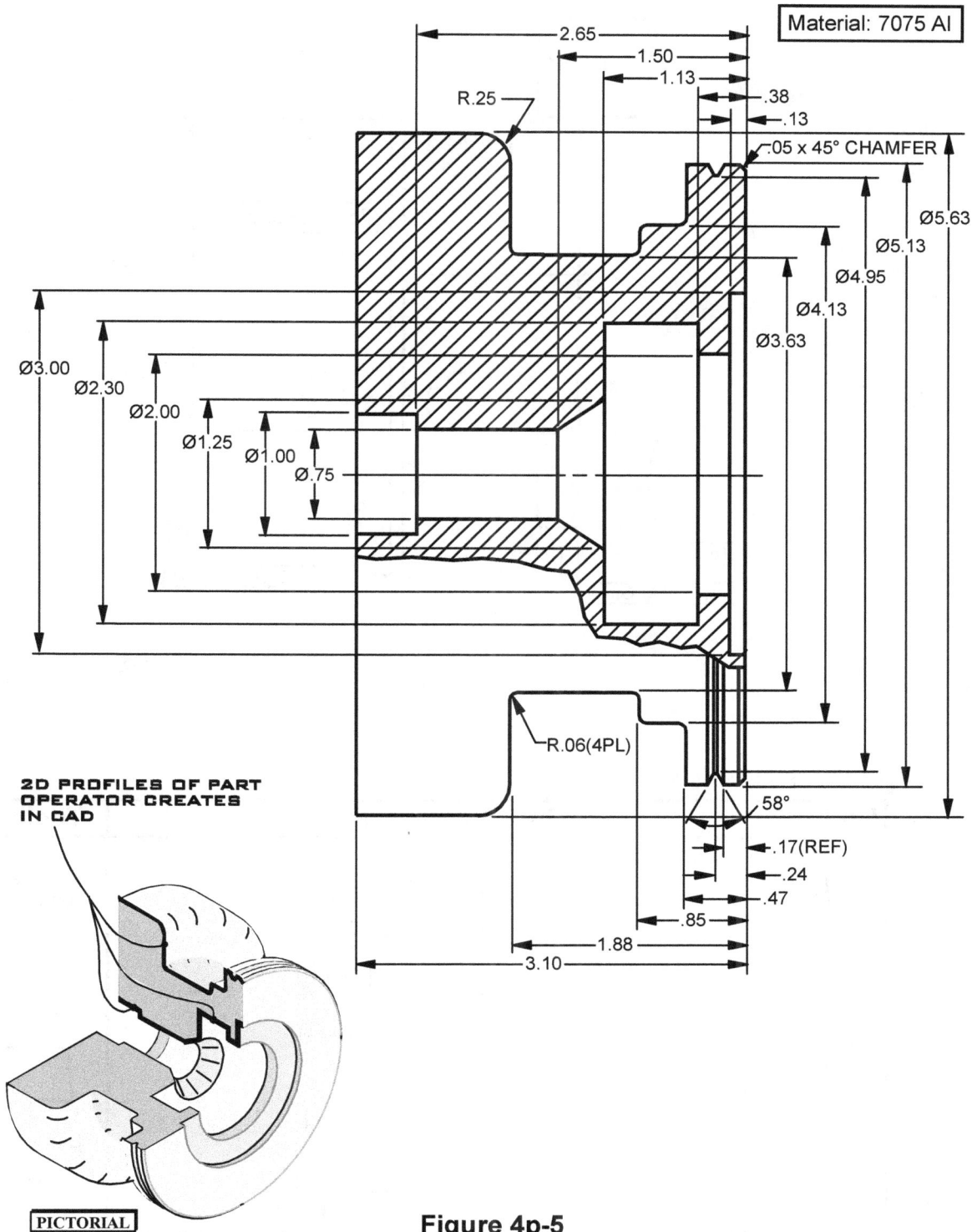

Material: 7075 Al

2.65
1.50
1.13
.38
.13
R.25
.05 x 45° CHAMFER
Ø5.63
Ø5.13
Ø4.95
Ø4.13
Ø3.63
Ø3.00
Ø2.30
Ø2.00
Ø1.25
Ø1.00
Ø.75

R.06(4PL)

2D PROFILES OF PART OPERATOR CREATES IN CAD

58°
.17(REF)
.24
.47
.85
1.88
3.10

PICTORIAL

Figure 4p-5

CHUCK

STOCK

Part Origin

0 OD MARGIN

Ø5.63 OD

0

GRIP LENGTH → .75 ←
LEFT MARGIN → 0
← 3.10 → ← .10 RIGHT MARGIN
LENGTH

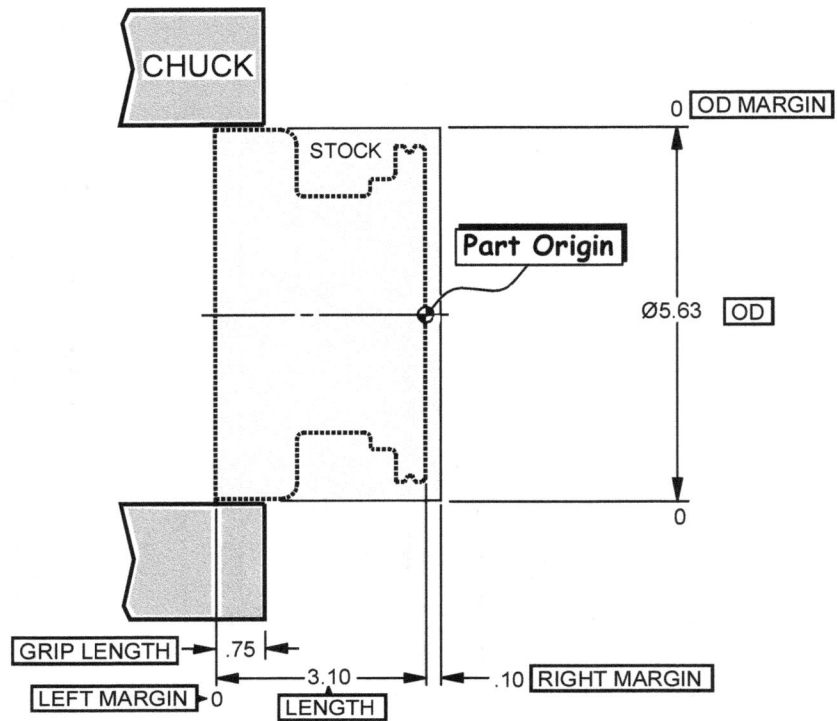

PROCESS PLAN 4P-5

No.	Operation	Tooling
1	Face End	1/32 TNR, RH OD TURNING TOOL
	Material Removed	R1/32
	→ ←.10	

PROCESS PLAN 4P-5(*continued*)

No.	Operation	Tooling
2	ROUGH TURN OD; LEAVE .01 IN X AND Z FOR FINISH CUT.	1/64 TNR, RH OD FINISHING TOOL

☑ Lead In/Out

| Lead In | Lead Out |

Adjust Contour
☐ Extend/shorten end of contour
Ammount 0.0
⦿ Extend
○ Shorten
☐ Add Line

Exit Vector
☑ Use exit vector
Fixed Direction
⦿ None
○ Tangent
○ Perpendicular
Angle: 45
Length: 0.1

Plunge Cutting
No Plunge Cutting allowed
⦿

R1/64

Material Removed

CHAIN 1 END PT ②

① CHAIN 1 START PT

.01 LEFT FOR FINISH CUT

PROCESS PLAN 4P-5(*continued*)

No.	Operation	Tooling
3	Finish OD Contour	1/64 TNR, RH OD FINISHING TOOL

Finish OD Contour

☑ Lead In/Out

Lead In | Lead Out

Adjust Contour
☐ Extend/shorten end of contour

Ammount 0.0   ⦿ Extend   ○ Shorten

☐ Add Line

Exit Vector
☑ Use exit vector

Fixed Direction
⦿ None
○ Tangent
○ Perpendicular

Angle: 45
Length: 0.1

Plunge Cutting
No Plunge Cutting allowed

CHAIN 1 END PT ②

Material Removed

① CHAIN 1 START PT

R1/64

PROCESS PLAN 4P-5(*continued*)

No.	Operation	Tooling
4	ROUGH AND FINISH OD GROOVE ; LEAVE .008 IN X AND Z FOR FINISHING.	R.01, W .25 OD GROOVING RIGHT GFG-250

PROCESS PLAN 4P-5(*continued*)

No.	Operation	Tooling
5	ROUGH AND FINISH OD GROOVE ; LEAVE .008 IN X AND Z FOR FINISHING. Groove Definition ○ 1 Point ○ 2 Points ○ 3 Lines ◉ Chain ○ Multiple chains ☑ Lead In Entry Vector ☑ Use Entry Vector Fixed Direction ○ None ○ Tangent ◉ Perpendicular Material Removed ② CHAIN 1 END PT ① CHAIN 1 START PT	R.01, W.125 OD GROOVING CENTER ⌀1 .75 .13 A DETAIL A R.01 .125 .5 .125 DOUBLE END (40° V) GC-4125
6	LATHE DRILL X .18 DEEP Depth -.18 ◉ Absolute ○ Incremental NO CHECK ☐ Drill tip compensation Material Removed	.25 DIA CENTER DRILL ⌀.25

PROCESS PLAN 4P-5(*continued*)

No.	Operation	Tooling
7	**LATHE DRILL THRU** Depth **-3.1** ◉ Absolute ○ Incremental Drill Cycle Parameters Cycle Peck drill ▼ CHECK ☑ Drill tip compensation Breakthrough ammount **.125** Material Removed	**3/4 DIA DRILL** 3.75 8.75 4 3.5 Ø.75 Removed
8	**LATHE DRILL X 1.13 DEEP** Depth **-1.13** ◉ Absolute ○ Incremental Material Removed NO CHECK ☐ Drill tip compensation	**2 DIA END MILL** 3.5 1.5 9 4 3.5 Ø2.0

PROCESS PLAN 4P-5(*continued*)

No.	Operation	Tooling
9	ROUGH AND FINISH OD GROOVE; LEAVE .01 IN X AND Z FOR FINISHING	R.005, W .15 ID GROOVING RIGHT HAND

Groove Definition

- ○ 1 Point
- ○ 2 Points
- ○ 3 Lines
- ◉ Chain
- ○ Multiple chains

☑ Lead In

Entry Vector

☑ Use Entry Vector

Fixed Direction
- ○ None
- ○ Tangent
- ◉ Perpendicular

.1   .175

.125

① CHAIN1 START PT

② CHAIN1 END PT

Material Removed

Tooling column:

.438   .375

A

.118
8

A08-NEL2

DETAIL A

.118

.625   .15

.195

R.005

NG-330L

4-6) Generate the 2D profiles of the part shown in Figure in 4p-6. Use PROCESS PLAN
4P-6 as an aid in generating the part program to produce the part.

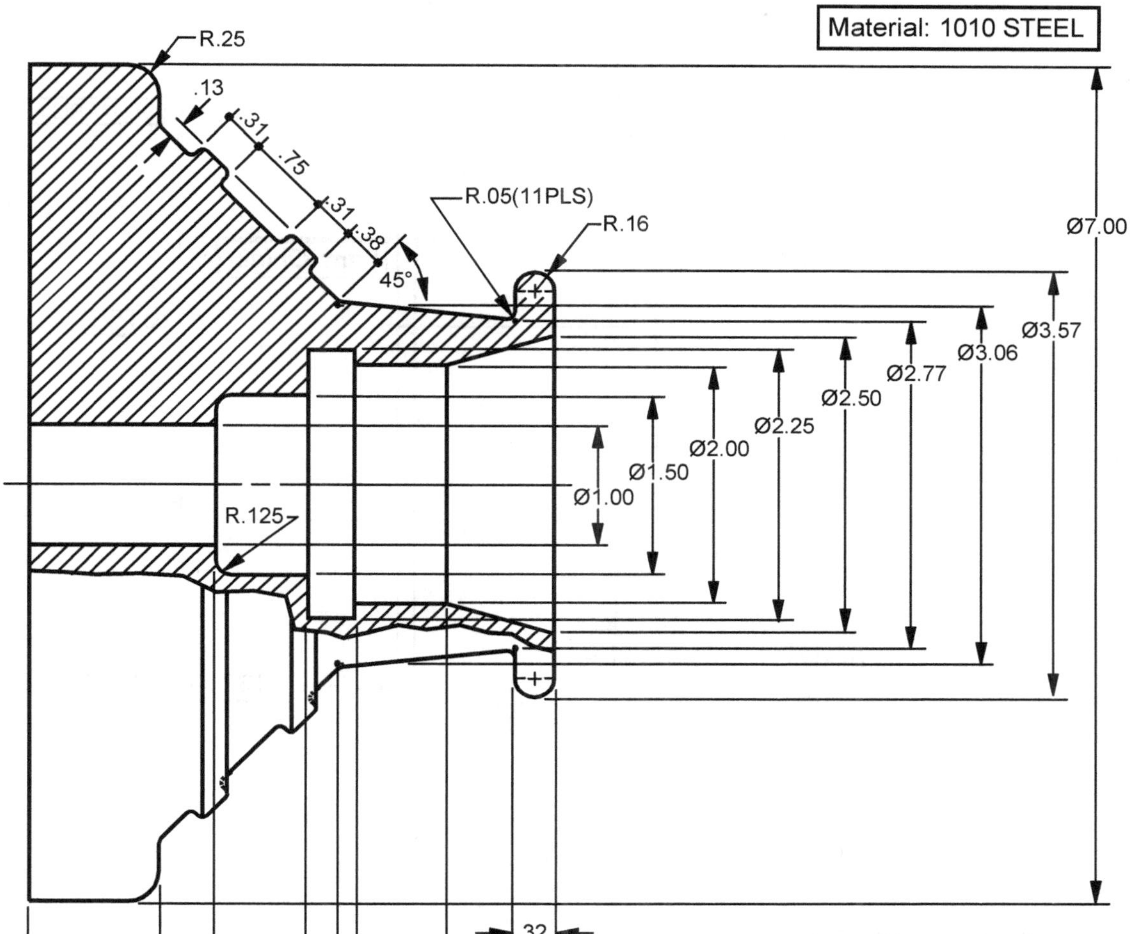

Material: 1010 STEEL

R.25
.13
.31
.75
.31
.38
45°
R.05(11PLS)
R.16
Ø7.00
Ø3.57
Ø3.06
Ø2.77
Ø2.50
Ø2.25
Ø2.00
Ø1.50
Ø1.00
R.125
.32
.88
1.63
1.73
2.00
2.75
3.25
4.30

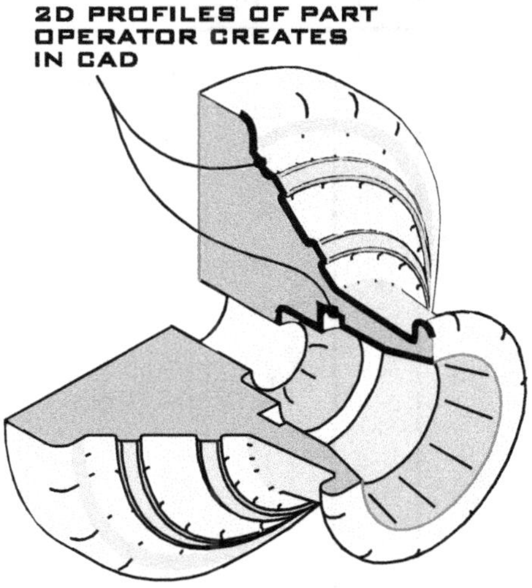

2D PROFILES OF PART
OPERATOR CREATES
IN CAD

PICTORIAL

Figure 4-p6

PROCESS PLAN 4P-6

No.	Operation	Tooling
1	FACE END Material Removed .10	1/32 TNR, RH OD TURNING TOOL R1/32

PROCESS PLAN 4P-6(*continued*)

No.	Operation	Tooling
2	ROUGH TURN FACE; LEAVE .01 IN X AND Z FOR FINISH CUT.	1/64 TNR, RH OD FINISHING TOOL

Rough Direction/Angle

Angle

0.0

Plunge parameters

☑ Lead In/Out

Lead In | Lead Out

Adjust Contour

☑ Extend/shorten end of contour

Ammount .05   ⦿ Extend   ○ Shorten

☐ Add Line

Exit Vector

☑ Use exit vector

Fixed Direction
○ None
○ Tangent
⦿ Perpendicular

Angle: 45

Length: 0.1

R1/64

CHAIN 1 START PT ②

Material Removed

.01 LEFT FOR FINISH CUT

① CHAIN 1 END PT

PROCESS PLAN 4P-6(*continued*)

No.	Operation	Tooling
3	FINISH OD CONTOUR	1/64 TNR, RH OD FINISHING TOOL

Plunge parameters

☑ Lead In/Out

| Lead In | Lead Out |

┌─ Adjust Contour ─────────────────
☑ Extend/shorten end of contour

Ammount **.05**   ● Extend
                    ○ Shorten

☐ Add Line

┌─ Exit Vector ─────
☑ Use exit vector
┌─ Fixed Direction ─
○ None
○ Tangent
● Perpendicular

Angle: 45

Length: 0.1

R1/64

② CHAIN 1 END PT

Material Removed

.01

① CHAIN 1 START PT

PROCESS PLAN 4P-6(*continued*)

No.	Operation	Tooling
4	ROUGH AND FINISH OD GROOVE ; LEAVE .01 IN X AND Z FOR FINISHING.	R.063, W .093, OD GROOVING RIGHT

Groove Definition

○ 1 Point
○ 2 Points
○ 3 Lines
◉ Chain
○ Multiple chains

☑ Lead In

Entry Vector
☑ Use Entry Vector

Fixed Direction
○ None
○ Tangent
◉ Perpendicular

Material Removed

① CHAIN 1 START PT
② CHAIN 1 END PT

.093

.75

A

1

DETAIL A

R.063

.5

.093

KENNAMETAL
GR-4125

PROCESS PLAN 4P-6(*continued*)

No.	Operation	Tooling
5	Rough and Finish OD Groove ; leave .01 in X and Z for finishing.	R.01, W.125 OD GROOVING CENTER

Groove Definition

- ○ 1 Point
- ○ 2 Points
- ○ 3 Lines
- ◉ Chain
- ○ Multiple chains

Tool Angle

Plunge Direction | Feed Direction

① -45

☑ Lead In

Entry Vector
☑ Use Entry Vector

Fixed Direction
- ○ None
- ○ Tangent
- ◉ Perpendicular

CHAIN 1 END PT

② Material Removed

CHAIN 1 START PT AT MIDPOINT OF LINE

①

DETAIL A

.093
.5
.125
R.01
.125

GC-4125

PROCESS PLAN 4P-6(*continued*)

No.	Operation	Tooling
6	ROUGH AND FINISH OD GROOVE ; LEAVE .01 IN X AND Z FOR FINISHING. CHAIN 1 END PT Material Removed ② ① CHAIN 1 START PT AT MIDPOINT OF LINE	R.01, W.125 OD GROOVING CENTER .125 A
7	LATHE DRILL X .18 DEEP Depth -.18 ⦿ Absolute ◯ Incremental NO CHECK ☐ Drill tip compensation Material Removed	.25 DIA CENTER DRILL Ø.25

PROCESS PLAN 4P-6(*continued*)

No.	Operation	Tooling
8	**LATHE DRILL THRU** Depth **-4.3** ⦿ Absolute ◯ Incremental Drill Cycle Parameters Cycle Peck drill ▼ **CHECK** ☑ Drill tip compensation Breakthrough ammount **.125** Material Removed	3/4 DIA DRILL 4 10 5 4 Ø1
9	**ROUGH BORE ID, LEAVE .01 IN X AND Z FOR FINISH CUT.** ☑ Lead In/Out Lead In Lead Out Plunge Cutting No Plunge Cutting allowed ◉ Adjust Contour ☑ Extend/shorten end of contour Ammount **.1** ⦿ Extend ◯ Shorten ☐ Add Line Exit Vector ☑ Use exit vector Fixed Direction ◯ None ◯ Tangent ⦿ Perpendicular Angle: 45 Length: 0.1 ① CHAIN 1 START PT ② CHAIN 1 END PT Material Removed .01 LEFT FOR FINISH CUT	1/64 TNR, ID ROUGH BORING TOOL ,.75 DIA R1/64 Ø.75

PROCESS PLAN 4P-6(*continued*)

No.	Operation	Tooling
10	**FINISH BORE ID CONTOUR** ☑ Lead In/Out Lead In | Lead Out ┌ Plunge Cutting ┐ No Plunge Cutting allowed ┌ Adjust Contour ┐ ☑ Extend/shorten end of contour Ammount .1 ● Extend ○ Shorten ☐ Add Line ┌ Exit Vector ┐ ☑ Use exit vector ┌ Fixed Direction ┐ ○ None ○ Tangent ● Perpendicular Angle: 45 Length: 0.1 ① CHAIN 1 START PT Material Removed ② CHAIN 1 END PT .01	**1/64 TNR, ID FINISH BORING TOOL ,.75 DIA** R1/64 Ø.75
11	**ROUGH AND FINISH ID GROOVE; LEAVE .01 IN X AND Z FOR FINISHING** ┌ Groove Definition ┐ ○ 1 Point ● 2 Points ○ 3 Lines ○ Chain ○ Multiple chains ① ② Material Removed ☑ Lead In ┌ Entry Vector ┐ ☑ Use Entry Vector ┌ Fixed Direction ┐ ○ None ○ Tangent ● Perpendicular	**R.01, W .25 ID GROOVING RIGHT HAND** .9 A Ø1 .144 8 DETAIL A .144 .5 .25 .25 R.01 GC-4250

4-7) Generate the 2D profiles of the part shown Figure 4-p7. Use PROCESS PLAN 4P-7 as an aid in generating the part program to produce the part.

Material: 7075 AL

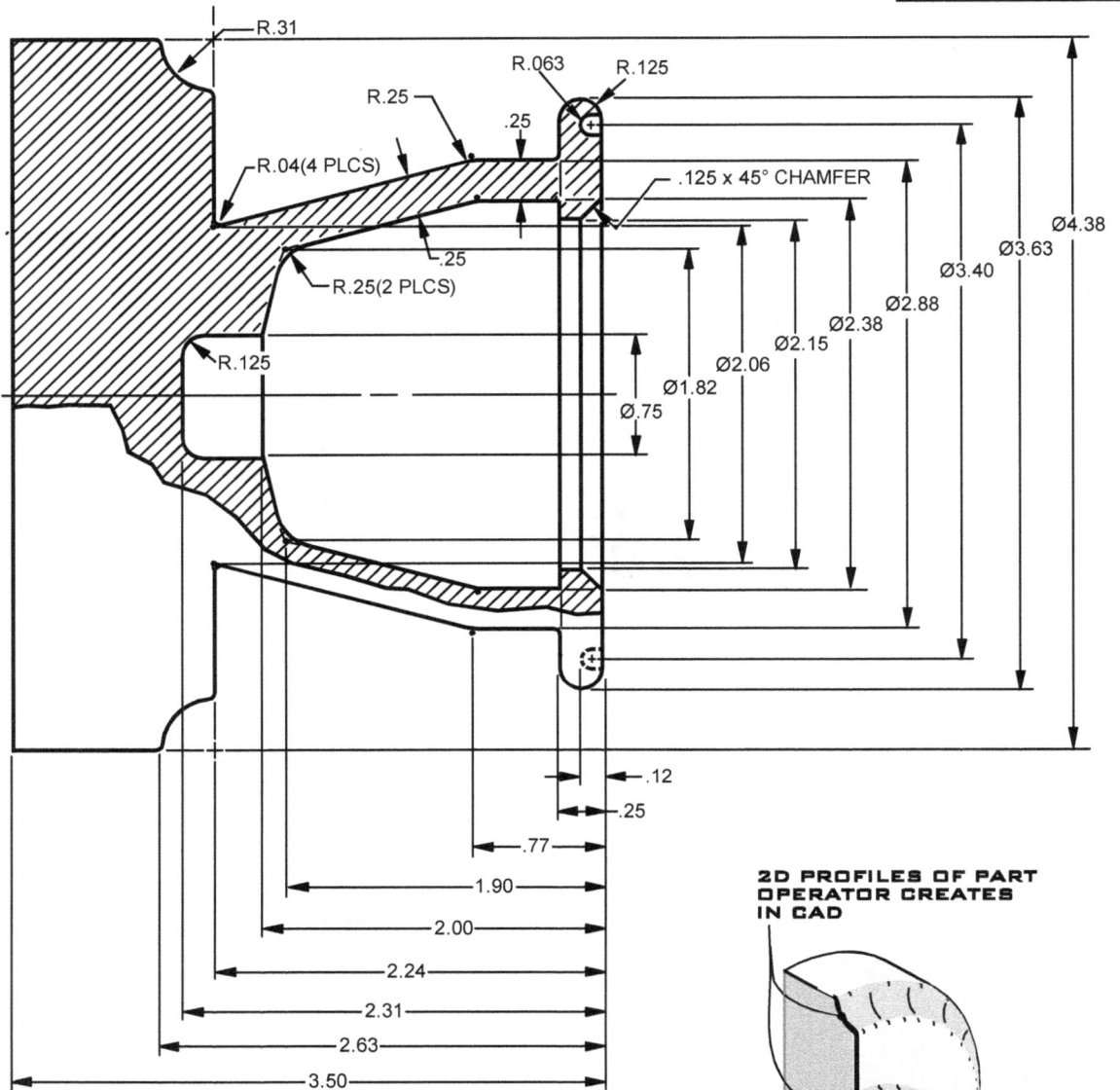

Figure 4p-7

PICTORIAL

2D PROFILES OF PART OPERATOR CREATES IN CAD

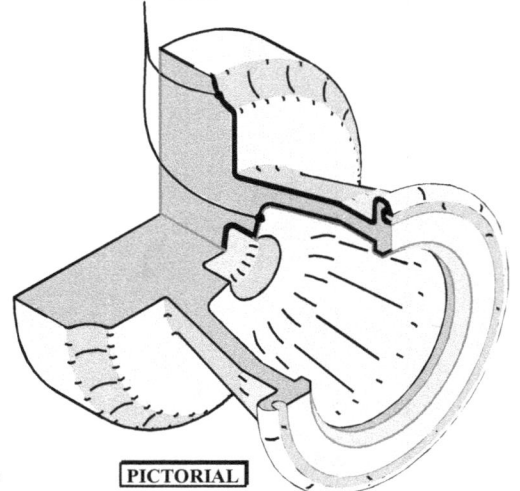

CHUCK

STOCK

0 | OD MARGIN |

| Part Origin |

Ø4.38 | OD |

0

| GRIP LENGTH | ← .75 →

| LEFT MARGIN | 0

← 3.50 →

.10 | RIGHT MARGIN |

| LENGTH |

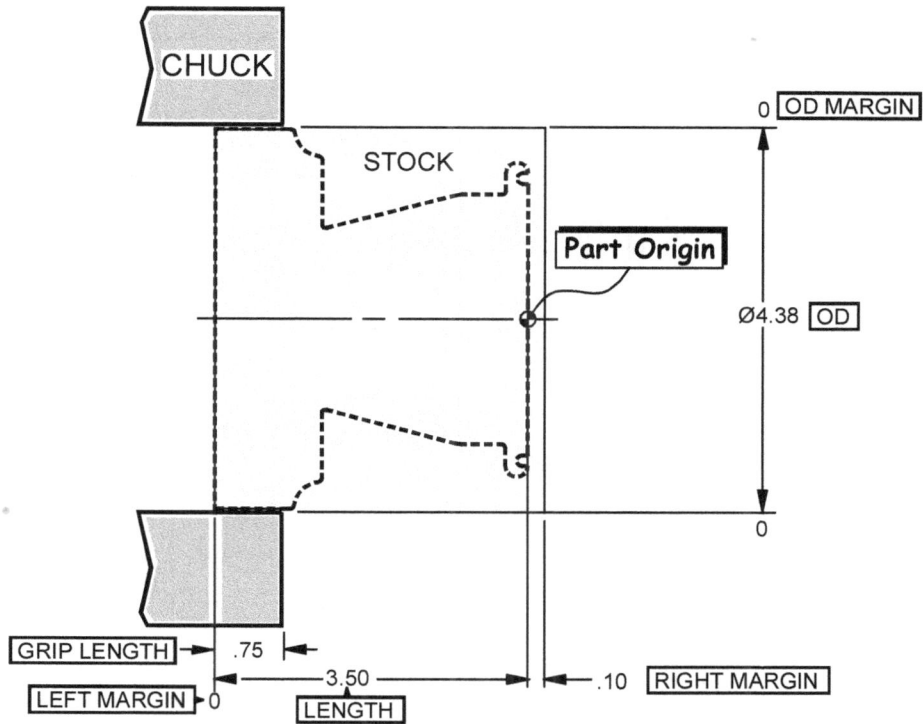

PROCESS PLAN 4P-7

No.	Operation	Tooling
1	FACE END Material Removed ← .10 →	1/32 TNR, RH OD TURNING TOOL R1/32

PROCESS PLAN 4P-7(*continued*)

No.	Operation	Tooling
2	ROUGH TURN OD; LEAVE .01 IN X AND Z FOR FINISH CUT.	1/64 TNR, RH OD FINISHING TOOL

☑ Lead In/Out

Plunge parameters

Lead In | Lead Out

┌─ Adjust Contour ─┐
☐ Extend/shorten end of contour

Ammount   0.0   ◉ Extend
                  ○ Shorten

☐ Add Line

┌─ Entry Vector ─┐
☑ Use entry vector

┌─ Fixed Direction ─┐
○ None
○ Tangent
◉ Perpendicular

Lead In | Lead Out

┌─ Adjust Contour ─┐
☑ Extend/shorten end of contour

Ammount   **.05**   ◉ Extend
                    ○ Shorten

☐ Add Line

┌─ Exit Vector ─┐
☑ Use exit vector

┌─ Fixed Direction ─┐
○ None
○ Tangent
◉ Perpendicular

─ R1/64

CHAIN1 END PT ②

Material Removed

CHAIN1 START PT ①

.01 LEFT FOR FINISH CUT

PROCESS PLAN 4P-7(*continued*)

No.	Operation	Tooling
3	ROUGH TURN OD, LEAVE .O1 IN X AND Z FOR SEMI-FINISH. SEMI-FINISH CONTOUR IN TWO PASSES USING A STEPOVER OF .OO5	1/64 TNR, LH OD FINISHING TOOL

Plunge parameters

☑ Lead In/Out

Lead In | Lead Out

Adjust Contour
☑ Extend/shorten end of contour
Ammount **.05** ● Extend ○ Shorten
☐ Add Line

Exit Vector
☑ Use exit vector
Fixed Direction
○ None
○ Tangent
● Perpendicular

☑ Semi Finish → **Semi Finish Parameters** ✕

Number of passes
2

Stepover
.005

Stock to leave in X:
0

Stock to leave in Z:
0

R1/64

Material Removed
② CHAIN1 END PT
① CHAIN1 START PT

.01 LEFT FOR FINISH CUT

PROCESS PLAN 4P-7(*continued*)

No.	Operation	Tooling
4	FINISH OD CONTOUR 	OD FINISHING TOOL R1/64

Inside the Operation panel, the following labels appear:

- Plunge parameters
- ☑ Lead In/Out
- Lead In | Lead Out
- Adjust Contour
  - ☑ Extend/shorten end of contour
  - Ammount **.05**   ● Extend   ○ Shorten
  - ☐ Add Line
- Exit Vector
  - ☑ Use exit vector
  - Fixed Direction
    - ○ None
    - ○ Tangent
    - ● Perpendicular

In the lower diagram:
- CHAIN 1 END PT ②
- CHAIN 1 START PT ①
- Material Removed
- .01

PROCESS PLAN 4P-7(*continued*)

No.	Operation	Tooling
5	**ROUGH FACE GROOVE**	**R.01, W.125** FACE GROOVING RIGHT

Operation 5:

Grooving Options

Groove Definition
- ● 1 Point
- ○ 2 Points
- ○ 3 Lines
- ○ Chain
- ○ Multiple chains

Point Selection
- ● Manual
- ○ Window

Groove shape parameters

Groove position

Height: .120

● Radius
0.00
○ Chamfer

Taper angle
0.00

● Radius
.0625
○ Chamfer

Radius ●
0.00
Chamfer ○

Taper angle
0.00

Radius ●
.0625
Chamfer ○

☐ Use tool width
Width: .125

Material Removed ①

Tooling (Operation 5):
- 1.5
- 1
- 6
- .125
- A
- NGEHR-24(I)
- DETAIL A
- R.01
- .5
- .125
- .125
- GC-4125

| 6 | **LATHE DRILL X .18 DEEP** | **.25 DIA CENTER DRILL** |

Depth: -.18
● Absolute ○ Incremental

NO CHECK
☐ Drill tip compensation

Material Removed

Ø.25

PROCESS PLAN 4P-7(*continued*)

No.	Operation	Tooling
	LATHE DRILL X 2.31 DEEP Depth [-2.31] ● Absolute ○ Incremental Drill Cycle Parameters Cycle [Peck drill ▼] [NO CHECK] □ Drill tip compensation Material Removed	5/8 DIA DRILL 4.75 .75 8.5 3 2.75 .625
8	LATHE DRILL X 2.31 DEEP Depth [-2.31] ● Absolute ○ Incremental [NO CHECK] □ Drill tip compensation Material Removed	3/4 DIA .125 CR BULL END MILL 4.5 8.5 1 3 2.5 R.125 .75

PROCESS PLAN 4P-7(*continued*)

No.	Operation	Tooling
9	LATHE DRILL X 1.93 DEEP Depth: **-1.90** ● Absolute ○ Incremental Drill Cycle Parameters Cycle: Peck drill ▼ NO CHECK □ Drill tip compensation Material Removed	1.5 DIA ENDMILL 3 2 9 4 3 ←1.5→
10	ROUGH BORE ID, LEAVE .01 IN X AND Z FOR FINISH CUT. ☑ Lead In/Out Plunge parameters Plunge Cutting No Plunge Cutting allowed ● Lead In | Lead Out Adjust Contour ☑ Extend/shorten end of contour Ammount **.1** ● Extend ○ Shorten □ Add Line Exit Vector ☑ Use exit vector Fixed Direction ○ None ○ Tangent ● Perpendicular ① CHAIN1 START PT CHAIN1 END PT ② .01 LEFT FOR FINISH CUT Material Removed	1/64 TNR, ID ROUGH BORING TOOL ,.5 DIA R1/64 Ø.5

PROCESS PLAN 4P-7(*continued*)

No.	Operation	Tooling
11	**FINISH BORE ID CONTOUR**	1/64 TNR, ID FINISH BORING TOOL ,.5 DIA

Plunge parameters

Plunge Cutting

No Plunge Cutting allowed

☑ Lead In/Out

Lead In | Lead Out

Adjust Contour
☑ Extend/shorten end of contour

Ammount .1   ⦿ Extend
               ○ Shorten

☐ Add Line

Exit Vector
☑ Use exit vector

Fixed Direction
○ None
○ Tangent
⦿ Perpendicular

R1/64

Ø.5

① CHAIN1 START PT

CHAIN1 END PT

.01

Material Removed

②

| 12 | **ROUGH AND FINISH ID GROOVE; LEAVE .01 IN X AND Z FOR FINISHING** | R.01, W .144 ID GROOVING RIGHT HAND |

Groove Definition
○ 1 Point
○ 2 Points
○ 3 Lines
⦿ Chain
○ Multiple chains

① CHAIN1 START PT

Material Removed

②

CHAIN1 END PT AT MIDPOINT OF LINE

☑ Lead In

Entry Vector
☑ Use Entry Vector

Fixed Direction
○ None
○ Tangent
⦿ Perpendicular

A

.9

1

.144

8

DETAIL A

.144

.5

.25

.25   R.01

CG-4250

4-8) Generate the 2D profiles of the part shown in Figure 4p-8. Use PROCESS PLAN 4P-8
as an aid in generating the part program to produce the part.

Material: 1030 STEEL

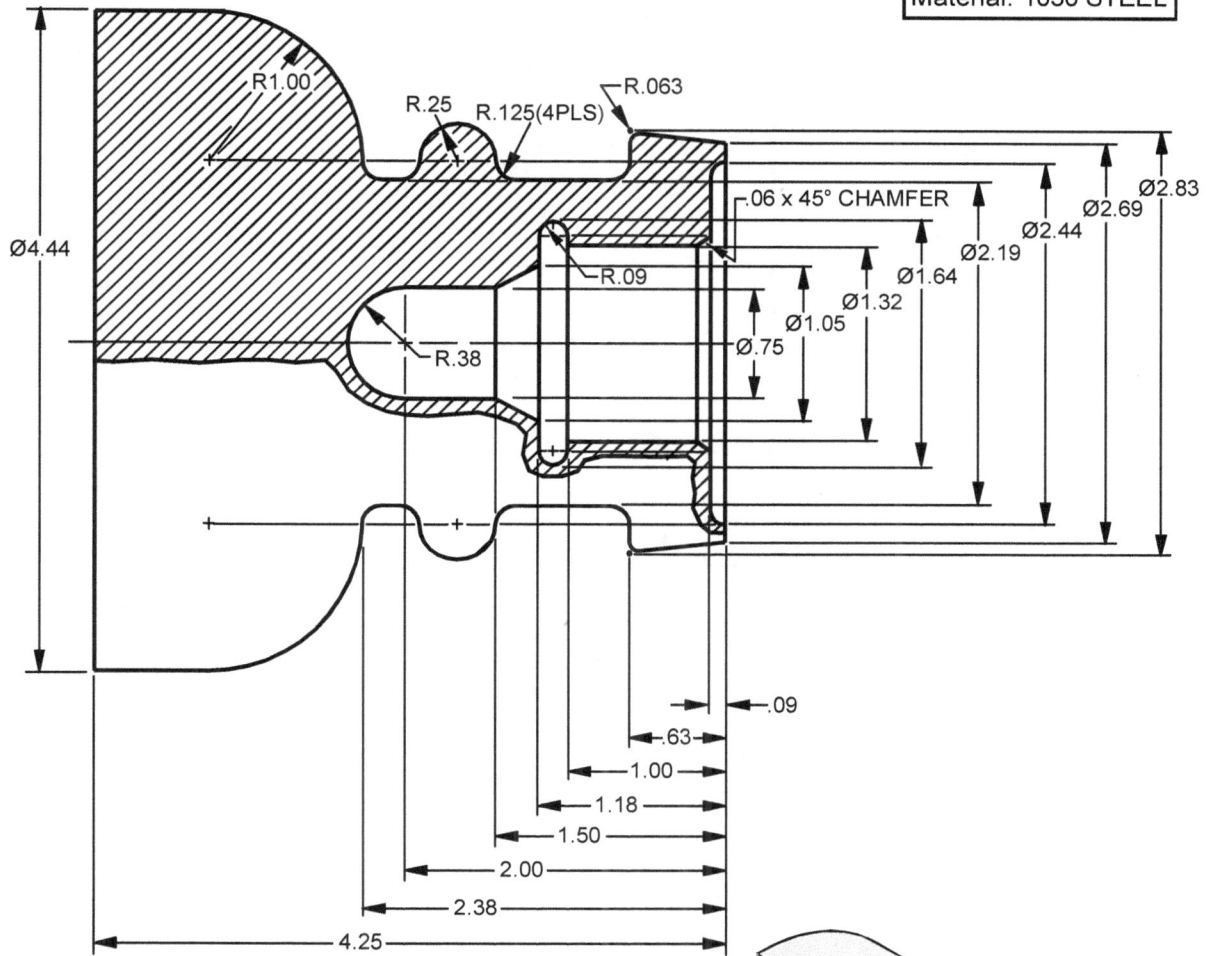

Figure 4p-8

2D PROFILES
OF PART
OPERATOR
CREATES
IN CAD

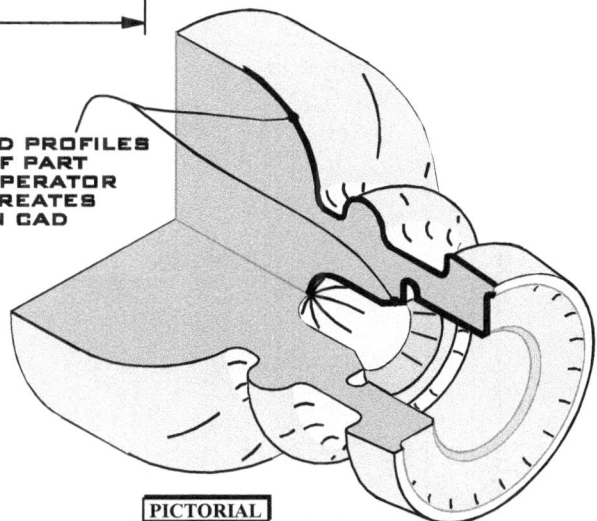

PICTORIAL

CHUCK

STOCK

0 OD MARGIN

Part Origin

Ø4.44 OD

0

GRIP LENGTH → .75 ←

LEFT MARGIN → 0

4.25

LENGTH

.10 RIGHT MARGIN

PROCESS PLAN 4P-8

No.	Operation	Tooling
1	FACE END Material Removed .10	1/32 TNR, RH OD TURNING TOOL R1/32

PROCESS PLAN 4P-8(*continued*)

No.	Operation	Tooling
2	ROUGH TURN OD; LEAVE .01 IN X AND Z FOR FINISH CUT.	1/32 TNR, RH OD TURNING TOOL

Plunge parameters

Plunge Cutting

No Plunge Cutting allowed

☑ Lead In/Out

Lead In | Lead Out

Adjust Contour
☑ Extend/shorten end of contour

Ammount **.05**   ● Extend
                 ○ Shorten

☐ Add Line

Exit Vector
☑ Use exit vector

Fixed Direction
○ None
○ Tangent
● Perpendicular

R1/32

CHAIN 1 END PT   ②

Material Removed

①

CHAIN 1 START PT

.01 LEFT FOR FINISH CUT

PROCESS PLAN 4P-8(*continued*)

No.	Operation	Tooling
3	Finish OD contour	1/64 TNR, RH OD FINISHING TOOL

Plunge parameters

Plunge Cutting

No Plunge Cutting allowed

☑ Lead In/Out

Lead In | Lead Out

Adjust Contour

☑ Extend/shorten end of contour

Ammount **.05**   ◉ Extend   ○ Shorten

☐ Add Line

Exit Vector

☑ Use exit vector

Fixed Direction
○ None
○ Tangent
◉ Perpendicular

R1/64

CHAIN 1 END PT ②

Material Removed

CHAIN 1 START PT ①

.01

PROCESS PLAN 4P-8(*continued*)

No.	Operation	Tooling
4	ROUGH AND FINISH OD GROOVE ; LEAVE .01 IN X AND Z FOR FINISHING.	R.01, W .125 OD GROOVING RIGHT

Groove Definition
- 1 Point
- 2 Points
- 3 Lines
- ● Chain
- Multiple chains

☑ Lead In

Entry Vector
- ☑ Use Entry Vector
- Fixed Direction
  - None
  - Tangent
  - ● Perpendicular

ARC-1

Material Removed

② LINE-1

① CHAIN 1 START PT

CHAIN 1 END PT AT THE INTERSECTION OF LINE-1 AND ARC-1

.093

A

.5

1

DETAIL A

.093

.5

.125

.125

R.01

GC-4125

PROCESS PLAN 4P-8(*continued*)

No.	Operation	Tooling
5	ROUGH AND FINISH FACE GROOVE; LEAVE .01 IN X AND Z FOR FINISHING ☑ Lead In/Out Lead In | Lead Out ┌ Adjust Contour ┐ ☐ Extend/shorten end of contour Ammount [0.0] ◉ Extend ○ Shorten ☐ [Add Line] ┌ Exit Vector ┐ ☑ Use exit vector ┌Fixed Direction┐ ○ None ○ Tangent ◉ Perpendicular ② CHAIN 1 END PT Material Removed ① CHAIN 1 START PT	R.01, W.125 FACE GROOVING RIGHT .75 .5 6 A .093 DETAIL A .093 .5 .125 .125 R.01 GC-4125
6	LATHE DRILL X .18 DEEP [Depth] [-.27] ◉ Absolute ○ Incremental NO CHECK ☐ Drill tip compensation Material Removed	.25 DIA CENTER DRILL Ø.25

PROCESS PLAN 4P-8(*continued*)

No.	Operation	Tooling
7	LATHE DRILL X 2.31 DEEP Depth **-2.31** ⦿ Absolute ◯ Incremental Drill Cycle Parameters — Cycle: Peck drill ▼ NO CHECK □ Drill tip compensation Material Removed	5/8 DIA DRILL 4.75 .75 8.5 3 2.75 .625
8	LATHE DRILL X 2.31 DEEP Depth **-2.38** ⦿ Absolute ◯ Incremental NO CHECK □ Drill tip compensation Material Removed	3/4 DIA BALL END MILL 4.5 1 8.5 3 2.5 .75
9	LATHE DRILL X 1.93 DEEP Depth **-1.18** ⦿ Absolute ◯ Incremental Drill Cycle Parameters — Cycle: Peck drill ▼ NO CHECK □ Drill tip compensation Material Removed	1- 5/16 DIA ENDMILL 3 2 9 4 3 1.313

PROCESS PLAN 4P-8(*continued*)

No.	Operation	Tooling
10	FINISH BORE ID CONTOUR	1/64 TNR, ID FINISH BORING TOOL ,.5 DIA

Plunge parameters

Plunge Cutting

No Plunge Cutting allowed

☑ Lead In/Out

Lead In | **Lead Out**

Adjust Contour

☑ Extend/shorten end of contour

Ammount [.1] ◉ Extend ○ Shorten

☐ Add Line

Exit Vector
☑ Use exit vector

Fixed Direction
○ None
○ Tangent
◉ Perpendicular

R1/64

17.5°   Ø.5

Toolpath Parameters | **Finish Parameters**

Finish stepover [.005]

Number of finish passes [2]

Stock to leave in X: [0]

Stock to leave in Z: [0]

① CHAIN 1 START PT

② CHAIN 1 END PT

Material Removed

PROCESS PLAN 4P-8(*continued*)

No.	Operation	Tooling
1 1	Rough and Finish ID groove; leave .01 in X and Z for finishing	ID GROOVING RIGHT HAND

Groove Definition

- ○ 1 Point
- ○ 2 Points
- ○ 3 Lines
- ◉ Chain
- ○ Multiple chains

☑ Lead In

Entry Vector
☑ Use Entry Vector
Fixed Direction
- ○ None
- ◉ Tangent
- ○ Perpendicular

CHAIN1 START PT ①

CHAIN1 END PT AT THE MIDPOINT OF LINE ②

Material Removed

.65   Ø.5

A

.25
6

DETAIL A
R.0625

.5

.25

.25

GR-4125R

4-9) Generate the 2D profiles of the part shown Figure 4p-9. Use PROCESS PLAN 4P-9
as an aid in generating the part program to produce the part.

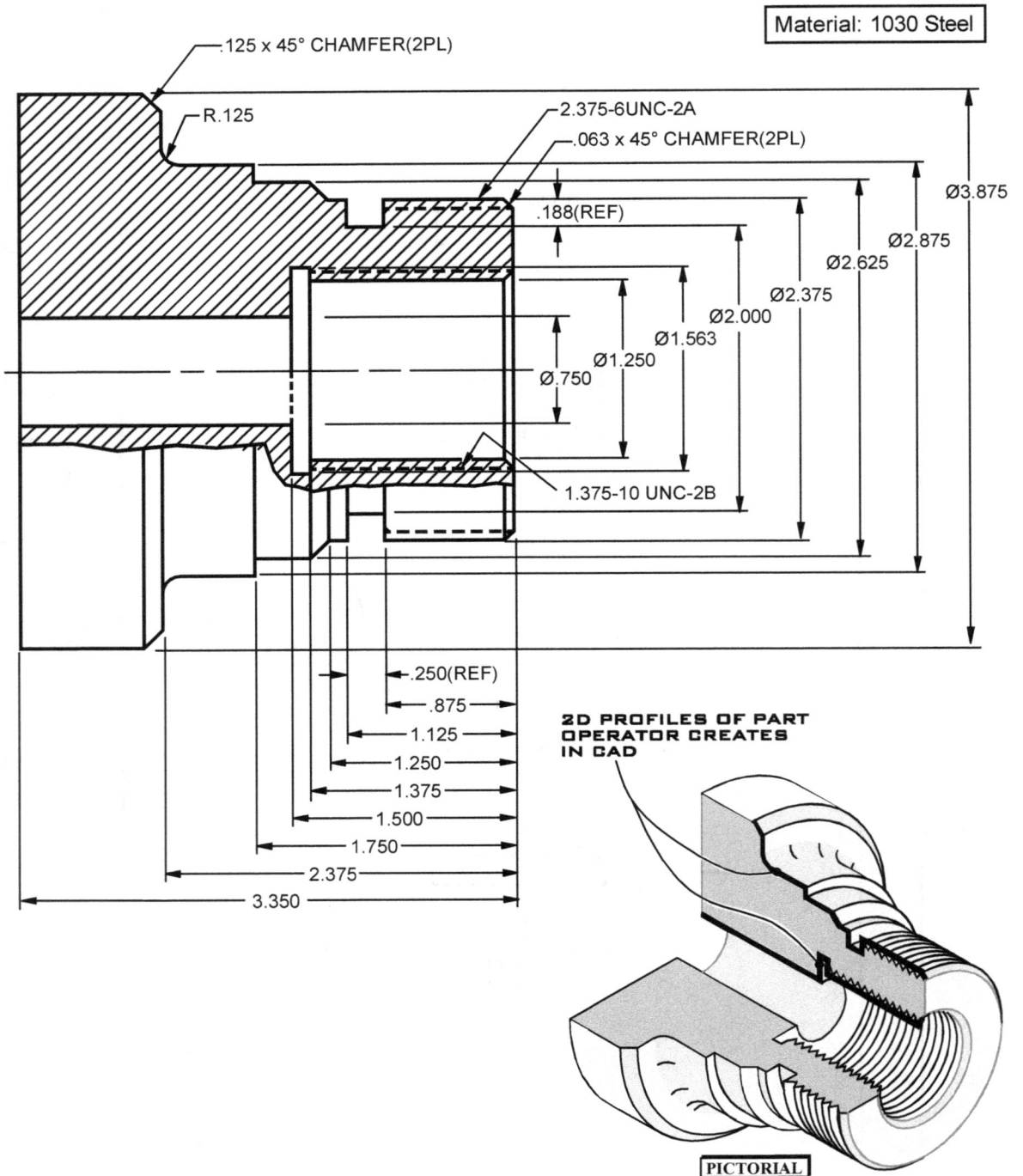

Material: 1030 Steel

.125 x 45° CHAMFER(2PL)

R.125

2.375-6UNC-2A

.063 x 45° CHAMFER(2PL)

.188(REF)

Ø3.875

Ø2.875

Ø2.625

Ø2.375

Ø2.000

Ø1.563

Ø1.250

Ø.750

1.375-10 UNC-2B

.250(REF)

.875

1.125

1.250

1.375

1.500

1.750

2.375

3.350

2D PROFILES OF PART OPERATOR CREATES IN CAD

PICTORIAL

Figure 4p-9

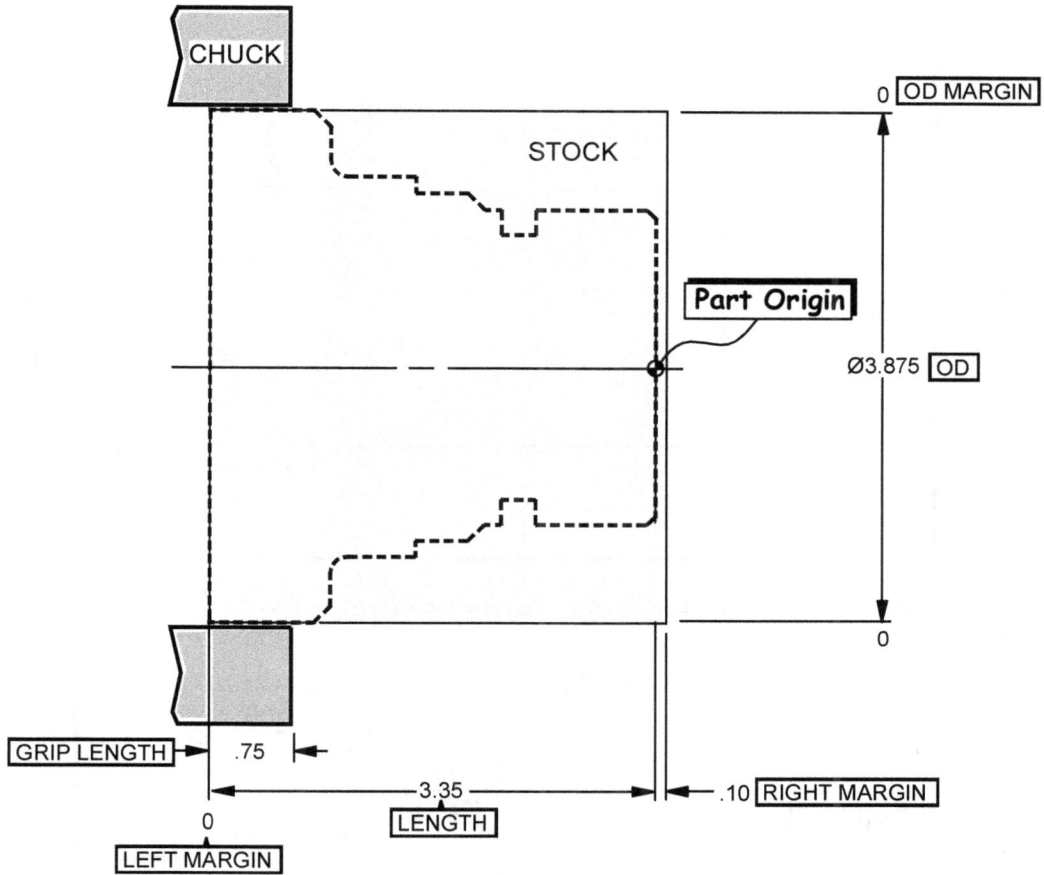

PROCESS PLAN 4P-9

No.	Operation	Tooling
1	FACE END	1/32 TNR, RH OD TURNING TOOL
2	ROUGH TURN OD, LEAVE .01 IN X AND Z FOR FINISHING.	

Operation 1 — FACE END

STOCK

Material Removed

.10

Tooling (Operation 1)

1/32 TNR, RH OD TURNING TOOL

R1/32

Operation 2 — ROUGH TURN OD, LEAVE .01 IN X AND Z FOR FINISHING.

CHAIN 1 END PT ②

Material Removed

Plunge Cutting

No Plunge Cutting allowed

.01 LEFT FOR FINISH CUT

① CHAIN 1 START PT

PROCESS PLAN 4P-9(*continued*)

No.	Operation	Tooling
3	**FINISH OD CONTOUR** Plunge Cutting — No Plunge Cutting allowed CHAIN 1 END PT ② Material Removed .01 CHAIN 1 START PT ①	1/64 TNR, RH OD FINISHING TOOL R1/64
4	**ROUGH OD GROOVE IN A SINGLE PASS** Groove Definition ○ 1 Point ● 2 Points ○ 3 Lines ○ Chain ○ Multiple chains Groove rough parameters Rough step: Percent of tool width ▼ **100** Stock to leave in X: **0** Stock to leave in Z: **0** Groove finish parameters ☐ Finish groove NO CHECK APPEARS Material Removed ① ②	R.01, W .25 OD GROOVING RIGHT A DETAIL A .144 .5 .250 .250 R.01 GCP-4250

PROCESS PLAN 4P-9(*continued*)

No.	Operation	Tooling
5	CUT 2.375-6 UNC OD THREAD	60°V,RH,OD THREADING TOOL

Thread Form
- Select from table
- Compute from formula
- Draw Thread

Thread Table

Thread Form: Unified - UN, UNC, UNF, UNEF ▼

Common diameter/lead combinations up to 4 inches

Basic major	Lead	Major diameter	Minor diameter	Comment ▼
2.3750	6.0000	2.3750	2.1946	2-3/8,6

Thread shape parameters

End Position	StartPosition
-.875	0.0

Thread cut parameters

NC code format: Canned ▼

Overcut
.06

DETAIL A

.195 → ← .117

.75

R.018

NA-3L6I

.06 ← → .875 →

Material Removed

PROCESS PLAN 4P-9(*continued*)

No.	Operation	Tooling
6	LATHE DRILL X .18 DEEP Depth: **-.18** ◉ Absolute ○ Incremental NO CHECK □ Drill tip compensation Material Removed	.25 DIA CENTER DRILL Ø.25
7	LATHE DRILL, THRU Depth: **-3.35** ◉ Absolute ○ Incremental Drill Cycle Parameters Cycle: Peck drill ▼ Ist peck: **.125** Subsequent peck: **.125** CHECK ☑ Drill tip compensation Breakthrough ammount: **.125** Material Removed	.5 DIA DRILL 2 4 9 4 3.5 Ø.75

PROCESS PLAN 4P-9(*continued*)

No.	Operation	Tooling
8	LATHE DRILL X 1.5 DEEP	.75 DIA END MILL
9	ROUGH ID GROOVE	R.01, W .125 ID GROOVING RIGHT HAND

Operation 8

LATHE DRILL X 1.5 DEEP

Depth　　-1.5

◉ Absolute　○ Incremental

—Material Removed

NO CHECK — Drill tip compensation

Tooling (Op 8): .75 DIA END MILL

2

3

1.5

8

3.5

3

Ø1.25

Operation 9

ROUGH ID GROOVE

Groove Definition
- ○ 1 Point
- ○ 2 Points
- ○ 3 Lines
- ◉ Chain
- ○ Multiple chains

Groove rough parameters

Rough step:
Percent of tool width ▼
100

Stock to leave in X: 0

Stock to leave in Z: 0

Groove finish parameters

☐ Finish groove

NO CHECK APPEARS

CHAIN 1 START PT ①

CHAIN 1 END PT ②

Material Removed

Tooling (Op 9): R.01, W .125 ID GROOVING RIGHT HAND

.5　　A　　Ø.625

.093

8

KGSPR-10-094

DETAIL A

.093

.5

.125

.125　　R.01

GC-4125

PROCESS PLAN 4P-9(*continued*)

No.	Operation	Tooling
10	FINISH ID CHAMFER Plunge Cutting No Plunge Cutting allowed CHAIN 1 START PT ① CHAIN 1 END PT ② — Material Removed	1/64 TNR, ID FINISH BORING TOOL ,.5 DIA R1/64 Ø.5

PROCESS PLAN 4P-9(*continued*)

No.	Operation	Tooling
11	**CUT 1.250-10 UNC ID THREAD**	60°V, RH, ID THREADING TOOL

Thread Form
- Select from table
- Compute from formula
- Draw Thread

Thread Table

Thread Form: Unified - UN, UNC, UNF, UNEF

Common diameter/lead combinations up to 4 inches

Basic major	Lead	Major diameter	Minor diameter	Comment
1.3750	10.0000	1.3750	1.2667	1-3/8,10

Thread cut parameters

NC code format: Canned

Thread shape parameters

End Position: **-1.375**

StartPosition: **0.0**

Overcut **.02**

.02 — 1.375

Material Removed

DETAIL A

.375

16 ER-10UN

.51 Ø.75

7

SIR 0750 P16

4-10) Generate the 2D profiles of the part shown in Figure 4p-10. Use PROCESS PLAN
4P-10 as an aid in generating the part program to produce the part

Material: 303 Stainless

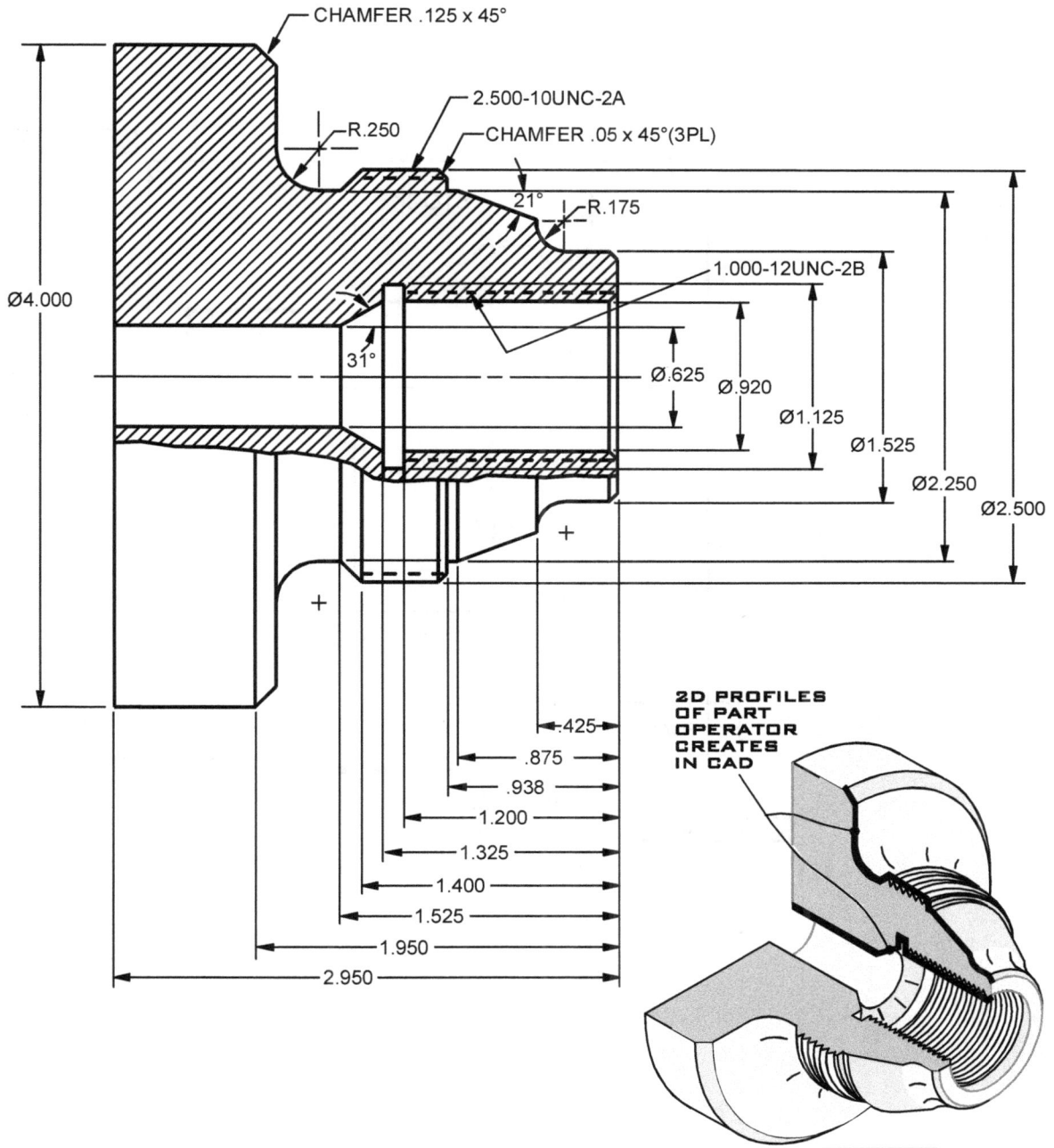

CHAMFER .125 x 45°

R.250

2.500-10UNC-2A

CHAMFER .05 x 45°(3PL)

21°

R.175

1.000-12UNC-2B

Ø4.000

31°

Ø.625

Ø.920

Ø1.125

Ø1.525

Ø2.250

Ø2.500

.425

.875

.938

1.200

1.325

1.400

1.525

1.950

2.950

2D PROFILES
OF PART
OPERATOR
CREATES
IN CAD

PICTORIAL

Figure 4p-10

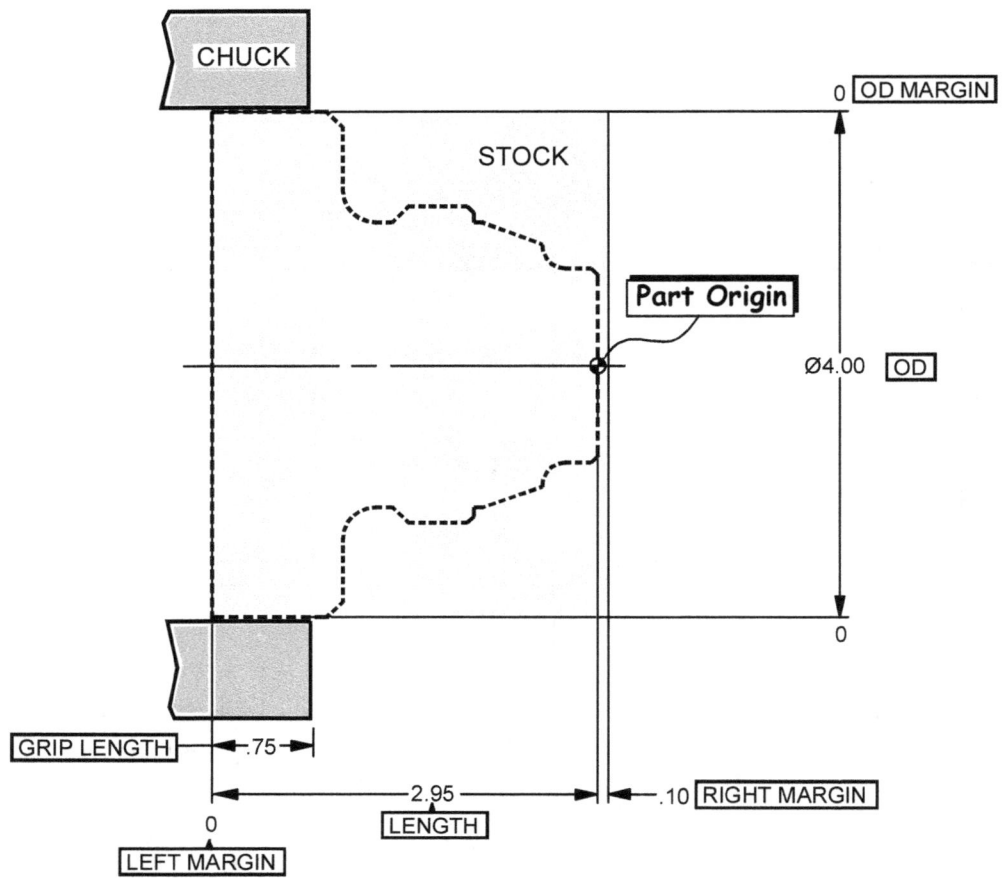

PROCESS PLAN 4P-10

No.	Operation	Tooling
1	FACE END STOCK Material Removed .10	1/32 TNR, RH OD TURNING TOOL R1/32
2	ROUGH TURN OD, LEAVE .01 IN X AND Z FOR FINISHING. Plunge Cutting — No Plunge Cutting allowed CHAIN1 END PT ② Material Removed .01 LEFT FOR FINISH CUT ① CHAIN1 START PT	
3	FINISH OD CONTOUR Plunge Cutting — No Plunge Cutting allowed CHAIN1 END PT ② Material Removed .01 ① CHAIN1 START PT	1/64 TNR, RH OD FINISHING TOOL R1/64

PROCESS PLAN 4P-10(*continued*)

No.	Operation	Tooling
4	ROUGH AND FINISH OD GROOVE ; LEAVE .01 IN X AND Z FOR FINISHING.	R.01, W .125 OD GROOVING RIGHT

Groove Definition
- ○ 1 Point
- ○ 2 Points
- ○ 3 Lines
- ● Chain
- ○ Multiple chains

☑ Lead In

Entry Vector
☑ Use Entry Vector
Fixed Direction
- ○ None
- ○ Tangent
- ● Perpendicular

② CHAIN1 END PT

Material Removed

① CHAIN1 START PT

.093
.5
1
A

DETAIL A
.093
.5
.125
.125
R.01

GC-4125

PROCESS PLAN 4P-10(*continued*)

No.	Operation	Tooling
5	CUT 2.375-6 UNC OD THREAD	60°V,RH,OD THREADING TOOL

Thread Form

Select from table

Thread Table

Thread Form: Unified - UN, UNC, UNF, UNEF ▼

Common diameter/lead combinations up to 4 inches

Basic major	Lead	Major diameter	Minor diameter	Comment ▼
2.5000	10.0000	2.5000	2.3917	2-1/2,10

Thread shape parameters

Thread cut parameters

NC code format: Canned ▼

Overcut **.15**

End Position **-1.4**

StartPosition **-.938**

Material Removed

.15   1.4   .938

DETAIL A

.375

16 ER 10UN

1

6

A

SER 1000 M16

PROCESS PLAN 4P-10(*continued*)

No.	Operation	Tooling
6	**LATHE DRILL X .18 DEEP** Depth `-.18` ◉ Absolute ○ Incremental Material Removed `NO CHECK` ☐ Drill tip compensation	**.25 DIA CENTER DRILL** ∅.25
7	**LATHE DRILL, THRU** Depth `-2.95` ◉ Absolute ○ Incremental ─ Drill Cycle Parameters ─ Cycle [Peck drill ▼] Ist peck `.125` Subsequent peck `.125` `CHECK` ☑ Drill tip compensation Breakthrough ammount `.125` Material Removed	**5/8 DIA DRILL** ∅1.88 4 8.5 1 3.5 2.75 ∅.625

PROCESS PLAN 4P-10(*continued*)

No.	Operation	Tooling
8	**ROUGH BORE ID, LEAVE .01 FOR FINISH CUT** ☑ Lead In/Out ↓ [Lead In] [Lead Out] ┌ Adjust Contour ┐ ☐ Extend/shorten end of contour Ammount [0.0] ◉ Extend / ○ Shorten ☐ [Add Line] ┌ Exit Vector ┐ ☑ Use exit vector ┌ Fixed Direction ┐ ○ None ○ Tangent ◉ Perpendicular Angle: [45] Length: [0.1] CHAIN 1 END PT ② CHAIN 1 START PT ① .01 LEFT FOR FINISH CUT — Material Removed ┌ Plunge Cutting ┐ No Plunge Cutting allowed ◉	5/64 TNR, ID ROUGH BORING TOOL ,.25 DIA R.5/64 A .188 Ø.25 **DETAIL A** Ø5/32 E(75 deg. diamond)
9	**FINISH BORE ID** ☑ Lead In/Out ↓ [Lead In] [Lead Out] ┌ Adjust Contour ┐ ☐ Extend/shorten end of contour Ammount [0.0] ◉ Extend / ○ Shorten ☐ [Add Line] ┌ Exit Vector ┐ ☑ Use exit vector ┌ Fixed Direction ┐ ○ None ○ Tangent ◉ Perpendicular Angle: [45] Length: [0.1] CHAIN 1 END PT ② CHAIN 1 START PT ① .01 — Material Removed ┌ Plunge Cutting ┐ No Plunge Cutting allowed ◉	5/64 TNR, ID FINISH BORING TOOL ,.25 DIA R1/64 A .188 Ø.25 Ø5/32 D(55 deg. diamond)

PROCESS PLAN 4P-10(*continued*)

No.	Operation	Tooling
10	ROUGH ID GROOVE	R.01, W .125 ID GROOVING RIGHT HAND

Groove Definition
- 1 Point
- 2 Points
- 3 Lines
- ● Chain
- Multiple chains

Groove rough parameters

Rough step:
Percent of tool width ▼
100

Stock to leave in X:
0

Stock to leave in Z:
0

Groove finish parameters

☐ Finish groove

NO CHECK APPEARS

② CHAIN 1 END PT

① CHAIN 1 START PT

Material Removed

Tooling:

.5   ⌀.625

A

.093

8

LATHE(IN) KGSPR-10-094

DETAIL A

.093

.5

.125

.125   R.01

LATHE(IN) GC-4125

PROCESS PLAN 4P-10(*continued*)

No.	Operation	Tooling
11	CUT 1-12 UNC ID THREAD	60°V, RH, ID THREADING TOOL

Operation (No. 11 — CUT 1-12 UNC ID THREAD):

Thread Form
- Select from table

Thread Table

Thread Form: Unified - UN, UNC, UNF, UNEF ▼

Common diameter/lead combinations up to 4 inches

Basic major	Lead	Major diameter	Minor diameter	Comment
1.0000	12.0000	1.0000	0.90981	1,12, UNF

Thread shape parameters

Thread cut parameters

NC code format: Canned ▼

Overcut
.06

End Position
-1.2

StartPosition
0.0

.06   1.2

Material Removed

Tooling:

⊙ ID

DETAIL A

.375

16 ER-12UN

—A   .45   Ø.62

7

SIR 0625 P16

4-11) Generate the 2D profiles of the part shown in Figure 4p-11. Use PROCESS PLAN 4P-11 as an aid in generating the part program to produce the part.

Material: 303 Stainless

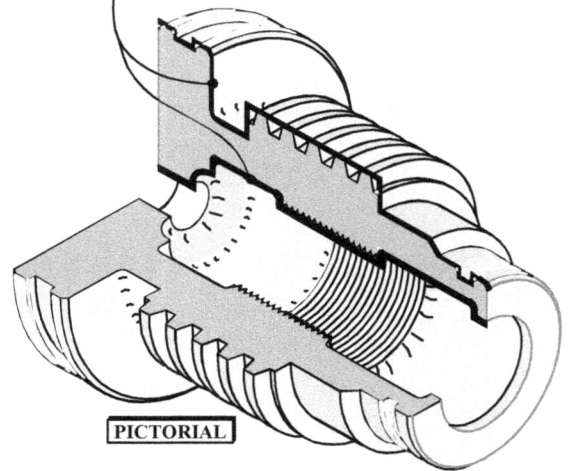

CHAMFER .040 x 45°(8PL)

R.050(5PL)   2-4 ACME   CHAMFER .100 x 45°

Ø2.875
Ø2.675
Ø2.000
Ø1.700
Ø1.300
Ø1.063
Ø1.500
Ø.883
Ø.750
Ø.375

R.125

.938-20UN
-2B

.097
.150
.300
.475
.560
1.000
1.960
2.285
2.650
2.800
3.000
3.150

2D PROFILES OF PART
OPERATOR CREATES
IN CAD

Figure 4p-11

PICTORIAL

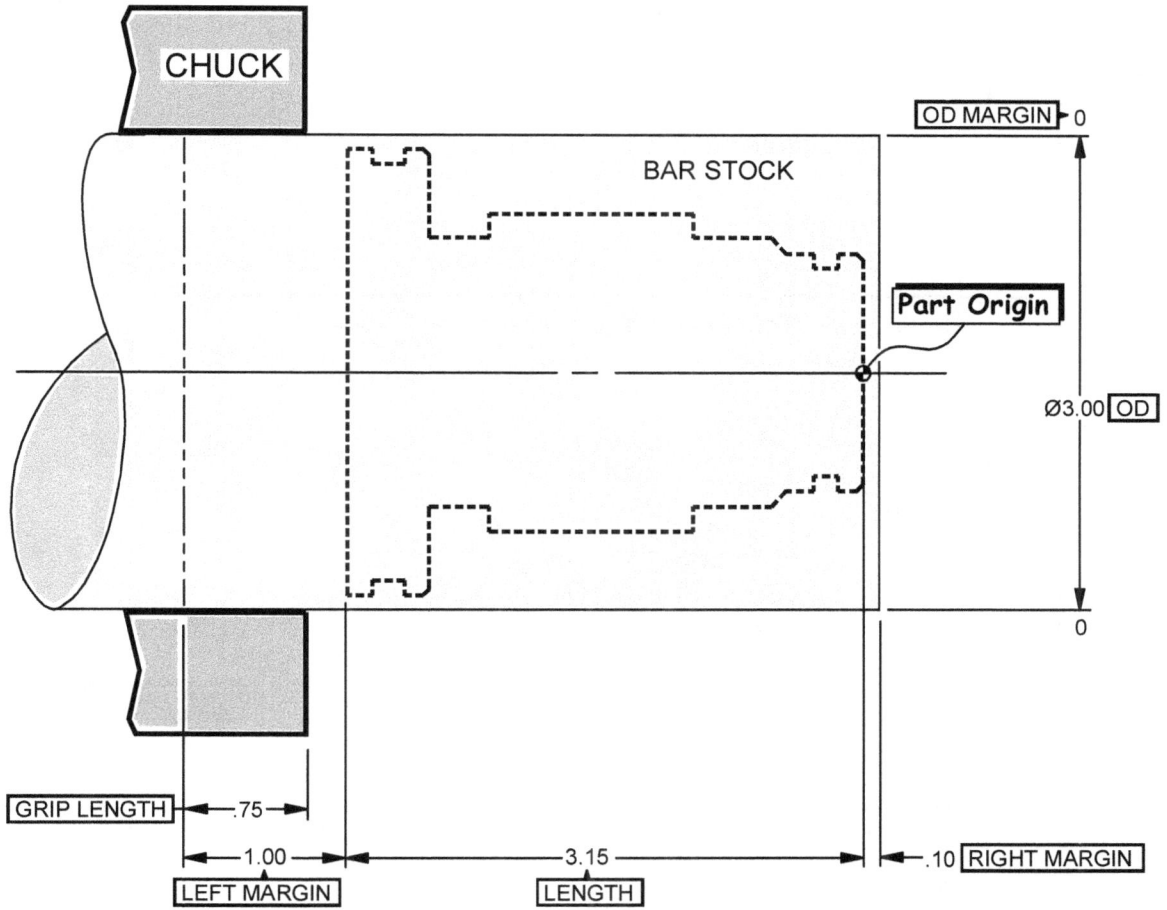

CHUCK

OD MARGIN ▸ 0

BAR STOCK

Part Origin

Ø3.00 OD

0

GRIP LENGTH |◄── .75 ──►|

1.00

3.15

.10 RIGHT MARGIN

LEFT MARGIN

LENGTH

PROCESS PLAN 4P-11

No.	Operation	Tooling
1	FACE END Material Removed — .10	1/32 TNR, RH OD TURNING TOOL R1/32
2	ROUGH TURN OD, LEAVE .01 IN X AND Z FOR FINISHING. Plunge Cutting No Plunge Cutting allowed .06 Material Removed CHAIN 1 END PT .01 LEFT FOR FINISH CUT CHAIN 1 START PT 	
3	FINISH OD CONTOUR Plunge Cutting No Plunge Cutting allowed .06 Material Removed CHAIN 1 END PT .01 CHAIN 1 START PT 	1/64 TNR, RH OD FINISHING TOOL R1/64

PROCESS PLAN 4P-11(*continued*)

No.	Operation	Tooling
4	ROUGH AND FINISH OD GROOVE; LEAVE .01 IN X AND Z FOR FINISHING	R.01, W .25 OD GROOVING RIGHT

Groove Definition

- ○ 1 Point
- ○ 2 Points
- ○ 3 Lines
- ◉ Chain
- ○ Multiple chains

Groove finish parameters

☑ Finish groove

☑ Lead In

Entry Vector
☑ Use Entry Vector

Fixed Direction
- ○ None
- ◉ Tangent
- ○ Perpendicular

② CHAIN 1 END PT

Material Removed

① CHAIN 1 START PT

DETAIL A

.144

.5

.250

.250

R.01

GCP-4250

PROCESS PLAN 4P-11(*continued*)

No.	Operation	Tooling
5	ROUGH AND FINISH OD GROOVE; LEAVE .01 IN X AND Z FOR FINISHING	R.01, W .125 OD GROOVING RIGHT

Toolpaths

Quick ➡ Groove

Quick Groove Options ☒

Groove Definition
- ○ 1 Point
- ◉ 2 Points
- ○ 3 Lines

Quick groove shape parameters

`0.04`
- ○ Radius
- ◉ 45 deg chamfer

`0.01`
- ◉ Radius
- ○ 45 deg chamfer

☑ Finish groove

☑ Lead In

Entry Vector
☑ Use Entry Vector

Fixed Direction
- ○ None
- ○ Tangent
- ◉ Perpendicular

Material Removed

Tooling dimensions: 1, 5, .093, A

DETAIL A
.075, .46, .093, .094, R.002

GS-094L

PROCESS PLAN 4P-11(*continued*)

No.	Operation	Tooling
6	CUT 2-4 ACME OD THREAD	ACME THREADING TOOL

Thread Form
Select from table

Thread Table

Thread Form | Acme General Purpose ▼

Common diameter/lead combinations up to 4 inches

Basic major	Lead	Major diameter	Minor diameter	Comment ▼
2.0000	4.0000	2.0000	1.7300	2, 4

Thread cut parameters

Thread shape parameters

NC code format: Canned ▼

End Position
-2.285

StartPosition
-1

Overcut
.125

DETAIL A

.202
.0875
1
.255

NA-4L4E

.125

2.285

1

Material Removed

PROCESS PLAN 4P-11(*continued*)

No.	Operation	Tooling
7	LATHE DRILL X .18 DEEP Depth **-.18** ● Absolute ○ Incremental NO CHECK — Drill tip compensation Material Removed	.25 DIA CENTER DRILL Ø.25
8	LATHE DRILL, THRU Depth **-3.15** ● Absolute ○ Incremental Drill Cycle Parameters Cycle: Peck drill Ist peck **.125** Subsequent peck **.125** CHECK ☑ Drill tip compensation Breakthrough ammount **.125** Material Removed 3.15	3/8 DIA DRILL Ø1.88 3 7.5 1 3.5 3 Ø.375

PROCESS PLAN 4P-11(*continued*)

No.	Operation	Tooling
9	**LATHE DRILL x 2.75 DEEP** — Depth **-2.75**, ◉ Absolute ○ Incremental. Drill Cycle Parameters — Cycle: Peck drill. Ist peck **.125**, Subsequent peck **.125**. NO CHECK, □ Drill tip compensation. Material Removed. 2.8	**3/4 DIA DRILL** — Ø2.25, 4, 1, 8.5, 3.5, 3, Ø.75
10	**LATHE DRILL x 2.8 DEEP** — Depth **-2.8**, ◉ Absolute ○ Incremental. Drill Cycle Parameters — Cycle: Drill/Counterbore. NO CHECK, □ Drill tip compensation. Material Removed. R.125. 2.85	**3/4 DIA BULL END MILL** — Ø2.25, 4, 8, 1, 3, 2.75, R.125, Ø.75
11	**LATHE DRILL x 2.65 DEEP** — Depth **-2.65**, ◉ Absolute ○ Incremental. Drill Cycle Parameters — Cycle: Peck drill. Ist peck **.125**, Subsequent peck **.125**. NO CHECK, □ Drill tip compensation. Material Removed. 2.65	**7/8 DIA END MILL** — Ø2.63, 4, 8, 1, 3, 2.5, Ø.875

PROCESS PLAN 4P-11(*continued*)

No.	Operation	Tooling
1 2	ROUGH AND FINISH ID GROOVE; LEAVE .01 IN X AND Z FOR FINISHING	R.01, W .125 ID GROOVING RIGHT HAND

Quick Groove Options

Groove Definition

○ 1 Point
◉ 2 Points
○ 3 Lines

Toolpaths

Quick ➡ Groove

Quick groove shape parameters

0.0   ○ Radius
        ◉ 45 deg chamfer

0.05   ◉ Radius
        ○ 45 deg chamfer

☑ Finish groove

☑ Lead In

─ Entry Vector ─
☑ Use Entry Vector
─ Fixed Direction ─
○ None
○ Tangent
◉ Perpendicular

④ ③ ② ①

Material Removed

.45   A   Ø.5

.093
6

DETAIL A
.093
.3125   .125
.125   R.01

GFG-125CC

PROCESS PLAN 4P-11(*continued*)

No.	Operation	Tooling
13	FINISH ID CHAMFER ☑ Lead In Entry Vector ☑ Use Entry Vector Fixed Direction ○ None ○ Tangent ◉ Perpendicular CHAIN 1 START PT ① ② Material Removed CHAIN 1 END PT	1/64 TNR, ID FINISH BORING TOOL ,.5 DIA R1/64 Ø.5

PROCESS PLAN 4P-11(*continued*)

No.	Operation	Tooling
14	CUT .938-20 UNC ID THREAD	60°V, RH, ID THREADING TOOL

Operation (No. 14 detail):

Thread Form — Select from table

Thread Table

Thread Form: Unified - UN, UNC, UNF, UNEF ▼

Common diameter/lead combinations up to 4 inches

Basic major	Lead	Major diameter	Minor diameter	Comment ▼
:	:	:	:	:
0.9375	20.0000	0.9375	0.8834	15/16,20,UNEF
:	:	:	:	:

Thread cut parameters

Thread shape parameters

NC code format: Canned ▼

Overcut **.06**

End Position	StartPosition
-1.96	**-1**

Material Removed

.06 1.96 1

Tooling (No. 14 detail):

⊙ ID

DETAIL A

.375

16 ER-20UN

A .42 Ø.62

7

SIR 0625 P16

PROCESS PLAN 4P-11(*continued*)

No.	Operation	Tooling
15	LATHE CUTOFF **Cutoff: Select boundary point:** Click ① the boundary point 	.01 TNR, RH OD CUTOFF TOOL 1.5 A DETAIL A .125 .125 R.01 .5 LATHE(IN) GC-4125

4-12) Generate the 2D profiles of the part shown Figure 4p-12. Use PROCESS PLAN
4P-12 as an aid in generating the part program to produce the part.

Material: 303 Stainless

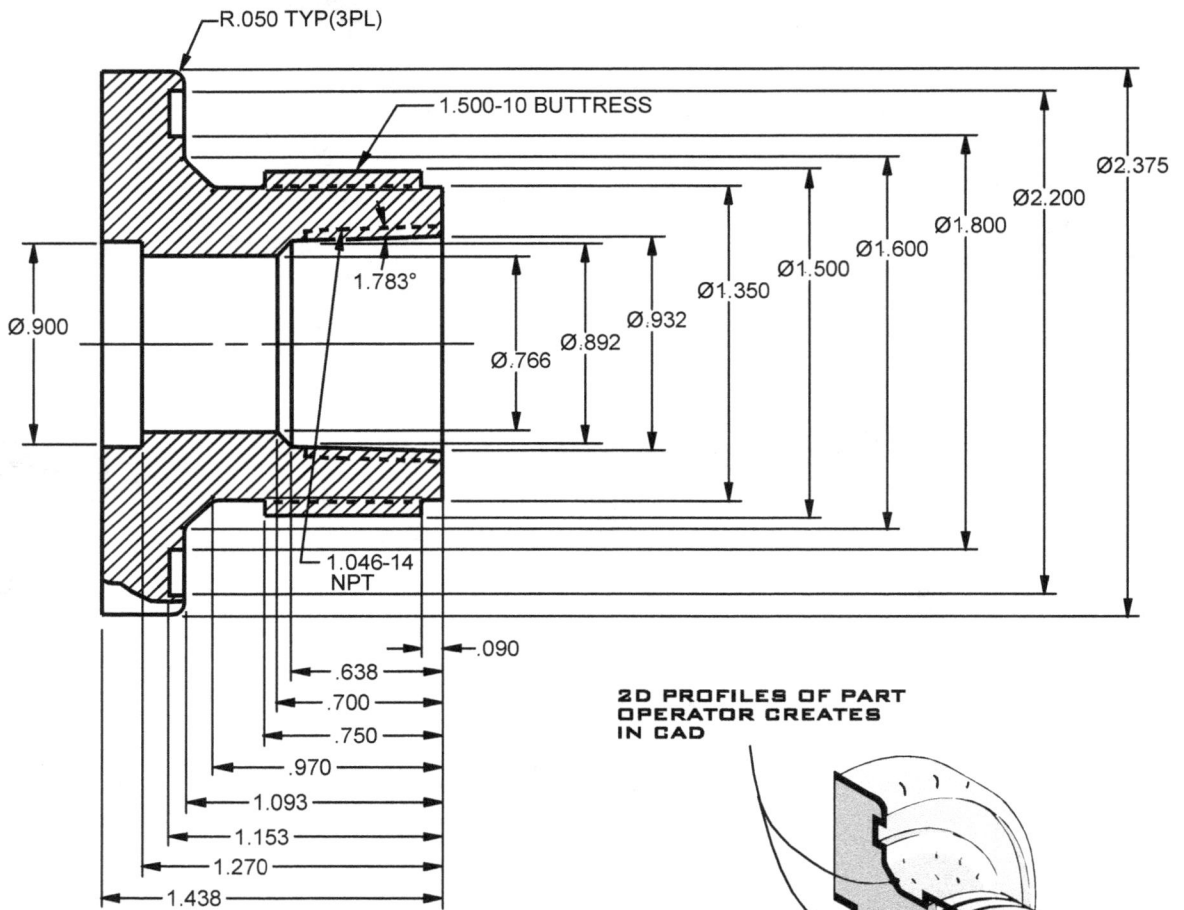

R.050 TYP(3PL)

1.500-10 BUTTRESS

Ø2.375

Ø2.200

Ø1.800

Ø1.600

Ø1.500

1.783°

Ø1.350

Ø.932

Ø.900

Ø.892

Ø.766

1.046-14
NPT

.090

.638

.700

.750

.970

1.093

1.153

1.270

1.438

**2D PROFILES OF PART
OPERATOR CREATES
IN CAD**

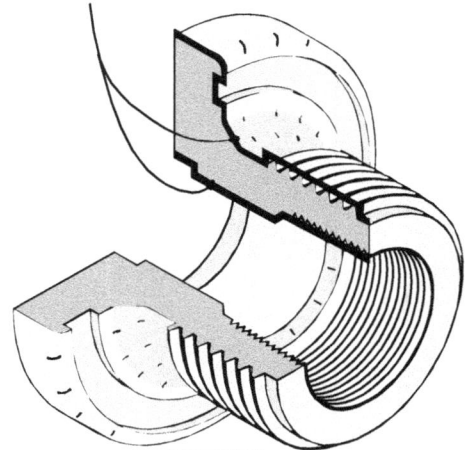

Figure4p-12   PICTORIAL

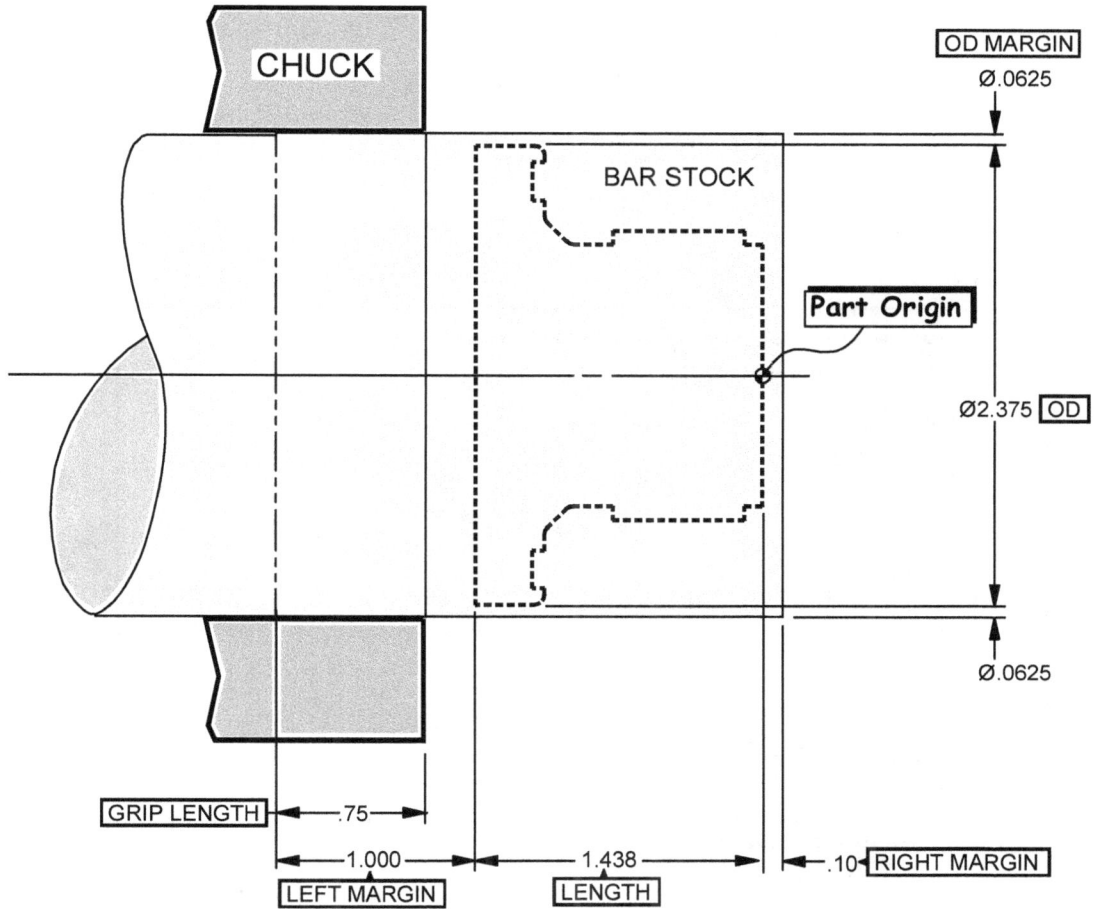

CHUCK

OD MARGIN
Ø.0625

BAR STOCK

Part Origin

Ø2.375 OD

Ø.0625

GRIP LENGTH .75

1.000
LEFT MARGIN

1.438
LENGTH

.10 RIGHT MARGIN

PROCESS PLAN 4P-12

No.	Operation	Tooling
1	FACE END BAR STOCK —Material Removed ←→ ←.10	1/32 TNR, RH OD TURNING TOOL R1/32
2	ROUGH TURN OD, LEAVE .01 IN X AND Z FOR FINISHING. Plunge Cutting No Plunge Cutting allowed ☑ Lead In/Out Lead In | Lead Out Adjust Contour ☑ Extend/shorten end of contour Ammount .125 ◉ Extend ○ Shorten .125 Material Removed ② CHAIN 1 END PT ① CHAIN 1 START PT .01 LEFT FOR FINISH CUT	
3	FINISH OD CONTOUR Plunge Cutting No Plunge Cutting allowed ☑ Lead In/Out Lead In | Lead Out Adjust Contour ☑ Extend/shorten end of contour Ammount .125 ◉ Extend ○ Shorten .125 Material Removed ② CHAIN 1 END PT ① CHAIN 1 START PT .01 LEFT FOR FINISH CUT	1/64 TNR, RH OD FINISHING TOOL R1/64

PROCESS PLAN 4P-12(*continued*)

No.	Operation	Tooling
4	ROUGH AND FINISH OD GROOVE ; LEAVE .O1 IN X AND Z FOR FINISHING.	R.O1, W .125 OD GROOVING RIGHT

Groove Definition
- ○ 1 Point
- ○ 2 Points
- ○ 3 Lines
- ◉ Chain
- ○ Multiple chains

☑ Lead In

Entry Vector
- ☑ Use Entry Vector

Fixed Direction
- ○ None
- ○ Tangent
- ◉ Perpendicular

Material Removed

② CHAIN 1 END PT

① CHAIN 1 START PT

.093

A

.5

1

DETAIL A

.093

.5

.125

.125

R.01

GC-4125

PROCESS PLAN 4P-12(*continued*)

No.	Operation	Tooling
5	ROUGH AND FINISH FACE GROOVE; LEAVE .01 IN X AND Z FOR FINISHING	R.01, W .125 ID GROOVING RIGHT HAND

Operation detail (No. 5):

Toolpaths

Quick ➡ Groove

Quick Groove Options ☒

Groove Definition
- ○ 1 Point
- ◉ 2 Points
- ○ 3 Lines

✓ ✗ ?

Quick groove shape parameters

0.0 — ○ Radius / ◉ 45 deg chamfer

0.01 — ◉ Radius / ○ 45 deg chamfer

☑ Finish groove

☑ Lead In

Entry Vector
☑ Use Entry Vector

Fixed Direction
- ○ None
- ◉ Tangent
- ○ Perpendicular

② ① Material Removed

Tooling detail:

1
5
A
1.38
.093

DETAIL A
.093
.5
.125
.125
R.01

GC-4125

PROCESS PLAN 4P-12(*continued*)

No.	Operation	Tooling
6	CUT 1-10 BUTRESS OD THREAD	BUTRESS THREADING TOOL

Thread Form — Select from table

Thread Table

Thread Form: Butress 7 degree/45 degree inch screw thread

Common diameter/lead combinations up to 4 inches

Basic major	Lead	Major diameter	Minor diameter	Comment
1.5000	10.0000	1.0000	1.3675	1-1/2, 10

Thread cut parameters

Thread shape parameters

NC code format: Canned

Overcut **.06**

End Position **-.75**   StartPosition **-.09**

.06   .75   .09

Material Removed

DETAIL A

.195 .117 1.25 .0125

BUTTRESS INSERT 10 THDS/IN

1 1.75 A

PROCESS PLAN 4P-12(*continued*)

No.	Operation	Tooling
7	LATHE DRILL X .18 DEEP Depth　　[**-.18**] ⦿ Absolute　○ Incremental [NO CHECK] ☐ Drill tip compensation Material Removed	**.25 DIA CENTER DRILL** Ø.25
8	LATHE DRILL X 1.563 DEEP Depth　　[**-1.563**] ⦿ Absolute　○ Incremental Drill Cycle Parameters Cycle [Peck drill ▼] Ist peck　[**.125**] Subsequent peck　[**.125**] [CHECK] ☑ Drill tip compensation Breakthrough ammount　[**.125**] Material Removed 1.563	**.766 (49/64) DIA DRILL** 6.25　Ø2.00 3 1 2.25　　2 Ø.766

PROCESS PLAN 4P-12(*continued*)

No.	Operation	Tooling
9	ROUGH AND FINISH ID GROOVE; LEAVE .01 IN X AND Z FOR FINISHING	R.007, W .013 ID GROOVING RIGHT HAND

Toolpaths

Quick ➡ Groove

Quick Groove Options ❌

Groove Definition

○ 1 Point
◉ 2 Points
○ 3 Lines

✓ ✗ ?

Quick groove shape parameters

`0.0` ○ Radius ◉ 45 deg chamfer

`0.007` ◉ Radius ○ 45 deg chamfer

☑ Finish groove

☑ Lead In

Entry Vector
☑ Use Entry Vector
Fixed Direction
○ None
◉ Tangent
○ Perpendicular

VHL 201 08 2

DETAIL A

.095
.4 A .5
.095
4.3

.5
.15
.13
R.007

VIPV .130E .007

.063 1.438

② Material Removed ①

PROCESS PLAN 4P-12(*continued*)

No.	Operation	Tooling
10	ROUGH BORE ID, LEAVE .01 IN X AND Z FOR SEMI-FINISH. SEMI-FINISH ID CONTOUR IN TWO PASSES USING A STEPOVER OF .05	1/64 TNR, ID FINISH BORING TOOL ,.5 DIA

☑ Lead In/Out

Lead In | Lead Out

Adjust Contour
☐ Extend/shorten end of contour

Ammount 0.0 ◉ Extend ○ Shorten

☐ Add Line

Exit Vector
☑ Use exit vector

Fixed Direction
○ None
◉ Tangent
○ Perpendicular

Angle: 45

Length: 0.1

Plunge Cutting

No Plunge Cutting allowed

◎

☑ Semi Finish ➡ **Semi Finish Parameters** ✖

Number of passes
2

Stepover
.005

Stock to leave in X:
0

Stock to leave in Z:
0

✓ ✖ ?

CHAIN 1 START PT
①
— Material Removed

② CHAIN 1 END PT

R1/64
17.5° Ø.5

PROCESS PLAN 4P-12(*continued*)

No.	Operation	Tooling
11	CUT .675-18 NPT ID THREAD	60°V,RH,ID THREADING TOOL

Thread Form

Select from table

Thread Table

Thread Form | American Pipe Threads - NPT |

Common diameter/lead combinations up to 4 inches

Basic major	Lead	Major diameter	Minor diameter	Comment
⋮	⋮	⋮	⋮	⋮
1.0500	14.0000	1.0460	0.9317	3/4,14, (1.050 OD)
⋮	⋮	⋮	⋮	⋮

Thread shape parameters

Thread cut parameters

NC code format: Canned

Overcut

0

Anticipated pulloff

.1

End Position

-.638

StartPosition

0

Thread orientation ID

Taper angle **-1.78333**

Material Removed

.1

.638

A   .45   .625

6

DETAIL A

.195

.117

.625

R.01

NAS=3R14I

PROCESS PLAN 4P-12(*continued*)

No.	Operation	Tooling
1 2	**LATHE CUTOFF** **Cutoff: Select boundary point:** ➤ Click ① the boundary point	.01 TNR, RH OD CUTOFF TOOL

Cutoff parameters

X Tangent point   .375

Clearance Cut   ☒

Entry ammount .1
X increment .01
Z increment .01

☐ Peck

✓   ✗   ?

Corner Geometry
○ None
○ Radius   0.003
◉ Chamfer   Parameters..
☑ Clearance Cut..

1.5

A─

DETAIL A

.125

.125   R.01

.5

GC-4125

Material Removed

.375

4-13) Generate the 2D profiles of the part shown in Figure 4p-13. Use PROCESS PLAN
      4P-13 as an aid in generating the part program to produce the part.

Material: 303 Stainless

1.25-5 ACME

2.38-11.5 NPT, .688 LONG

.04 x 45°
CHAMFER

1.66-11.5 NPT, 1 LONG

.08 x 45°
CHAMFER(2PL)

1.78°(2PL)

.175

R.04(4PL)

R.25

.19

Ø2.375

Ø2.332

Ø1.438

Ø1.063

Ø2.600

Ø2.300

Ø1.960

Ø1.676

Ø1.598

1.125

1.250

1.730

1.980

2.170

2.250

3.000

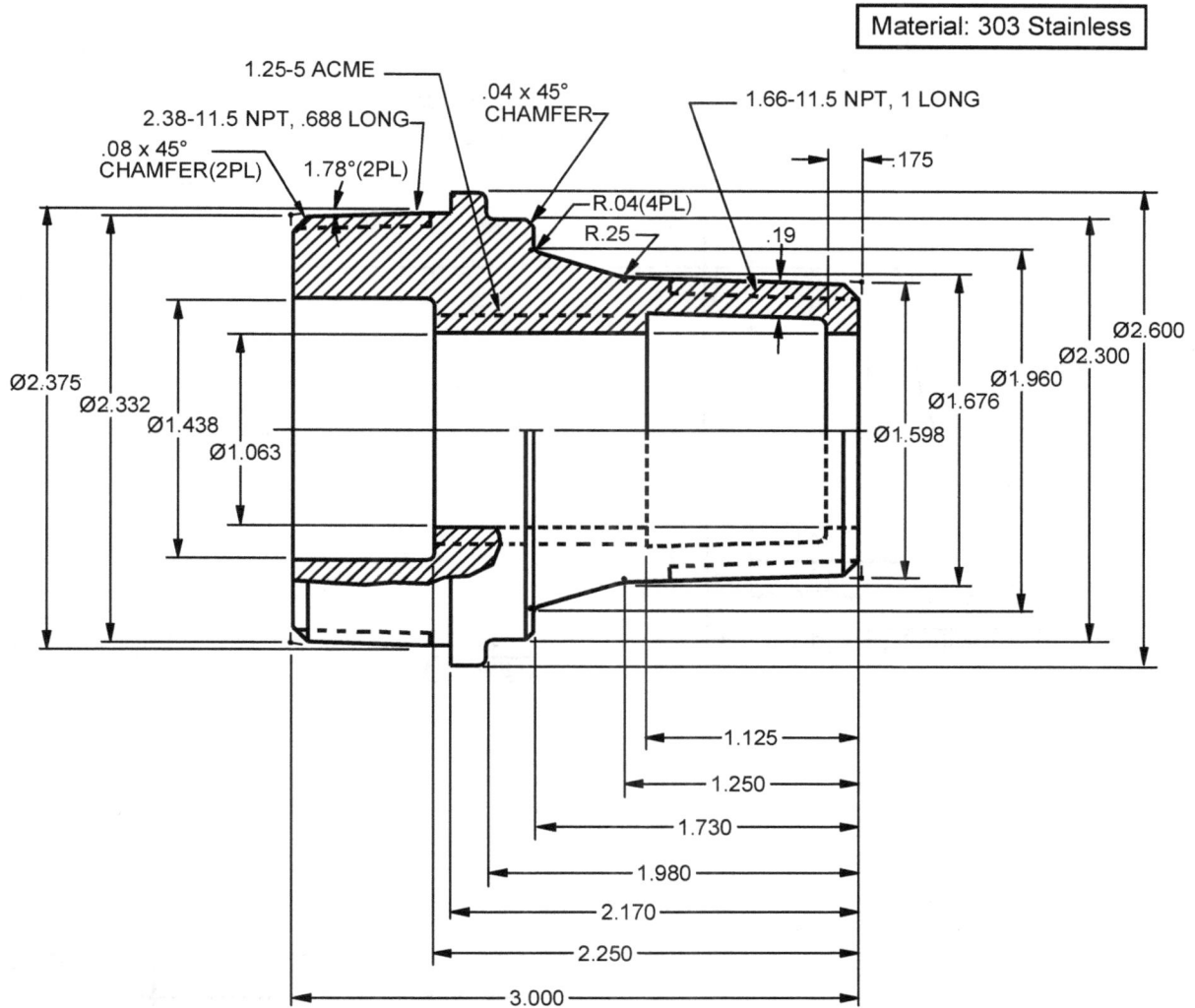

2D PROFILES
OF PART
OPERATOR
CREATES
IN CAD

PICTORIAL

Figure 4p-13

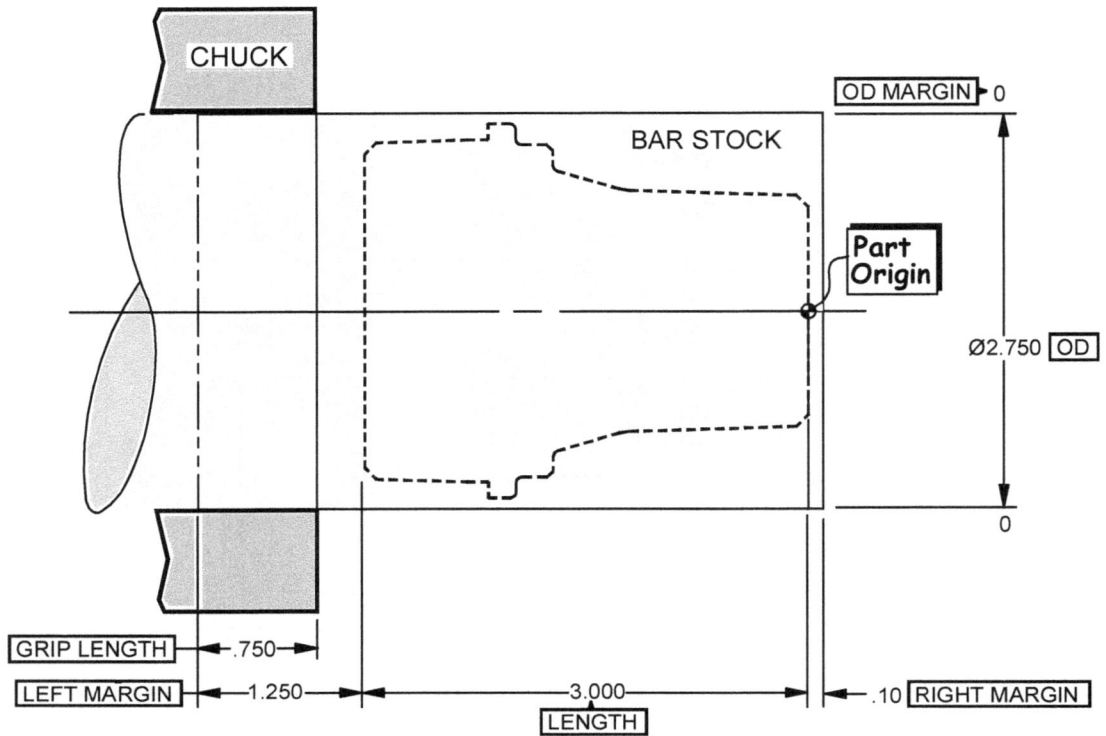

CHUCK

OD MARGIN ▸ 0

BAR STOCK

Part Origin

Ø2.750 OD

0

GRIP LENGTH ◂.750▸

LEFT MARGIN ◂1.250▸ ◂3.000▸ ◂.10 RIGHT MARGIN

LENGTH

PROCESS PLAN 4P-13

No.	Operation	Tooling
1	FACE END BAR STOCK Material Removed ◂ ▸.10	1/32 TNR, RH OD TURNING TOOL R1/32

PROCESS PLAN 4P-13(*continued*)

No.	Operation	Tooling
2	ROUGH TURN OD, LEAVE .01 IN X AND Z FOR SEMI-FINISH. SEMI-FINISH ID CONTOUR IN TWO PASSES USING A STEPOVER OF .05	1/64 TNR, RH OD FINISHING TOOL

☑ Lead In/Out

Lead In | Lead Out

Adjust Contour
☐ Extend/shorten end of contour
Ammount [0.0] ◉ Extend ◯ Shorten
☐ Add Line

Exit Vector
☑ Use exit vector
Fixed Direction
◯ None
◯ Tangent
◉ Perpendicular
Angle: 45
Length: 0.1

Plunge Cutting
No Plunge Cutting allowed

☑ Semi Finish → **Semi Finish Parameters** ✖

Number of passes: **2**
Stepover: **.005**
Stock to leave in X: **0**
Stock to leave in Z: **0**

←.40
Material Removed

② CHAIN 1 END PT

① CHAIN 1 START PT

R1/64

PROCESS PLAN 4P-13(*continued*)

No.	Operation	Tooling
3	ROUGH and FINISH OD GROOVE ; LEAVE .01 IN X AND Z FOR FINISHING.	R.01, W .187 OD GROOVING RIGHT DETAIL A GC-4187

PROCESS PLAN 4P-13(*continued*)

No.	Operation	Tooling
4	CUT 2.375-11.5 NPT ID THREAD	60°V, RH, OD THREADING TOOL

Thread Form

Select from table

Thread Table

Thread Form | American Pipe Threads - NPT |

Common diameter/lead combinations up to 4 inches

Basic major	Lead	Major diameter	Minor diameter	Comment
2.3750	11.5000	2.3386	2.1995	2, 11.5(2.375OD)

Thread shape parameters

Thread cut parameters

NC code format: | Canned |

Overcut

0

Anticipated pulloff

0

End Position	StartPosition
-2.25	**-3**

Thread orientation | ID |

Taper angle | **-1.78333** |

Material Removed

2.25

3

DETAIL A

.195

.117

1

R.012

1

1.25

A

VALENITE
16 ER 12UN

PROCESS PLAN 4P-13(*continued*)

No.	Operation	Tooling
5	Cut 1.66-11.5 NPT ID thread	60°V,RH,OD THREADING TOOL

Thread Form — Select from table

Thread Table

Thread Form   American Pipe Threads - NPT   ▼

Common diameter/lead combinations up to 4 inches

Basic major	Lead	Major diameter	Minor diameter	Comment
1.6600	11.5000	1.6267	1.4876	1 1/4,11.5(1.660OD)

Thread shape parameters

Thread cut parameters

NC code format: Canned   ▼

Overcut

`0`

Anticipated pulloff

`.1`

End Position
`-1.0`

StartPosition
`0`

Thread orientation ID ▼

Taper angle **1.78333**

DETAIL A

.195

.117

1

R.012

16 ER 12UN

.1

Material Removed

1.0

PROCESS PLAN 4P-13(*continued*)

No.	Operation	Tooling
6	LATHE DRILL X .18 DEEP Depth **-.18** ⦿ Absolute ◯ Incremental NO CHECK ☐ Drill tip compensation Material Removed	.25 DIA CENTER DRILL ⌀.25
7	LATHE DRILL X 3.125 DEEP Depth **-3.0** ⦿ Absolute ◯ Incremental Drill Cycle Parameters Cycle Peck drill ▼ Ist peck **.125** Subsequent peck **.125** CHECK ☑ Drill tip compensation Breakthrough ammount **.125** Material Removed 3.125	1 1/16 DIA DRILL ⌀3.188 3.5 2 4.5 4 ⌀1.063

PROCESS PLAN 4P-13(*continued*)

No.	Operation	Tooling
8	ROUGH AND FINISH ID GROOVE ; LEAVE .01 IN X AND Z FOR FINISHING.	R.01, W .125 ID GROOVING RIGHT

Groove Definition
- ○ 1 Point
- ○ 2 Points
- ○ 3 Lines
- ● Chain
- ○ Multiple chains

☑ Lead In

Entry Vector
- ☑ Use Entry Vector

Fixed Direction
- ○ None
- ○ Tangent
- ● Perpendicular

.1

② CHAIN1 END PT

① CHAIN1 START PT

Material Removed

.56   A   Ø.75
.093
8

DETAIL A
.093
.313   .125
.125   R.01

GC-4125

PROCESS PLAN 4P-13(*continued*)

No.	Operation	Tooling
9	ROUGH AND FINISH ID GROOVE ; LEAVE .01 IN X AND Z FOR FINISHING.	R.01, W .125 ID GROOVING RIGHT

Groove Definition

- ○ 1 Point
- ○ 2 Points
- ○ 3 Lines
- ◉ Chain
- ○ Multiple chains

☑ Lead In

Entry Vector

☑ Use Entry Vector

Fixed Direction

- ○ None
- ○ Tangent
- ◉ Perpendicular

② CHAIN 1 END PT

Material Removed

① CHAIN 1 START PT

.56 — A — Ø.75

.093

8

DETAIL A

.093

.313 .125

.125 R.01

GFG-125CW

PROCESS PLAN 4P-13(*continued*)

No.	Operation	Tooling
10	**CUT 1.25-5 ACME ID THREAD**	**ACME ID THREADING TOOL**

Thread Form — Select from table

Thread Table

Thread Form: Acme General Purpose

Common diameter/lead combinations up to 4 inches

Basic major	Lead	Major diameter	Minor diameter	Comment
1.2500	5.0000	1.2700	1.0500	1-1/4, 5

Thread cut parameters

Thread shape parameters

NC code format: Canned

Overcut .1

End Position **-2.25**

StartPosition **-1.125**

.1 — 2.25 — 1.125

Material Removed

A10-NER2

DETAIL A

.117
.07414
.75
.195

LT-22NR-5ACME

Ø.625
.5
A
10
⊙ ID

PROCESS PLAN 4P-13(*continued*)

No.	Operation	Tooling
1 1	**LATHE CUTOFF** **Cutoff: Select boundary point:** ➤ Click ① the boundary point	.01 TNR, RH OD CUTOFF TOOL

Cutoff parameters

X Tangent point `.5`

Clearance Cut

Entry ammount `.1`
X increment `.01`
Z increment `.01`

☐ Peck

✔ ✖ ?

Corner Geometry
- ○ None
- ○ Radius `0.003`
- ◉ Chamfer [Parameters..]
- ☑ [Clearance Cut..]

Material Removed ①

1.5

A

DETAIL A

.125
.125 R.01
.5

GC-4125

.5

4-14) Generate the 2D profiles of the part shown in Figures 4p-14 and 4p-15. Use
PROCESS PLAN 4P-14 as an aid in generating the part program to produce the part.

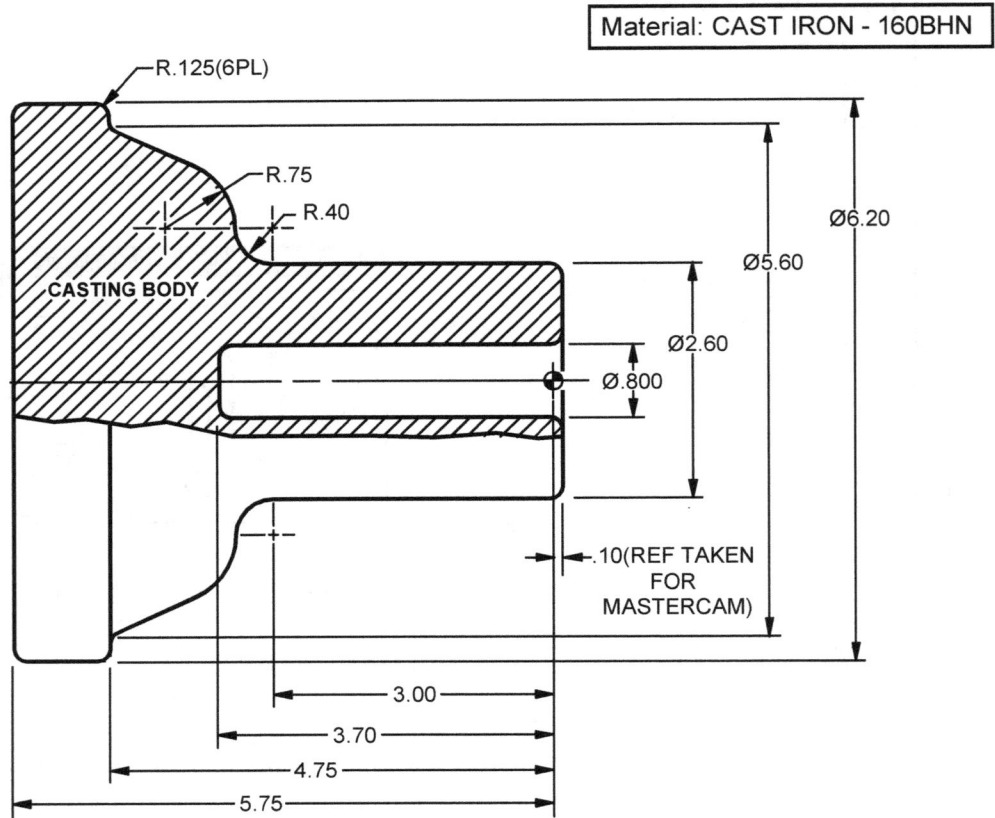

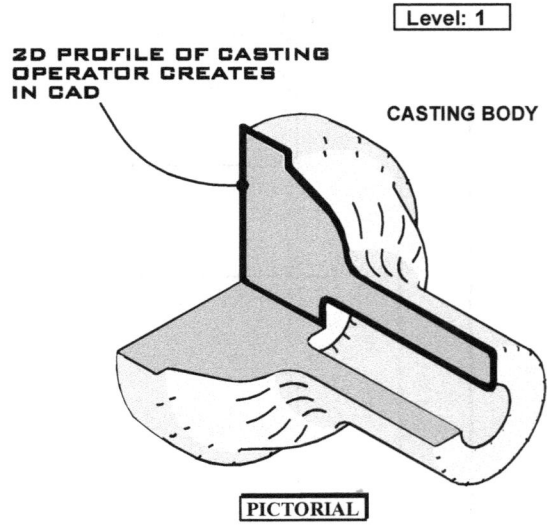

Material: CAST IRON - 160BHN

R.125(6PL)
R.75
R.40
CASTING BODY
Ø6.20
Ø5.60
Ø2.60
Ø.800
.10(REF TAKEN FOR MASTERCAM)
3.00
3.70
4.75
5.75

Level: 1

2D PROFILE OF CASTING OPERATOR CREATES IN CAD

CASTING BODY

PICTORIAL

Figure 4p-14

CASTING OUTLINE

.10(REF FROM CASTING)

R.65

R.50

R.09(2PL)

21.8°

Ø6.20

Ø5.38

Ø2.40

Ø2.15

FINISHED PART BODY

Ø1.80

Ø1.60

R.125

Ø1.20

Ø1.00

.10(REF TAKEN FOR MASTERCAM)

.35

.50

.80

1.20

2.25

2.50

3.00

3.80

4.75

Level: 2

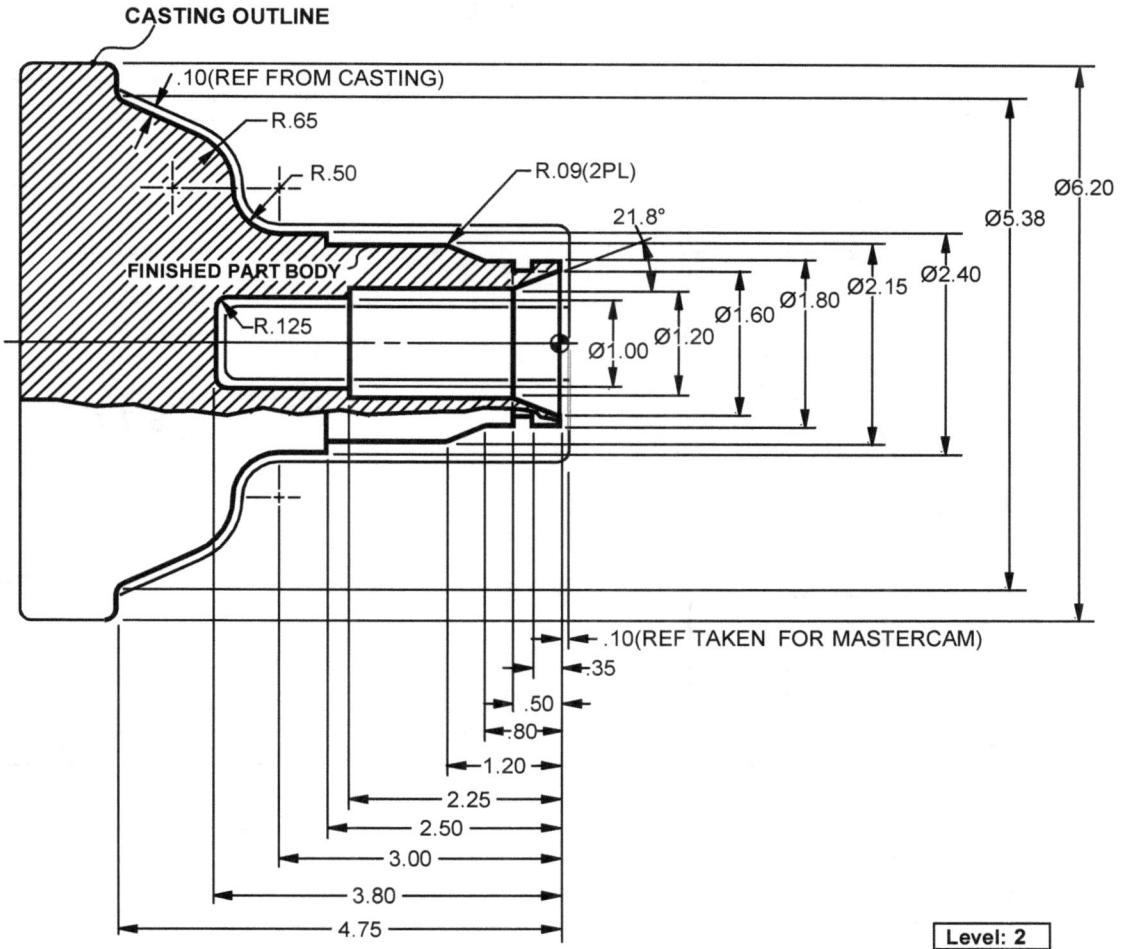

2D PROFILES OF FINISHED
PART OPERATOR CREATES
IN CAD

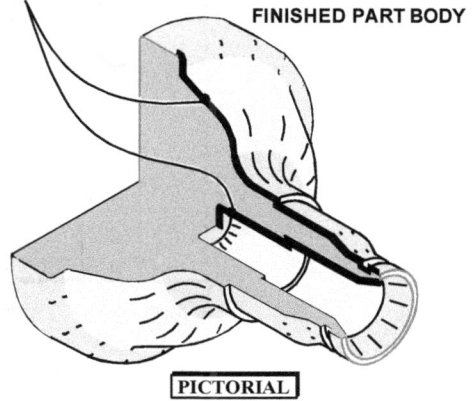

FINISHED PART BODY

PICTORIAL

Figure4p-15

Level: 1

Name: Stock:(Left Spindle)

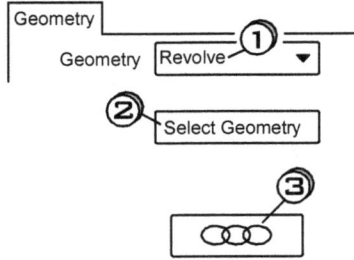

Name: Chuck Jaws:(Left Spindle)

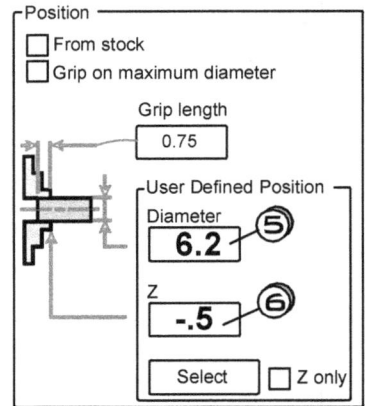

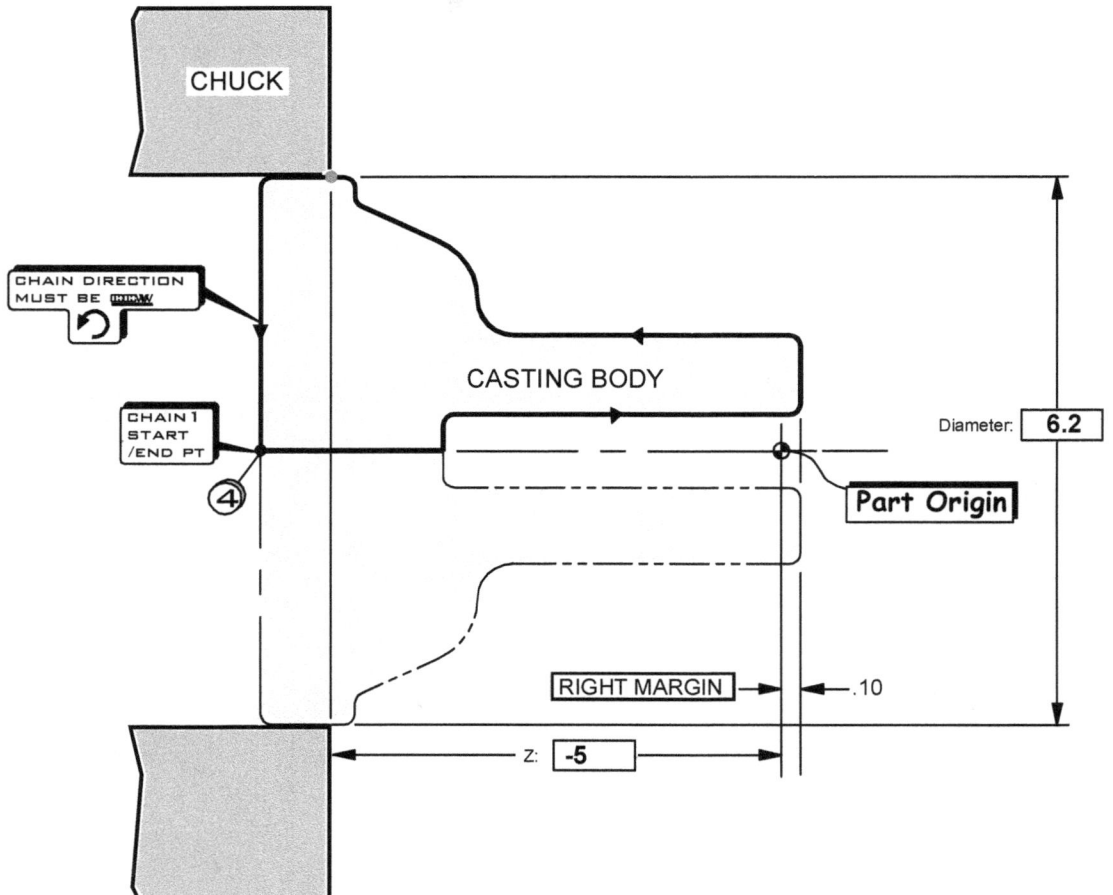

Geometry

Geometry Revolve ① ▼

② Select Geometry

③

Position
☐ From stock
☐ Grip on maximum diameter

Grip length
0.75

User Defined Position

Diameter ⑤
6.2

Z ⑥
-.5

Select   ☐ Z only

CHUCK

CHAIN DIRECTION
MUST BE CCW

CHAIN 1
START
/END PT

④

CASTING BODY

Diameter: **6.2**

Part Origin

RIGHT MARGIN ──► ◄── .10

Z: **-5**

PROCESS PLAN 4P-14

No.	Operation	Tooling
1	**FACE END**	1/32 TNR, RH OD TURNING TOOL R1/32
2	**ROUGH TURN OD, LEAVE .01 IN X AND Z FOR FINISHING.**	
3	**FINISH OD CONTOUR**	1/64 TNR, RH OD FINISHING TOOL R1/64

PROCESS PLAN 4P-14(*continued*)

No.	Operation	Tooling
4	Rough and finish OD groove; leave .01 in X and Z for finishing	R.01, W .125 OD GROOVING RIGHT

Toolpaths

Quick ➡ Groove

Quick Groove Options

Groove Definition
- ○ 1 Point
- ◉ 2 Points
- ○ 3 Lines

Quick groove shape parameters

0.0　○ Radius　◉ 45 deg chamfer

0.01　◉ Radius　○ 45 deg chamfer

☑ Finish groove

☑ Lead In

Entry Vector
☑ Use Entry Vector

Fixed Direction
- ○ None
- ◉ Tangent
- ○ Perpendicular

Material Removed ① ②

DETAIL A
.093
.5
.125
.125
R.01

GC-4125

1
5
.10
A

PROCESS PLAN 4P-14(*continued*)

No.	Operation	Tooling
5	**LATHE DRILL X 2.85 DEEP** Depth: **-3.8** ⦿ Absolute ◯ Incremental Drill Cycle Parameters Cycle Drill/Counterbore ▾ Material Removed R.125 3.80	**1 DIA BULL END MILL** 3 12 5.5 2 4.5 4 R.125 1
6	**ROUGH BORE ID, LEAVE .01 FOR FINISH CUT** ☑ Lead In/Out Lead In | Lead Out — Adjust Contour — ☐ Extend/shorten end of contour Ammount 0.0 ⦿ Extend ◯ Shorten ☐ Add Line — Exit Vector — ☑ Use exit vector Fixed Direction ◯ None ◯ Tangent ⦿ Perpendicular Angle: 45 Length: 0.1 — Plunge Cutting — No Plunge Cutting allowed ⦿ ① CHAIN 1 START PT ② CHAIN 1 END PT .01 LEFT FOR FINISH CUT Material Removed	**5/64 TNR, ID ROUGH BORING TOOL ,.25 DIA** R1/32 A .5 Ø.75 <u>DETAIL A</u> Ø5/16 E(75 deg. diamond)

PROCESS PLAN 4P-14(*continued*)

No.	Operation	Tooling
7	Finish Bore ID	5/64 TNR, ID FINISH BORING TOOL ,.25 DIA

Operation (No. 7):

☑ Lead In/Out

↓

Lead In | Lead Out

Adjust Contour

☐ Extend/shorten end of contour

Ammount 0.0   ◉ Extend ○ Shorten

☐ Add Line

Exit Vector

☑ Use exit vector

Fixed Direction
- ○ None
- ○ Tangent
- ◉ Perpendicular

Angle: 45

Length: 0.1

Plunge Cutting

No Plunge Cutting allowed

Tooling:

R1/32
A
.5
Ø.75

Ø5/16

D(55 deg. diamond)

① CHAIN 1 START PT

CHAIN 1 END PT ②

.01

Material Removed

4-15) Generate the 2D profiles of the part shown in Figures 4p-16 and 4p-17. Use
      PROCESS PLAN 4P-15 as an aid in generating the part program to produce the part.

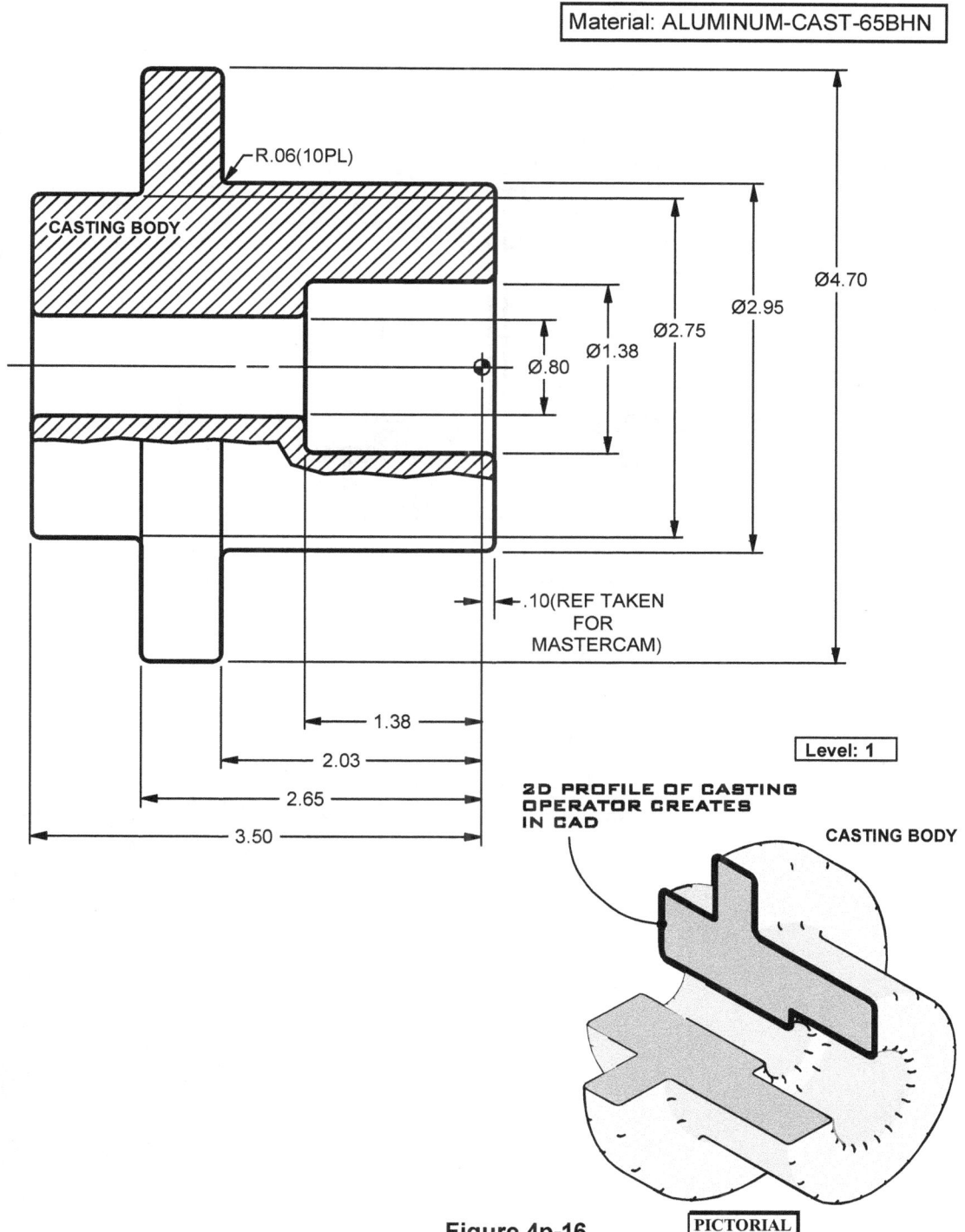

Material: ALUMINUM-CAST-65BHN

R.06(10PL)

CASTING BODY

Ø4.70

Ø2.95

Ø2.75

Ø1.38

Ø.80

.10(REF TAKEN
FOR
MASTERCAM)

1.38

2.03

2.65

3.50

Level: 1

2D PROFILE OF CASTING
OPERATOR CREATES
IN CAD

CASTING BODY

PICTORIAL

Figure 4p-16

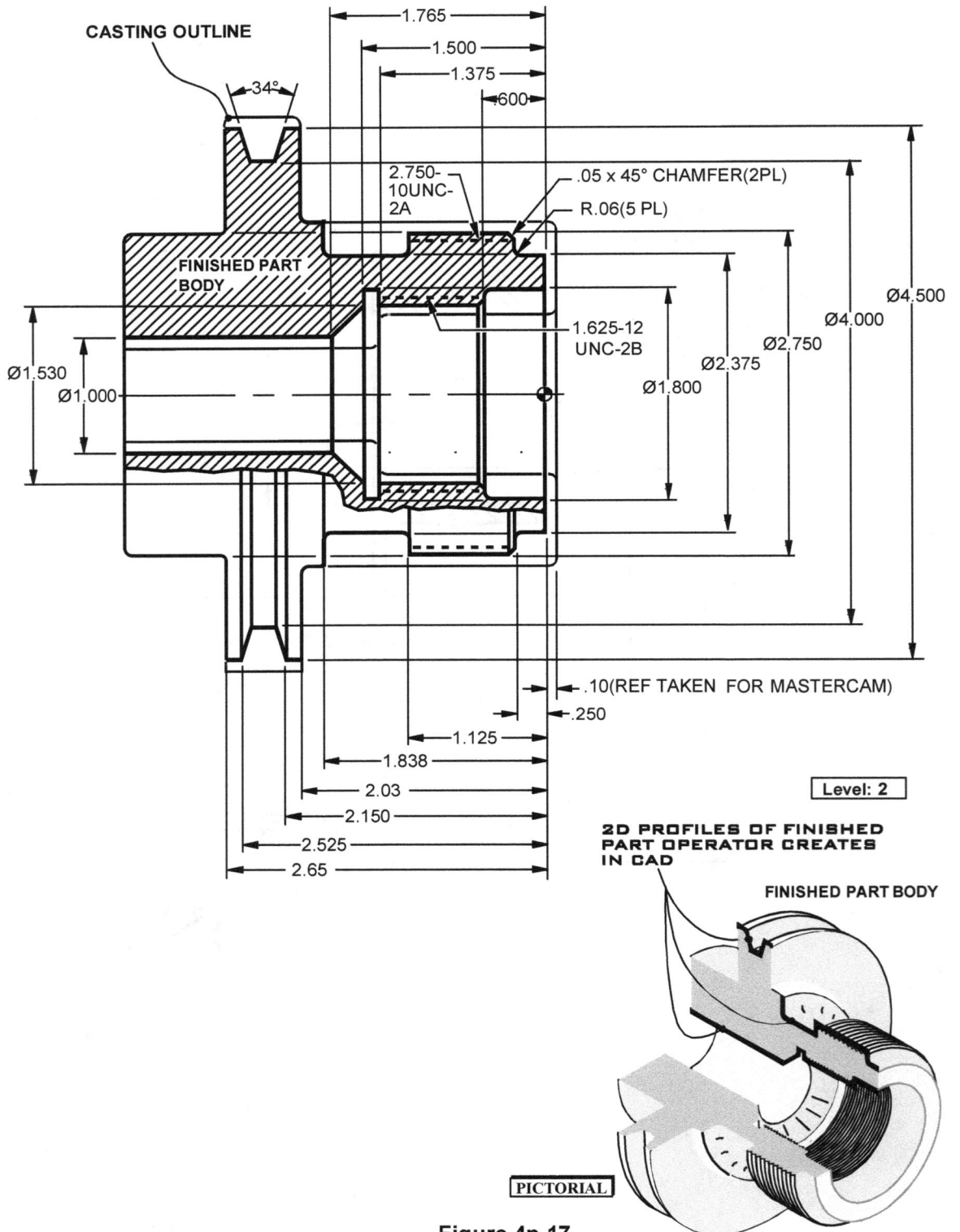

CASTING OUTLINE

34°

1.765
1.500
1.375
.600

2.750-
10UNC-
2A

.05 x 45° CHAMFER(2PL)

R.06(5 PL)

FINISHED PART
BODY

Ø4.500
Ø4.000

Ø1.530
Ø1.000

1.625-12
UNC-2B

Ø2.750
Ø2.375
Ø1.800

.10(REF TAKEN FOR MASTERCAM)
.250

1.125
1.838
2.03
2.150
2.525
2.65

Level: 2

2D PROFILES OF FINISHED
PART OPERATOR CREATES
IN CAD

FINISHED PART BODY

PICTORIAL

Figure 4p-17

Level: 1

Stock:(Left Spindle)

Name: Chuck Jaws:(Left Spindle)

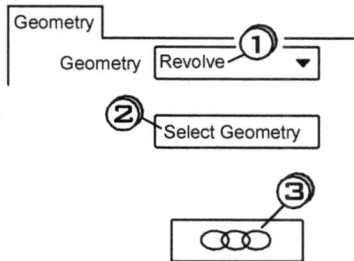

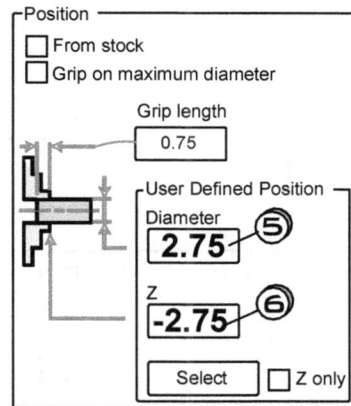

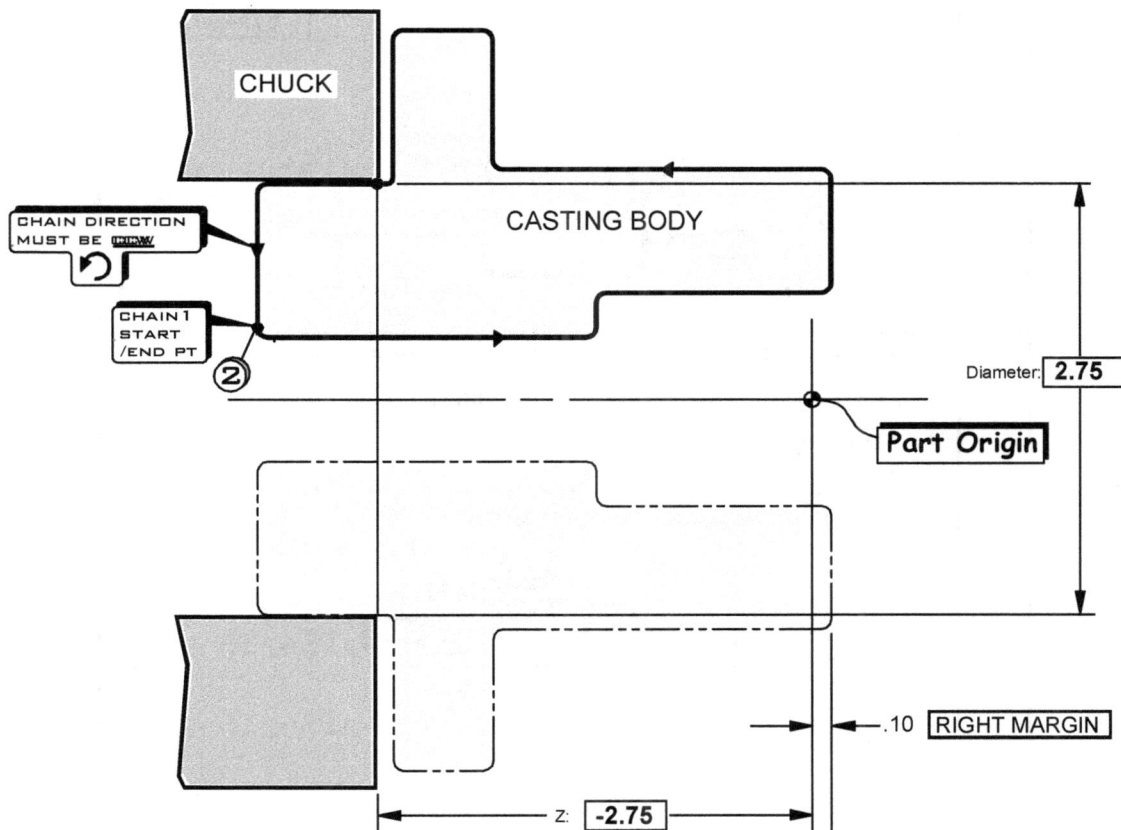

Geometry

Geometry Revolve ① ▼

② Select Geometry

③ ⬭⬭⬭

Position
☐ From stock
☐ Grip on maximum diameter

Grip length
0.75

User Defined Position
Diameter ⑤
2.75

Z ⑥
-2.75

Select ☐ Z only

CHUCK

CASTING BODY

CHAIN DIRECTION
MUST BE CCW
↺

CHAIN 1
START
/END PT
②

Diameter: **2.75**

Part Origin

.10 → RIGHT MARGIN

Z: **-2.75**

PROCESS PLAN 4P-15

No.	Operation	Tooling
1	FACE END Material Removed ← → .10	1/32 TNR, RH OD TURNING TOOL R1/32
2	ROUGH TURN OD, LEAVE .01 IN X AND Z FOR FINISHING. Level: 2 Plunge Cutting — No Plunge Cutting allowed ⦿ ☑ Lead Out Entry Vector — ☑ Use Entry Vector Fixed Direction — ○ None ⦿ Tangent ○ Perpendicular Material Removed ② CHAIN1 END PT .01 LEFT FOR FINISH CUT ① CHAIN1 START PT	
3	ROUGH TURN OD, LEAVE .01 IN X AND Z FOR FINISHING. Plunge Cutting — No Plunge Cutting allowed ⦿ ☑ Lead Out Entry Vector — ☑ Use Entry Vector Fixed Direction — ○ None ○ Tangent ⦿ Perpendicular .01 LEFT FOR FINISH CUT Material Removed ② CHAIN1 END PT ① CHAIN1 START PT	

PROCESS PLAN 4P-15

No.	Operation	Tooling
4	**FINISH OD CONTOUR** Plunge Cutting — No Plunge Cutting allowed ☑ Lead Out Entry Vector ☑ Use Entry Vector Fixed Direction ○ None ◉ Tangent ○ Perpendicular Material Removed ② CHAIN 1 END PT .01 ① CHAIN 1 START PT	1/64 TNR, RH OD FINISHING TOOL R1/64
5	**FINISH OD CONTOUR** Plunge Cutting — No Plunge Cutting allowed ☑ Lead Out Entry Vector ☑ Use Entry Vector Fixed Direction ○ None ○ Tangent ◉ Perpendicular Removed .01 Material Removed ② CHAIN 1 END PT ① CHAIN 1 START PT	

PROCESS PLAN 4P-15(*continued*)

No.	Operation	Tooling
6	ROUGH AND FINISH OD GROOVE ; LEAVE .01 IN X AND Z FOR FINISHING.	R.01, W .375 ID GROOVING RIGHT

PROCESS PLAN 4P-15(*continued*)

No.	Operation	Tooling
7	ROUGH AND FINISH OD GROOVE ; LEAVE .01 IN X AND Z FOR FINISHING. **Groove Definition** ○ 1 Point ○ 2 Points ○ 3 Lines ◉ Chain ○ Multiple chains ☑ Lead In **Entry Vector** ☑ Use Entry Vector **Fixed Direction** ○ None ◉ Tangent ○ Perpendicular Material Removed ② ① CHAIN 1 END PT · CHAIN 1 START PT	R.01, W .187 OD GROOVING RIGHT .144 A .75 1 DETAIL A .144 .5 .187 .187 · R.01 CG - 4187

PROCESS PLAN 4P-15(*continued*)

No.	Operation	Tooling
8	CUT 2.75-10 UNC OD THREAD	60°V,RH,OD THREADING TOOL

No. 8 — CUT 2.75-10 UNC OD THREAD

Thread Form — Select from table

Thread Table

Thread Form: Unified - UN, UNC, UNF, UNEF ▼

Common diameter/lead combinations up to 4 inches

Basic major	Lead	Major diameter	Minor diameter	Comment ▼
2.7500	10.0000	2.7500	2.6417	2-3/4,10

Thread cut parameters

Thread shape parameters

NC code format: Canned ▼

Overcut **.15**

End Position	StartPosition
-1.125	**-.25**

.15 — 1.125 — .25

Material Removed

Tooling: 60°V,RH,OD THREADING TOOL

DETAIL A

.375

16 ER 10UN

1

6

A

SER 1000 M16

PROCESS PLAN 4P-15(*continued*)

No.	Operation	Tooling
9	ROUGH BORE ID, LEAVE .01 FOR FINISH CUT	5/64 TNR, ID ROUGH BORING TOOL ,.25 DIA

☑ Lead In/Out

Lead In | **Lead Out**

Adjust Contour
☐ Extend/shorten end of contour
Ammount 0.0   ◉ Extend   ○ Shorten
☐ Add Line

Exit Vector
☑ Use exit vector
Fixed Direction
○ None
◉ Tangent
○ Perpendicular
Angle: 45
Length: 0.1

Plunge Cutting
No Plunge Cutting allowed

R1/32 A .5 Ø.75

DETAIL A
Ø5/16
E(75 deg. diamond)

CHAIN 1 START PT ①
CHAIN 1 END PT ②
Material Removed
.01 LEFT FOR FINISH CUT

PROCESS PLAN 4P-15(*continued*)

No.	Operation	Tooling
10	**FINISH BORE ID** ☑ [Lead In/Out] [Lead In] [Lead Out] **Adjust Contour** ☐ Extend/shorten end of contour Ammount [0.0] ◉ Extend ○ Shorten ☐ [Add Line] **Exit Vector** ☑ Use exit vector Fixed Direction ○ None ◉ Tangent ○ Perpendicular Angle: [45] Length: [0.1] Plunge Cutting — No Plunge Cutting allowed ◉ CHAIN 1 START PT ① CHAIN 1 END PT ② Material Removed .01 LEFT FOR FINISH CUT	**5/64 TNR, ID FINISH BORING TOOL ,.25 DIA** R1/32 A .5 Ø.75 Ø5/16 D(55 deg. diamond)
12	**ROUGH ID GROOVE** [Toolpaths] Quick ➤ Groove **Quick Groove Options** ✕ Groove Definition ○ 1 Point ◉ 2 Points ○ 3 Lines [✓] [✗] [?] ② ① Material Removed	**R.01, W.125 ID GROOVING RIGHT HAND** .5 A Ø.625 .093 8 LATHE(IN) KGSPR-10-094 DETAIL A .093 .5 .125 .125 R.01 GC-4125

PROCESS PLAN 4P-15(*continued*)

No.	Operation	Tooling
8	CUT 1.625-12 UNC ID THREAD	60°V,RH,ID THREADING TOOL

Thread Form — Select from table

Thread Table

Thread Form: Unified - UN, UNC, UNF, UNEF

Common diameter/lead combinations up to 4 inches

Basic major	Lead	Major diameter	Minor diameter	Comment
1.6250	12.0000	1.6250	1.5348	1-5/8,12

Thread cut parameters

Thread shape parameters

NC code format: Canned

Overcut **.06**

End Position **-1.375**   StartPosition **-.5**

DETAIL A

.375

16 ER-12UN

—A .51 Ø.75

7

SIR 0750 P16

Material Removed

.06   1.375   .5

4-16) Generate the 2D profiles of the part shown Figure 4p-18. Use PROCESS
PLAN 4P16 as an aid in generating the part program to produce the part.

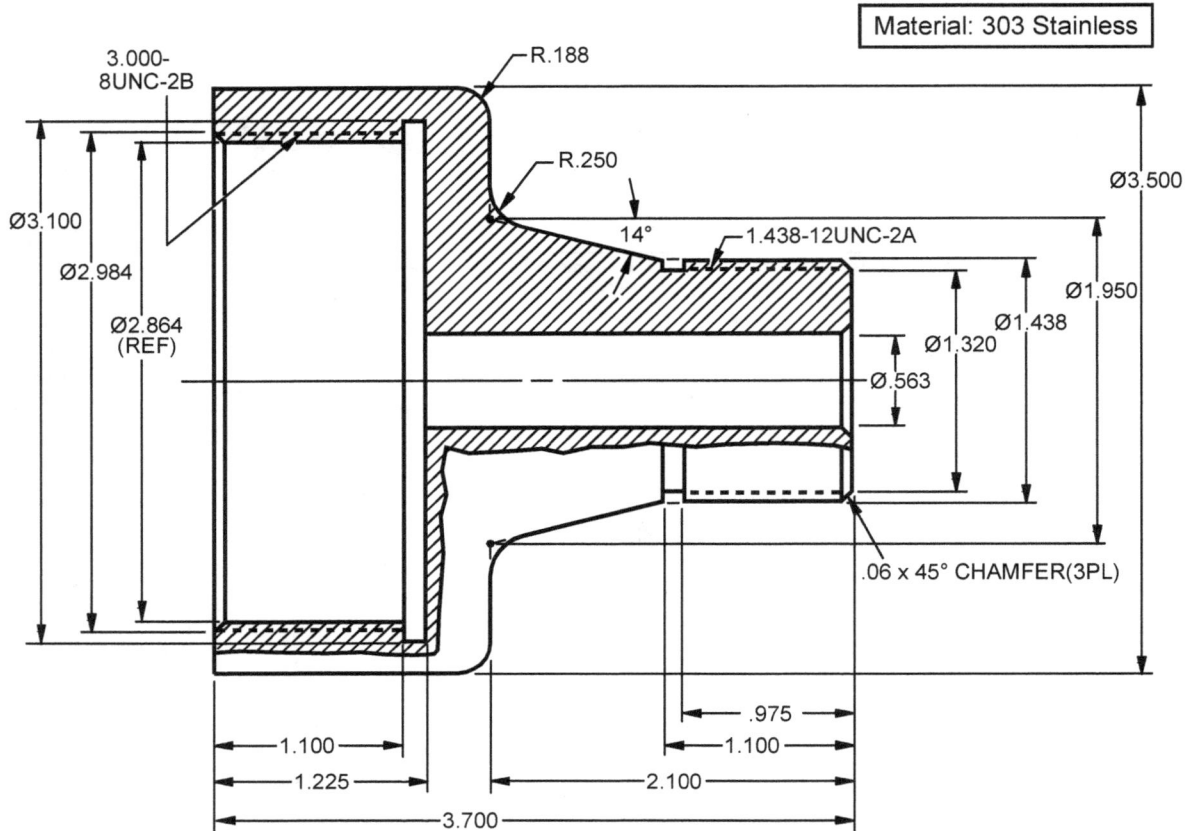

Material: 303 Stainless

3.000-
8UNC-2B

R.188

R.250

Ø3.500

Ø3.100

14°   1.438-12UNC-2A

Ø2.984

Ø1.950

Ø2.864
(REF)

Ø1.438

Ø1.320

Ø.563

.06 x 45° CHAMFER(3PL)

.975

1.100

1.100

1.225

2.100

3.700

2D PROFILES OF FINISHED
PART OPERATOR CREATES
IN CAD

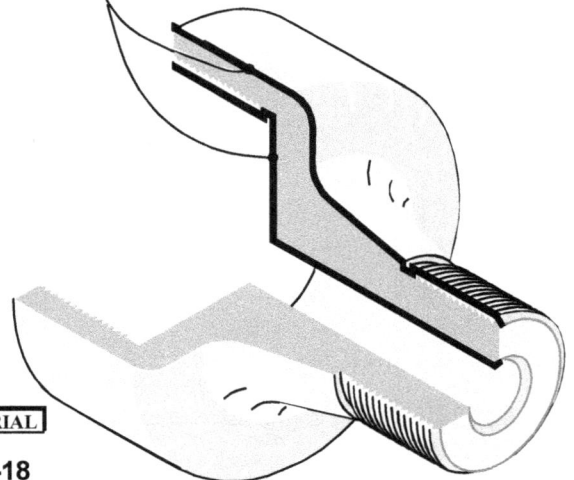

PICTORIAL

Figure 4p-18

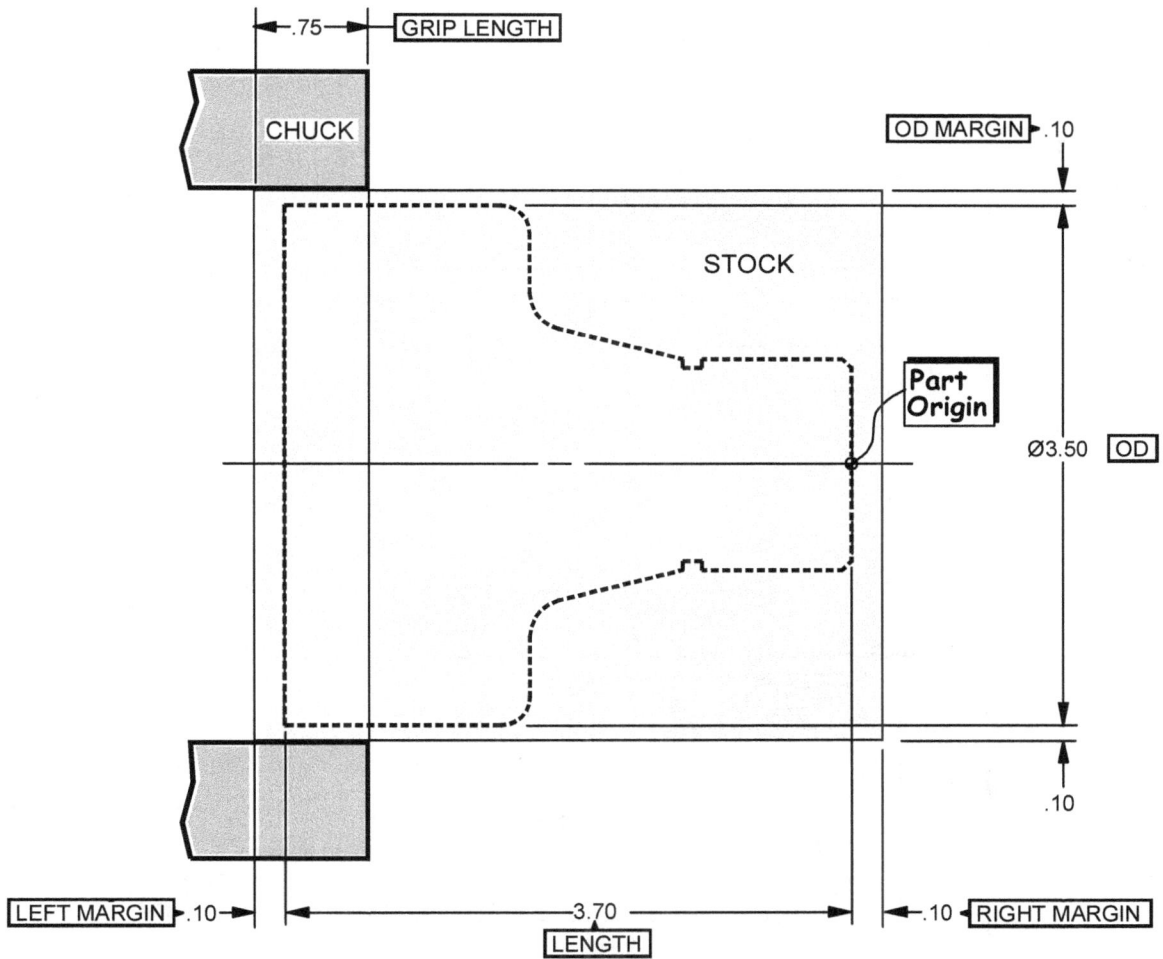

PROCESS PLAN 4P-16

No.	Operation	Tooling
1	FACE END —Material Removed .20	1/32 TNR, RH OD TURNING TOOL R1/32
2	ROUGH TURN OD, LEAVE .01 IN X AND Z FOR FINISHING. Plunge Cutting No Plunge Cutting allowed ☑ Lead Out Entry Vector ☑ Use Entry Vector Fixed Direction ○ None ○ Tangent ⦿ Perpendicular CHUCK .06 3.09 Break line Material Removed CHAIN 1 END PT ② ① .01 LEFT FOR FINISH CUT CHAIN 1 START PT	

PROCESS PLAN 4P-16(*continued*)

No.	Operation	Tooling
3	**FINISH OD CONTOUR** Plunge Cutting No Plunge Cutting allowed ☑ Lead Out Entry Vector ☑ Use Entry Vector Fixed Direction ○ None ○ Tangent ⦿ Perpendicular CHUCK .06 ← 3.09 → CHAIN 1 END PT ② Material Removed ① CHAIN 1 START PT .01	**1/64 TNR, RH OD FINISHING** R1/64
4	**ROUGH OD GROOVE** Toolpaths Quick ➡ Groove **Quick Groove Options** ✕ Groove Definition ○ 1 Point ⦿ 2 Points ○ 3 Lines ✓ ✕ ? ☑ Groove rough Groove step ○ Step ammount [.1] ⦿ Step % of tool width [**100**] ⦿ Stock to leave [**0**] Material Removed ① ②	**R.01, W .125 OD GROOVING RIGHT** .01 A .5 ← 1 → <u>DETAIL</u> A .093 .5 .125 .125 ← → R.01 GC-4125

PROCESS PLAN 4P-16(*continued*)

No.	Operation	Tooling
5	CUT 1.5-12 UNC OD THREAD	60°V,RH,OD THREADING TOOL

Thread Form

Select from table

Thread Table

Thread Form: Unified - UN, UNC, UNF, UNEF ▼

Common diameter/lead combinations up to 4 inches

Basic major	Lead	Major diameter	Minor diameter	Comment ▼
1.4375	12.0000	1.4375	1.3473	1-7/16,12,UN

Thread cut parameters

Thread shape parameters

NC code format: Canned ▼

Overcut
.15

End Position
-.975

StartPosition
0

.15 — .975

Material Removed

SER 1000 M16

DETAIL A

.375

16 ER 12UN

PROCESS PLAN 4P-16(*continued*)

No.	Operation	Tooling
6	**LATHE DRILL X .18 DEEP** Depth: **-.17** ● Absolute ○ Incremental Material Removed NO CHECK ☐ Drill tip compensation	**.25 DIA CENTER DRILL** Ø.25
7	**LATHE DRILL X 2.6 DEEP** Depth: **-2.475** ● Absolute ○ Incremental Drill Cycle Parameters Cycle Peck drill ▼ Ist peck: **.125** Subsequent peck: **.125** CHECK ☑ Drill tip compensation Breakthrough ammount: **.125** Material Removed .125 2.475	**9/16 DIA DRILL** 1.75 3.5 8 1.25 3.25 2.75 Ø.5625

PROCESS PLAN 4P-16(*continued*)

No.	Operation	Tooling
8	CHAMFER ID	5/64 TNR, ID FINISH BORING TOOL ,.25 DIA

Operation 8 (Chamfer ID)

☑ Lead In/Out

Lead In / **Lead Out**

— Adjust Contour —
☐ Extend/shorten end of contour

Ammount [0.0]   ◉ Extend   ○ Shorten

☐ Add Line

— Exit Vector —
☑ Use exit vector

— Fixed Direction —
○ None
◉ Tangent
○ Perpendicular

Angle: [45]
Length: [0.1]

Plunge Cutting

No Plunge Cutting allowed

CHAIN 1 START PT ①
CHAIN 1 END PT
②

Material Removed

Tooling (Operation 8):
— R1/64
— A
.6
Ø.5
Ø1/4
D(55 deg. diamond)

Operation 9 — LATHE FLIP STOCK

Level: 2

.07
.06

CREATE OFFSET LINE 1 AT Z LOCATION AND TRIM IT TO THE X LOCATION WHERE **THE CHUCK GRIPS THE PART**

CHUCK

CREATE STOCK POLAR LINE-ORIG POS

LINE 1

LINE 2

CREATE OFFSET LINE 2 AT Z LOCATION AND TRIM IT TO THE DESIRED X LOCATION **WHERE THE CHUCK IS TO GRIP THE FLIPPED PART**

CREATE STOCK POLAR LINE- TRANS POS

.75
3.8
.10

PROCESS PLAN 4P-16(*continued*)

No.	Operation	Tooling

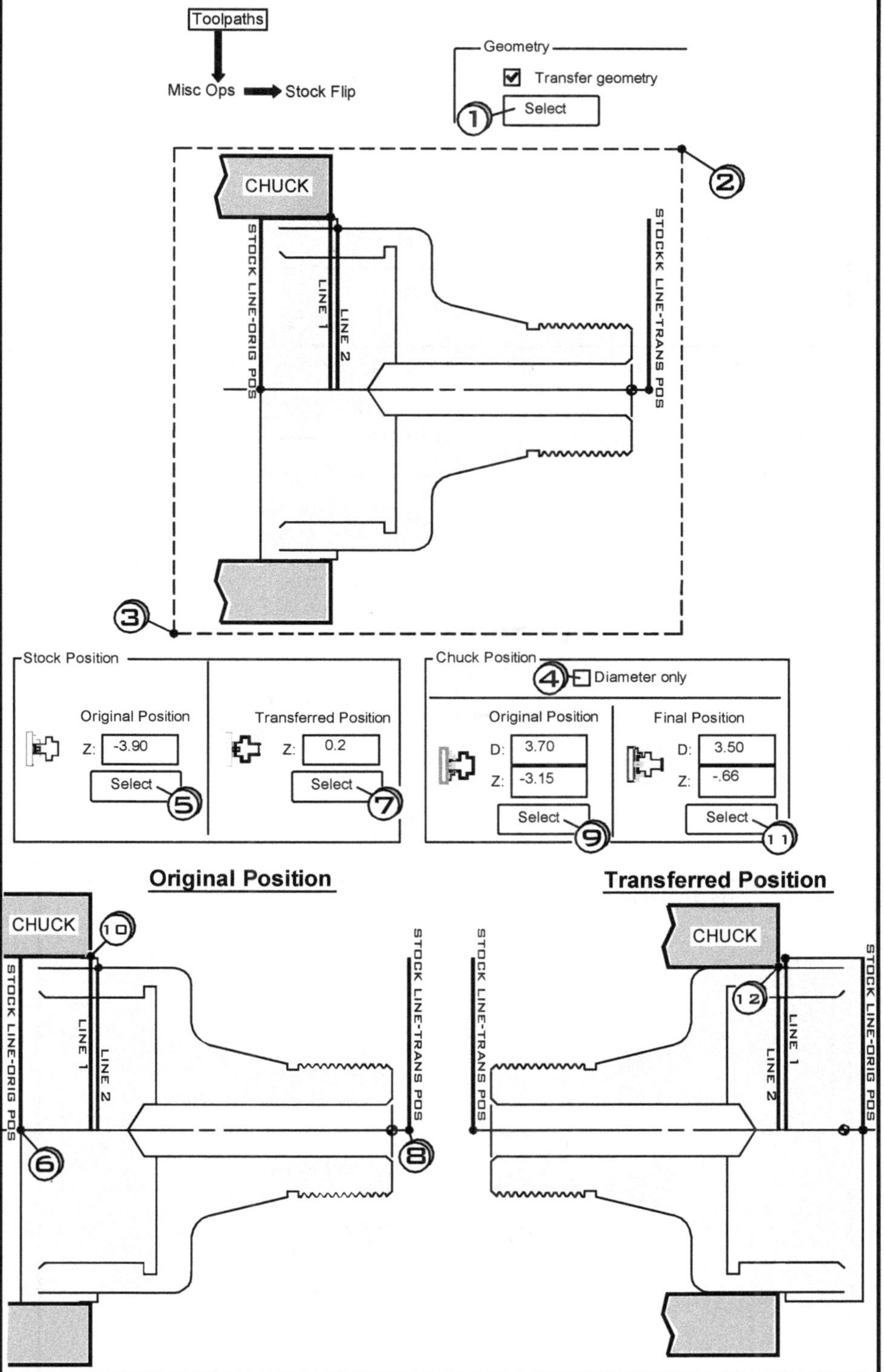

PROCESS PLAN 4P-16(*continued*)

No.	Operation	Tooling
10	**Face End**	1/32 TNR, RH OD TURNING TOOL
11	**Rough Turn OD, leave .01 in X and Z for finishing.**	
12	**Finish OD Contour**	1/64 TNR, RH OD FINISHING

Operation 10:

CHUCK — Material Removed — .10

Operation 11:

Plunge Cutting — No Plunge Cutting allowed

☑ Lead Out

Adjust Contour — ☑ Extend/shorten end of contour — Ammount **0.01** — ⊙ Extend ○ Shorten

Entry Vector — ☑ Use Entry Vector — Fixed Direction — ○ None ○ Tangent ⊙ Perpendicular

CHUCK — Material Removed — ① — CHAIN1 START PT — CHAIN1 END PT — ② — .01 LEFT FOR FINISH CUT

Operation 12:

Plunge Cutting — No Plunge Cutting allowed

☑ Lead Out

Adjust Contour — ☑ Extend/shorten end of contour — Ammount **0.03** — ⊙ Extend ○ Shorten

Entry Vector — ☑ Use Entry Vector — Fixed Direction — ○ None ○ Tangent ⊙ Perpendicular

CHUCK — Material Removed — ① — CHAIN1 START PT — CHAIN1 END PT — ② — .01 LEFT FOR FINISH CUT

R1/32　　R1/64

PROCESS PLAN 4P-16(*continued*)

No.	Operation	Tooling
1 3	**LATHE DRILL X .18 DEEP** Depth [-.17] ◉ Absolute ○ Incremental NO CHECK ☐ Drill tip compensation Material Removed	**.25 DIA CENTER DRILL** Ø.25
1 4	**LATHE DRILL X 1.3 DEEP** Depth [-1.3] ◉ Absolute ○ Incremental Drill Cycle Parameters Cycle [Peck drill ▼] Ist peck [.125] Subsequent peck [.125] CHECK ☑ Drill tip compensation Breakthrough ammount [.125] Material Removed 1.3	**1.75 DIA DRILL** Ø2 3.75 1.25 8.75 3.75 3.75 Ø1.75

PROCESS PLAN 4P-16(*continued*)

No.	Operation	Tooling
15	**LATHE DRILL X 1.225 DEEP** Depth **-1.225** ◉ Absolute ○ Incremental Drill Cycle Parameters Cycle Drill/Counterbore ▼ NO CHECK □ Drill tip compensation Material Removed ←1.225→	**2 DIA END MILL** ←Ø3→ 3 9 2 4 4 Ø2
16	**ROUGH BORE ID, LEAVE .01 FOR FINISH CUT** ☑ Lead Out Adjust Contour ☑ Extend/shorten end of contour Ammount **0.03** ◉ Extend ○ Shorten Entry Vector ☑ Use Entry Vector Fixed Direction ○ None ◉ Tangent ○ Perpendicular Plunge Cutting No Plunge Cutting allowed ◉ .01 LEFT FOR FINISH CUT ① CHAIN 1 START PT ② CHAIN 1 END PT **AT MIDPOINT OF LINE** Material Removed	**ROUGH BORING TOOL ,.25 DIA** R1/32 A .5 Ø.75 DETAIL A Ø5/16 E(75 deg. diamond)

PROCESS PLAN 4P-16(*continued*)

No.	Operation	Tooling
17	**FINISH BORE ID** ☑ Lead Out ─ Adjust Contour ─ ☑ Extend/shorten end of contour Ammount [0.03] ◉ Extend ○ Shorten ─ Exit Vector ─ ☑ Use Entry Vector ─Fixed Direction─ ○ None ◉ Tangent ○ Perpendicular ─ Plunge Cutting ─ No Plunge Cutting allowed .01 ① CHAIN1 START PT Material Removed ② CHAIN1 END PT AT MIDPOINT OF LINE	5/64 TNR, ID FINISH BORING TOOL ,.25 DIA R1/32 A .5 Ø.75 Ø5/16 D(55 deg. diamond)
18	**ROUGH ID GROOVE** ─Groove Definition─ ○ 1 Point ○ 2 Points ○ 3 Lines ◉ Chain ○ Multiple chains ① CHAIN1 START PT Material Removed ② CHAIN1 END PT	R.01, W.125 ID GROOVING RIGHT HAND .7 A 1 .075 8 DETAIL A .075 .46 .094 .094 R.002 LATHE(IN) GS-094L

PROCESS PLAN 4P-16(*continued*)

No.	Operation	Tooling
19	CUT 3 - 8 UNC ID THREAD	60°V, RH, ID THREADING TOOL

Thread Form

Select from table

Thread Table

Thread Form: Unified - UN, UNC, UNF, UNEF ▼

Common diameter/lead combinations up to 4 inches

Basic major	Lead	Major diameter	Minor diameter	Comment ▼
⋮	⋮	⋮	⋮	⋮
3.0000	8.0000	3.0000	2.8647	3, 8
⋮	⋮	⋮	⋮	⋮

Thread cut parameters

Thread shape parameters

NC code format: Canned ▼

Overcut
.05

End Position
-1.1

StartPosition
0

Material Removed

.05 ← → ← 1.1 →

Tooling column:

.688 Ø1

A

8

A16-NEL2

DETAIL A

.195

.117

1

R.0180

LT-16ER-8UN

4-17) Generate the 2D profile of the part shown in Figure 4p-19. Use PROCESS
PLAN 4P-17 as an aid in generating the part program to produce the part.

Material: 303 Stainless

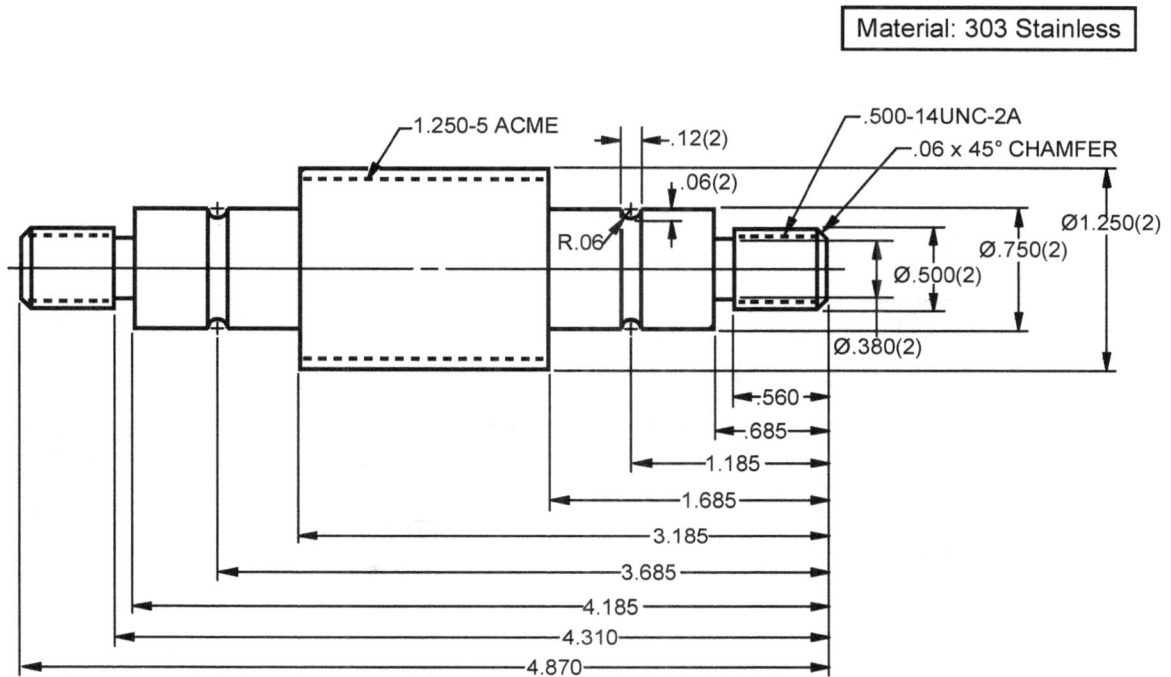

1.250-5 ACME

.12(2)

.06(2)

.500-14UNC-2A

.06 x 45° CHAMFER

R.06

Ø1.250(2)

Ø.750(2)

Ø.500(2)

Ø.380(2)

.560

.685

1.185

1.685

3.185

3.685

4.185

4.310

4.870

2D PROFILE OF FINISHED
PART OPERATOR CREATES
IN CAD

PICTORIAL

Figure 4p-19

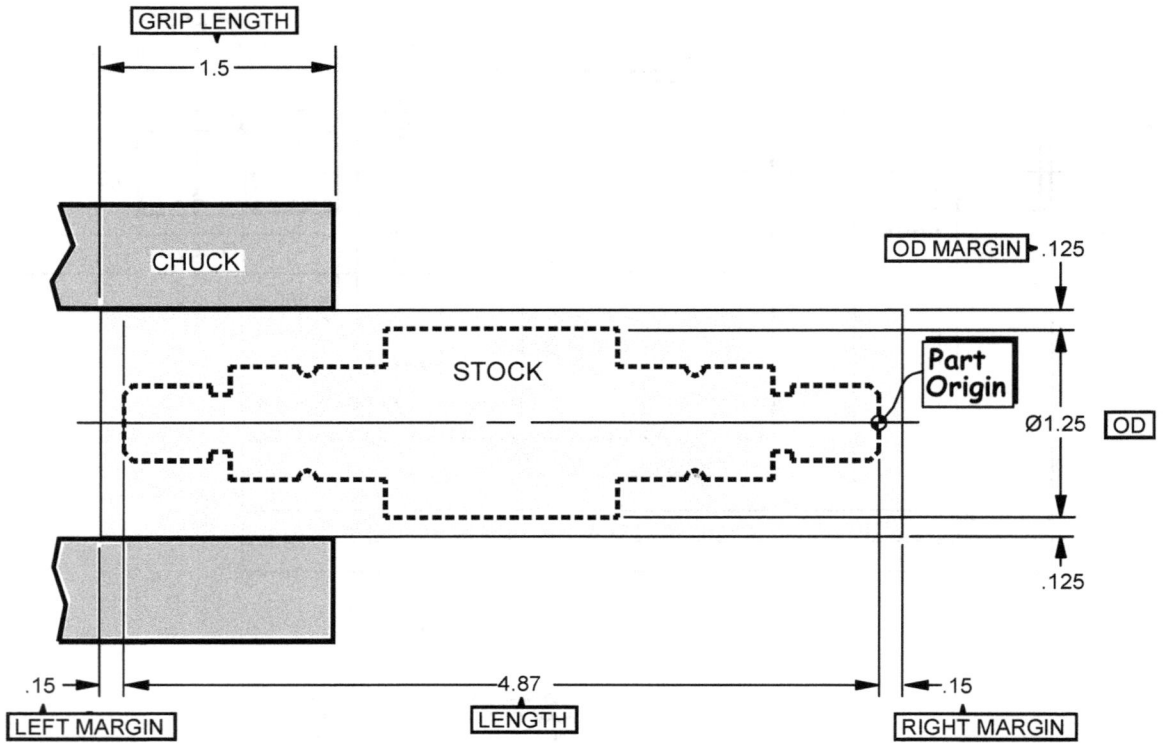

PROCESS PLAN 4P-17

No.	Operation	Tooling
1	FACE END Material Removed .15	1/32 TNR, RH OD TURNING TOOL R1/32
2	ROUGH TURN OD, LEAVE .01 IN X AND Z FOR FINISHING. ☑ Lead Out Adjust Contour ☑ Extend/shorten end of contour Ammount **.125** ⦿ Extend ○ Shorten Exit Vector ☑ Use Entry Vector Fixed Direction ○ None ○ Tangent ⦿ Perpendicular Plunge Cutting No Plunge Cutting allowed .125 CHUCK Material Removed ② CHAIN 1 END PT .01 LEFT FOR FINISH CUT ① CHAIN 1 START PT	

PROCESS PLAN 4P-17(*continued*)

No.	Operation	Tooling
3	**FINISH OD CONTOUR** Plunge Cutting No Plunge Cutting allowed ☑ Lead Out Adjust Contour ☑ Extend/shorten end of contour Ammount **.125** ◉ Extend ○ Shorten Exit Vector ☑ Use Entry Vector Fixed Direction: ○ None ○ Tangent ◉ Perpendicular .125 CHUCK Material Removed .01 CHAIN 1 END PT ② CHAIN 1 START PT ①	1/64 TNR, RH OD FINISHING R1/64
4	**ROUGH OD GROOVE** **Quick Groove Options** Groove Definition: ◉ 1 Point ○ 2 Points ○ 3 Lines ☑ Groove rough Groove step ○ Step ammount .1 ◉ Step % of tool width **100** ◉ Stock to leave **0** **Quick groove shape parameters** Height **.125** Width **.125** **0.00** ○ Radius ◉ 45 deg chamfer **.0625** ◉ Radius ○ 45 deg chamfer CREATE LINE ① Material Removed .0625	R.063, W .093, OD GROOVING RIGHT .093 .75 A 1 **DETAIL A** R.063 .5 .093 GR-4125

PROCESS PLAN 4P-17(*continued*)

No.	Operation	Tooling
5	ROUGH OD GROOVE	R.01, W .125 OD GROOVING RIGHT

Groove Definition
- ○ 1 Point
- ○ 2 Points
- ○ 3 Lines
- ◉ Chain
- ○ Multiple chains

Material Removed

② CHAIN 1 END PT

① CHAIN 1 START PT

1

5

.075

A

DETAIL A
.075
.46
.094
.094
R.002

GS-094L

PROCESS PLAN 4P-17(*continued*)

No.	Operation	Tooling
6	Cut .50-14 unc OD thread	60°V,RH,OD THREADING TOOL

Operation detail (No. 6):

Thread Form — Select from table

Thread Table

Thread Form: Unified - UN, UNC, UNF, UNEF ▼

Common diameter/lead combinations up to 4 inches

Basic major	Lead	Major diameter	Minor diameter	Comment ▼
⋮	⋮	⋮	⋮	⋮
0.5000	14.0000	0.5000	0.4227	1/2,14
⋮	⋮	⋮	⋮	⋮

Thread shape parameters

Thread cut parameters

NC code format: Canned ▼

Overcut .05

End Position	StartPosition
-.56	0

.05 → | ← .560 →

Material Removed

Tooling detail (No. 6):

DETAIL A

.375

16 ER 14UN

1

6

A

SER 1000 M16

PROCESS PLAN 4P-17(*continued*)

No.	Operation	Tooling
7	LATHE FLIP STOCK	

Level: 2

CREATE OFFSET LINE 1 AT Z LOCATION AND TRIM IT TO THE X LOCATION WHERE **THE CHUCK GRIPS THE PART**

CREATE OFFSET LINE 2 AT Z LOCATION AND TRIM IT TO THE X LOCATION WHERE **THE CHUCK GRIPS THE FLIPPED PART**

CREATE STOCK POLAR LINE-ORIG POS

CREATE STOCK POLAR LINE-TRANS POS

CHUCK

LINE 1

LINE 2

.21

.20

1.5

5.02

.15

Toolpaths

Misc Ops ➡ Stock Flip

Geometry

☑ Transfer geometry

① Select

②

③

STOCK LINE-ORIG POS

STOCK LINE-TRANS POS

CHUCK

LINE 1

LINE 2

PROCESS PLAN 4P-17(*continued*)

No.	Operation	Tooling
7	LATHE FLIP STOCK	

Original Position

Transferred Position

PROCESS PLAN 4P-17(*continued*)

No.	Operation	Tooling
8	FACE END Material Removed .15	1/32 TNR, RH OD TURNING TOOL R1/32
9	ROUGH TURN OD, LEAVE .01 IN X AND Z FOR FINISHING. Plunge Cutting No Plunge Cutting allowed ☑ Lead Out Exit Vector ☑ Use Entry Vector Fixed Direction ○ None ○ Tangent ◉ Perpendicular Material Removed CHUCK CHAIN 1 END PT ② CHAIN 1 START PT ① .01 LEFT FOR FINISH CUT	

PROCESS PLAN 4P-17(*continued*)

No.	Operation	Tooling
3	**FINISH OD CONTOUR** 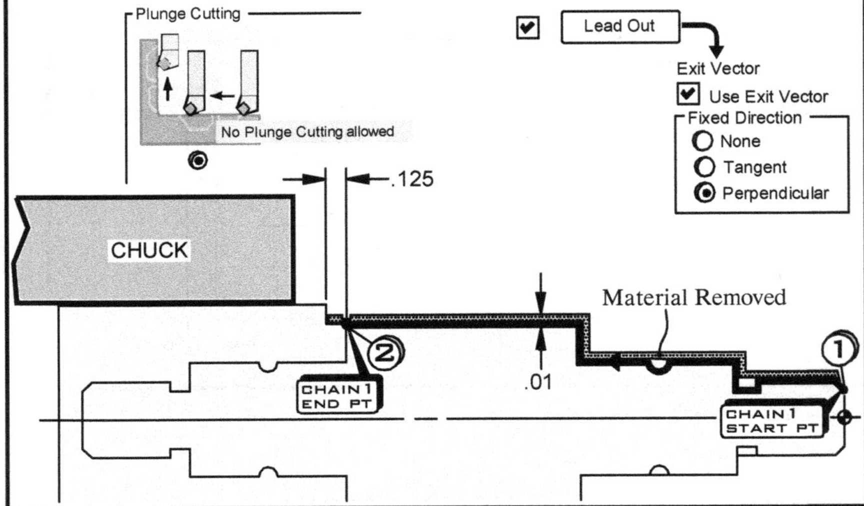	1/64 TNR, RH OD FINISHING 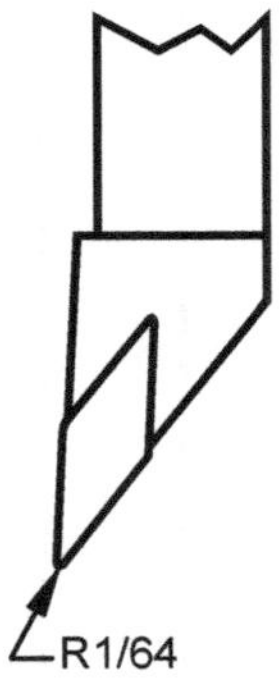 R1/64
4	**ROUGH OD GROOVE** 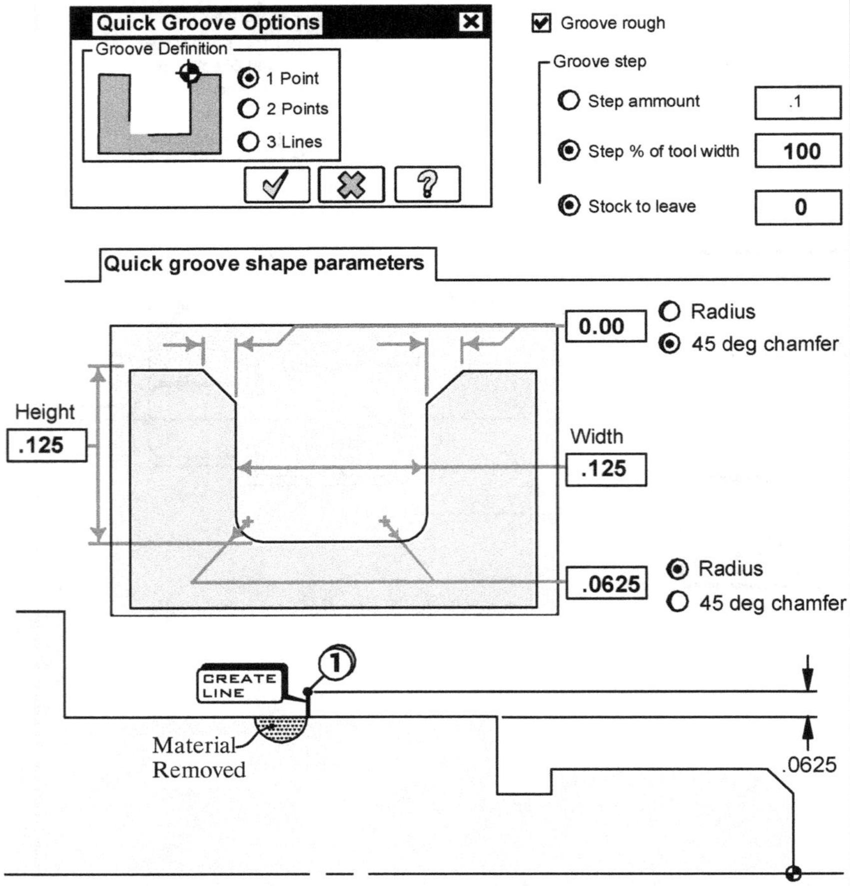	R.063, W .093, OD GROOVING RIGHT 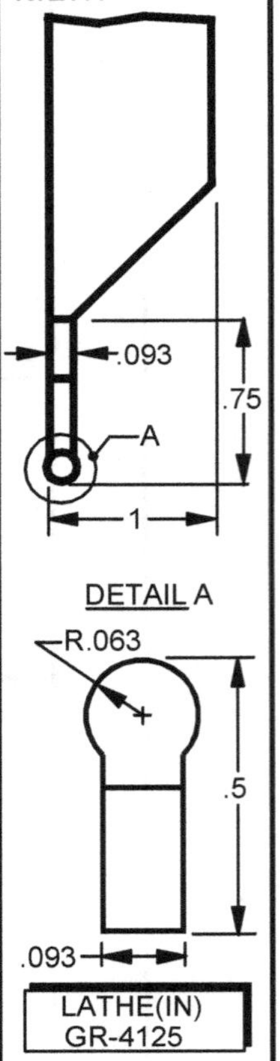 .093 .75 1 DETAIL A R.063 .5 .093 LATHE(IN) GR-4125

Operation 3 diagram labels:
- Plunge Cutting — No Plunge Cutting allowed
- ☑ Lead Out — Exit Vector — ☑ Use Exit Vector — Fixed Direction: ○ None ○ Tangent ◉ Perpendicular
- .125
- CHUCK
- Material Removed
- .01
- ② CHAIN 1 END PT
- ① CHAIN 1 START PT

Operation 4 diagram labels:
- **Quick Groove Options** ✕
- Groove Definition: ◉ 1 Point ○ 2 Points ○ 3 Lines
- Groove rough — Groove step
- ○ Step ammount [.1]
- ◉ Step % of tool width [100]
- ◉ Stock to leave [0]
- **Quick groove shape parameters**
- Height [.125]
- Width [.125]
- [0.00] ○ Radius ◉ 45 deg chamfer
- [.0625] ◉ Radius ○ 45 deg chamfer
- CREATE LINE ①
- Material Removed
- .0625

PROCESS PLAN 4P-17(*continued*)

No.	Operation	Tooling
12	ROUGH OD GROOVE	R.01, W .125 OD GROOVING RIGHT

Groove Definition
- ○ 1 Point
- ○ 2 Points
- ○ 3 Lines
- ◉ Chain
- ○ Multiple chains

Material Removed

② CHAIN 1 END PT

① CHAIN 1 START PT

1

5

.075

A

DETAIL A

.075

.46

.094

.094

R.002

GS-094L

PROCESS PLAN 4P-17(*continued*)

No.	Operation	Tooling
1 3	CUT .50-14 UNC OD THREAD	60°V,RH,OD THREADING TOOL

Thread Form — Select from table

Thread Table

Thread Form | Unified - UN, UNC, UNF, UNEF ▼

Common diameter/lead combinations up to 4 inches

Basic major	Lead	Major diameter	Minor diameter	Comment ▼
⋮	⋮	⋮	⋮	⋮
0.5000	14.0000	0.5000	0.4227	1/2,14
⋮	⋮	⋮	⋮	⋮

Thread cut parameters

Thread shape parameters

NC code format: | Canned ▼

Overcut **.05**

End Position **-.56**   StartPosition **0**

DETAIL A

.375

16 ER 14UN

1

6

A

SER 1000 M16

.05 ← | → .560

Material Removed

PROCESS PLAN 4P-17(*continued*)

No.	Operation	Tooling
14	Cut 1.25-5 acme OD thread	ACME THREADING TOOL

Thread Form — Select from table

Thread Table

Thread Form | Acme General Purpose ▼

Common diameter/lead combinations up to 4 inches

Basic major	Lead	Major diameter	Minor diameter	Comment ▼
1.2500	5.0000	1.2500	1.0300	1-1/4, 5

Thread cut parameters

Thread shape parameters

NC code format: Canned ▼

Overcut .06

End Position **-3.185**   StartPosition **-1.685**

Material Removed

1.685
3.185
.215

DETAIL A

.117 .07414 1 .195

LT-22NR-5ACME

1 1.25

4-18) Generate the 2D profiles of the part shown in Figures 4p-20 and 4p-21. Use
PROCESS PLAN 4P-18 as an aid in generating the part program to produce the part.

Material: IRON-CAST-220BHN

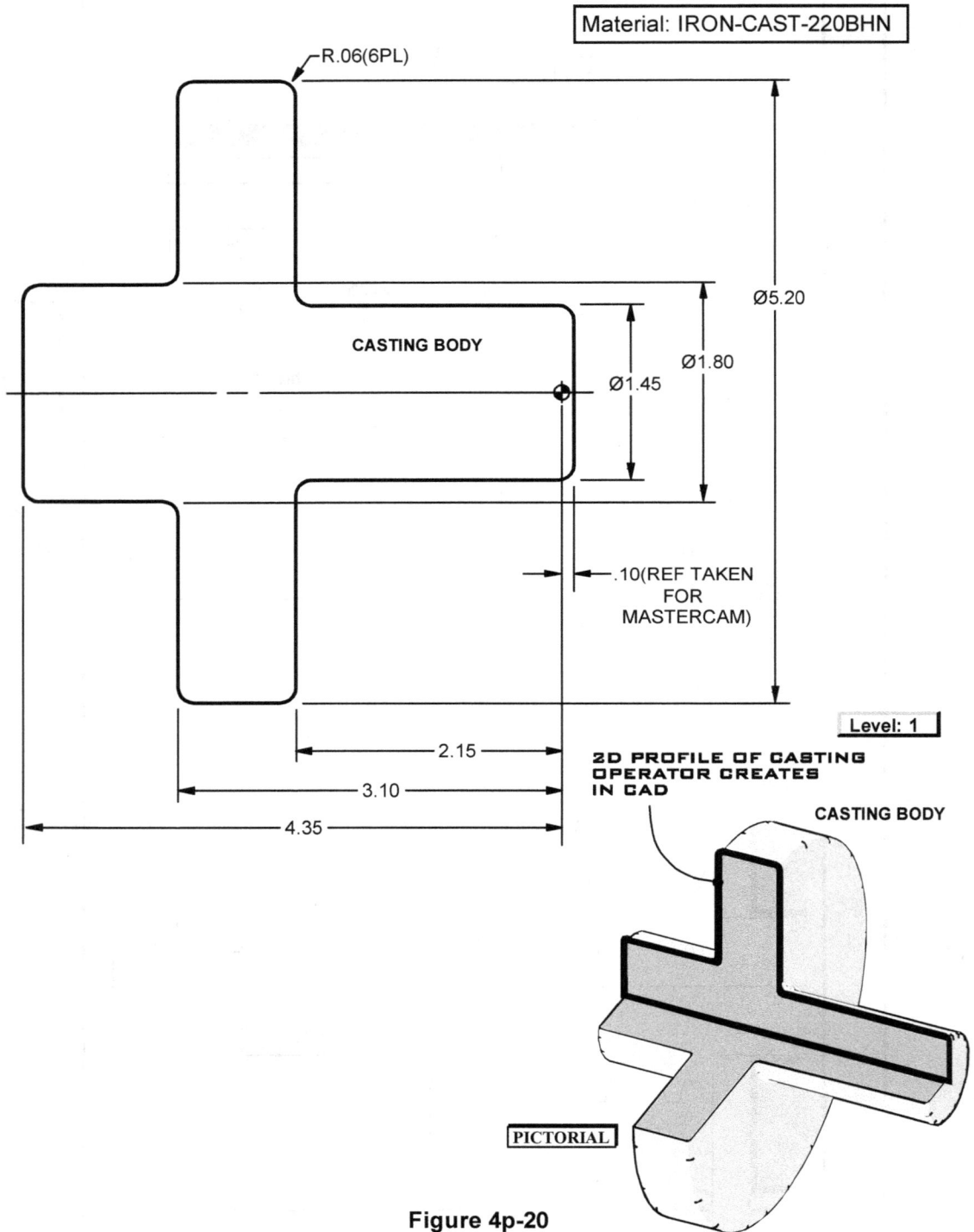

R.06(6PL)

CASTING BODY

Ø5.20

Ø1.80

Ø1.45

.10(REF TAKEN
FOR
MASTERCAM)

Level: 1

2.15

3.10

4.35

2D PROFILE OF CASTING
OPERATOR CREATES
IN CAD

CASTING BODY

PICTORIAL

Figure 4p-20

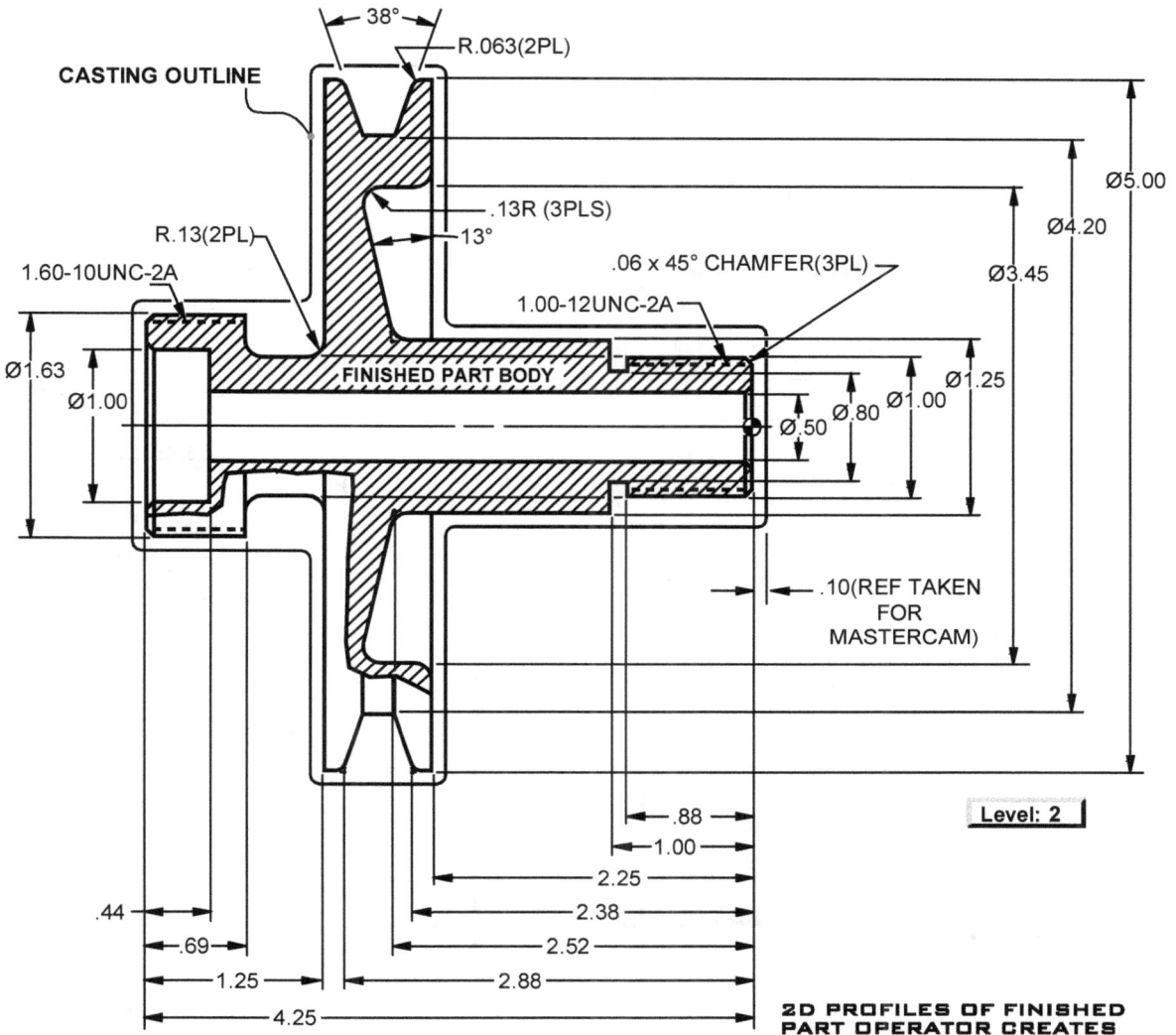

38°

R.063(2PL)

CASTING OUTLINE

.13R (3PLS)

R.13 (2PL)

13°

1.60-10UNC-2A

.06 x 45° CHAMFER(3PL)

1.00-12UNC-2A

Ø1.63

Ø1.00

FINISHED PART BODY

Ø.50

Ø.80

Ø1.00

Ø1.25

Ø5.00

Ø4.20

Ø3.45

.10(REF TAKEN FOR MASTERCAM)

Level: 2

.88

1.00

2.25

2.38

2.52

2.88

.44

.69

1.25

4.25

2D PROFILES OF FINISHED PART OPERATOR CREATES IN CAD

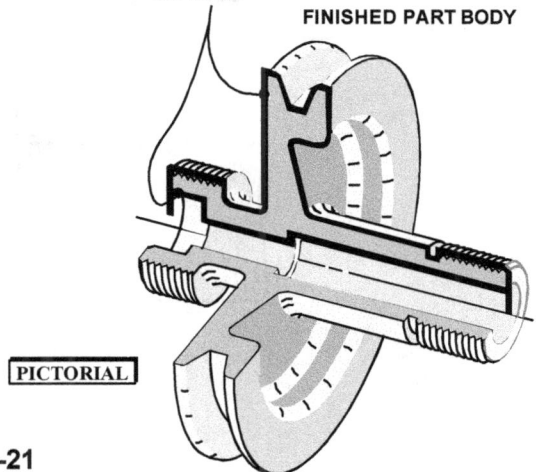

FINISHED PART BODY

PICTORIAL

Figure 4p-21

Level: 1

Stock:(Left Spindle)   Name: Chuck Jaws:(Left Spindle)

Geometry

Geometry   Revolve ─①─ ▼

②─ Select Geometry

③
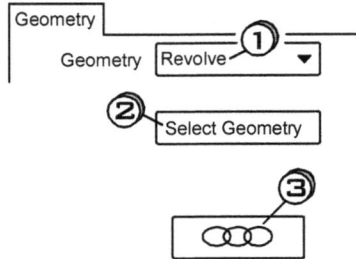

Position

☐ From stock
☐ Grip on maximum diameter

Grip length
0.75

User Defined Position

Diameter ⑤
1.8

Z ⑥
-3.35

Select   ☐ Z only

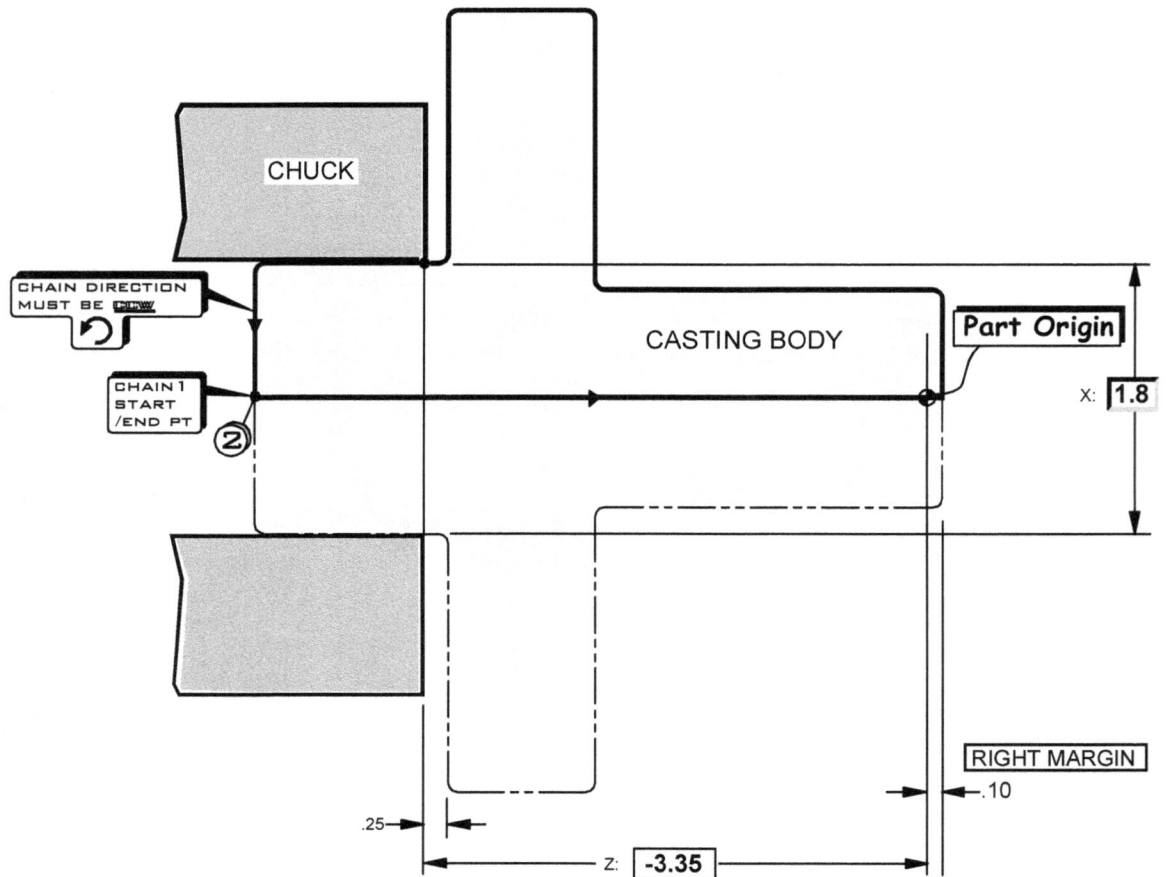

CHUCK

CHAIN DIRECTION
MUST BE CCW

CHAIN 1
START
/END PT
②

CASTING BODY

Part Origin

X: **1.8**

RIGHT MARGIN
.10

.25

Z: **-3.35**

PROCESS PLAN 4P-18

No.	Operation	Tooling
1	**Finish OD Contour**	**1/64 TNR, RH OD FINISHING**

Finish OD Contour

Level: 2

Plunge Cutting ─────────

No Plunge Cutting allowed

☑ Lead Out

Adjust Contour
☑ Extend/shorten end of contour
Ammount .15   ⊙ Extend   ○ Shorten

Exit Vector
☑ Use Exit Vector
Fixed Direction
○ None
○ Tangent
⊙ Perpendicular

Finish parameters

Finish stepover .005
Number of finish passes 3
Stock to leave in X: 0
Stock to leave in Z: 0

.15

② CHAIN 1 END PT

CHUCK

Material Removed

① CHAIN 1 START PT

R1/64

1/64 TNR, RH OD FINISHING

PROCESS PLAN 4P-18(*continued*)

No.	Operation	Tooling
2	Rough and Finish OD Groove ; leave .01 in X and Z for finishing.	R.01, W .125 OD GROOVING RIGHT

PROCESS PLAN 4P-18(*continued*)

No.	Operation	Tooling
3	ROUGH AND FINISH OD GROOVE ; LEAVE .01 IN X AND Z FOR FINISHING.	R.01, W .125 OD GROOVING RIGHT

Groove Definition
- ○ 1 Point
- ○ 2 Points
- ○ 3 Lines
- ◉ Chain
- ○ Multiple chains

☑ Lead In

Entry Vector
☑ Use Entry Vector

Fixed Direction
- ○ None
- ○ Tangent
- ◉ Perpendicular

CHUCK

② CHAIN 1 END PT

— Material Removed

① CHAIN 1 START PT

A24-NEL3

1

14

A

.144

1.5

DETAIL A

R.125

.5

.144

GR-4125

PROCESS PLAN 4P-18(*continued*)

No.	Operation	Tooling
4	ROUGH OD GROOVE	R.01, W .125 OD GROOVING RIGHT

Groove Definition
- ○ 1 Point
- ○ 2 Points
- ○ 3 Lines
- ◉ Chain
- ○ Multiple chains

Material Removed

CHAIN 1 END PT
CHAIN 1 START PT

DETAIL A

GS-094L

PROCESS PLAN 4P-18(*continued*)

No.	Operation	Tooling
5	CUT 1-12 UNC OD THREAD	60°V,RH,OD THREADING TOOL

Operation content:

Thread Form — Select from table

Thread Table

Thread Form | Unified - UN, UNC, UNF, UNEF ▼

Common diameter/lead combinations up to 4 inches

Basic major	Lead	Major diameter	Minor diameter	Comment ▼
⋮	⋮	⋮	⋮	⋮
1.0000	12.0000	1.0000	0.9098	1,12,UNF
⋮	⋮	⋮	⋮	⋮

Thread cut parameters

Thread shape parameters

NC code format: Canned ▼

Overcut .06

End Position -.88

StartPosition 0

.06   .88   Material Removed

Tooling content:

DETAIL A

.375

16 ER 12UN

1

6

A

SER 1000 M16

PROCESS PLAN 4P-18(*continued*)

No.	Operation	Tooling
6	**LATHE DRILL X .18 DEEP** Depth **-.17** ● Absolute ○ Incremental NO CHECK — Drill tip compensation	.125 DIA CENTER DRILL Ø.25
7	**LATHE DRILL X 4.475 DEEP** Depth **-4.35** ● Absolute ○ Incremental Drill Cycle Parameters Cycle: Peck drill Ist peck **.125** Subsequent peck **.125** CHECK ✔ Drill tip compensation Breakthrough ammount **.125** CHUCK Material Removed .125 4.35	1/2 DIA DRILL Ø2 3.75 9.75 1.25 .25 4.5 3.75 Ø.5

PROCESS PLAN 4P-18(*continued*)

No.	Operation	Tooling
8	**FINISH ID** ☑ Lead Out Exit Vector ☑ Use Exit Vector Fixed Direction ○ None ◉ Tangent ○ Perpendicular Plunge Cutting No Plunge Cutting allowed CHAIN1 START PT ① CHAIN1 END PT ② Material Removed	5/64 TNR, ID FINISH BORING TOOL ,.25 DIA R1/64 A .375 15.7° (end clr). Ø.5 Ø1/4 D(55 deg. diamond)
9	**LATHE FLIP STOCK** Level: 3 .25 .25 CREATE OFFSET LINE 1 AT Z LOCATION AND AND TRIM IT TO THE DESIRED X LOCATION WHERE THE CHUCK GRIPS THE PART CREATE OFFSET LINE 2 AT Z LOCATION AND AND TRIM IT TO THE DESIRED X LOCATION WHERE THE CHUCK IS TO GRIP THE FLIPPED PART LINE 1 LINE 2 CHUCK Casting outline	

PROCESS PLAN 4P18(*continued*)

No.	Operation	Tooling
	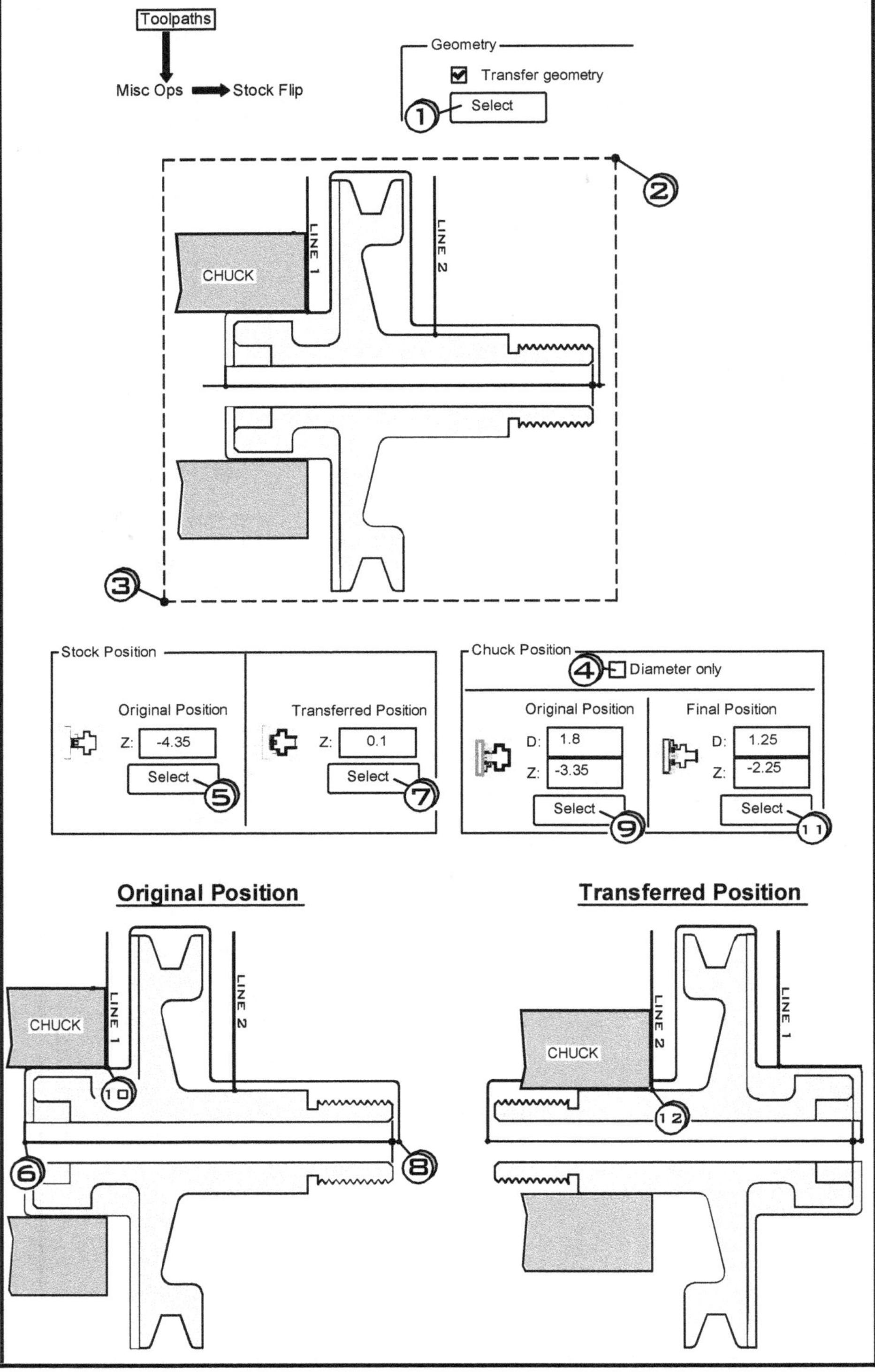	

PROCESS PLAN 4P-18(*continued*)

No.	Operation	Tooling
10	**FINISH OD CONTOUR** Plunge Cutting — *No Plunge Cutting allowed* ☑ Lead Out Exit Vector ☑ Use Exit Vector Fixed Direction ○ None ○ Tangent ◉ Perpendicular Finish parameters Finish stepover **.005** Number of finish passes **2** Stock to leave in X: **0** Stock to leave in Z: **0** CHAIN 1 END PT ② Material Removed CHUCK CHAIN 1 START PT ①	1/64 TNR, RH OD FINISHING R1/64

PROCESS PLAN 4P-18(*continued*)

No.	Operation	Tooling
11	ROUGH AND FINISH OD GROOVE ; LEAVE .01 IN X AND Z FOR FINISHING.	R.01, W .125 OD GROOVING RIGHT

Groove Definition
○ 1 Point
○ 2 Points
○ 3 Lines
◉ Chain
○ Multiple chains

CHUCK

CHAIN 1 END PT ②

CHAIN 1 START PT ①

Material Removed

DETAIL A — .093, .5, .125, .125, R.01

GC-4125

1, 5, .10, A

PROCESS PLAN 4P-18(*continued*)

No.	Operation	Tooling
12	CUT 1.625-10 UNC OD THREAD	60°V,RH,OD THREADING TOOL

Operation (No. 12):

Thread Form — Select from table

Thread Table

Thread Form: Unified - UN, UNC, UNF, UNEF ▼

Common diameter/lead combinations up to 4 inches

Basic major	Lead	Major diameter	Minor diameter	Comment ▼
⋮	⋮	⋮	⋮	⋮
1.6250	10.0000	1.6250	1.5167	1-5/8,10
⋮	⋮	⋮	⋮	⋮

Thread cut parameters

Thread shape parameters

NC code format: Canned ▼

Overcut **.06**

End Position **-.69**   StartPosition **0**

.06 → | ← .69 →
Material Removed

Tooling:

← 1 →

6

A

SER 1000 M16

DETAIL A

.375

16 ER 10UN

PROCESS PLAN 4P-18(*continued*)

No.	Operation	Tooling
13	LATHE DRILL X .44 DEEP Depth: **-.44** ● Absolute ○ Incremental Drill Cycle Parameters Cycle: Peck drill ▼ 1st peck: **.125** Subsequent peck: **.125** NO CHECK → □ Drill tip compensation Material Removed .44	1 DIA END MILL Ø3 3 9 2 4 3.5 Ø1
14	FINISH ID ☑ Lead Out → Exit Vector ☑ Use Exit Vector Fixed Direction ○ None ● Tangent ○ Perpendicular Plunge Cutting ● No Plunge Cutting allowed CHAIN1 START PT ① CHAIN1 END PT ② Material Removed	5/64 TNR, ID FINISH BORING TOOL , .25 DIA R1/64 A .375 15.7° (end clr). Ø.5 Ø1/4 D(55 deg. diamond)

4-19) Generate the 2D profile of the part shown Figure 4p-22. Use PROCESS
PLAN 4P-19 as an aid in generating the part program to produce the part.

Material: 303 Stainless

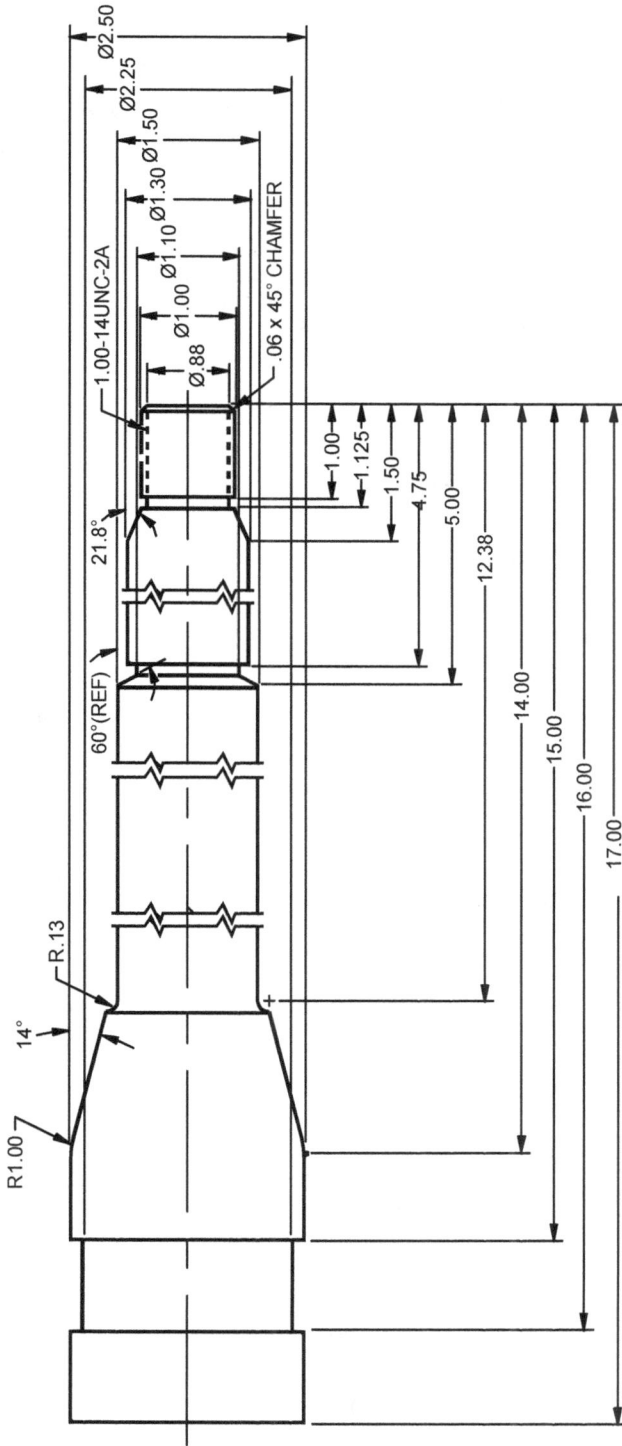

Figure 4p-22

2D PROFILE OF FINISHED
PART OPERATOR CREATES
IN CAD

PICTORIAL

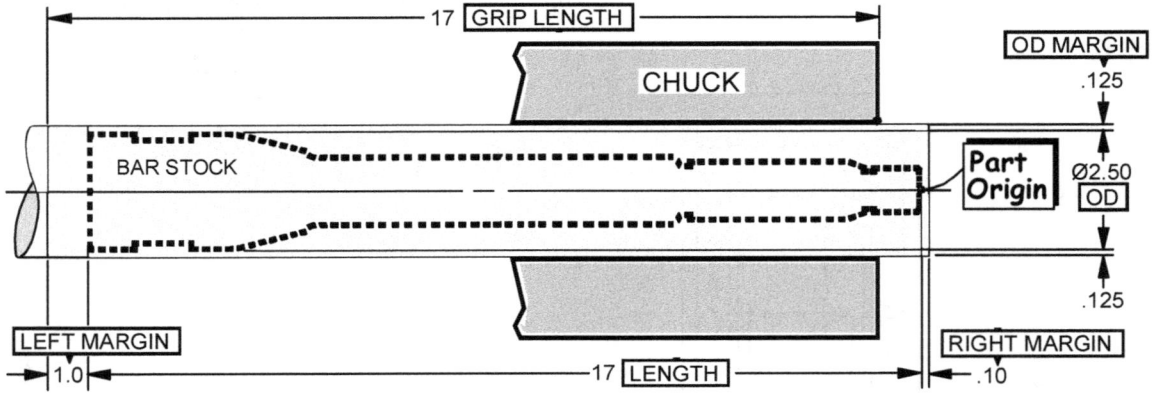

Tailstock Dimensions and Retract Position

Level: 2

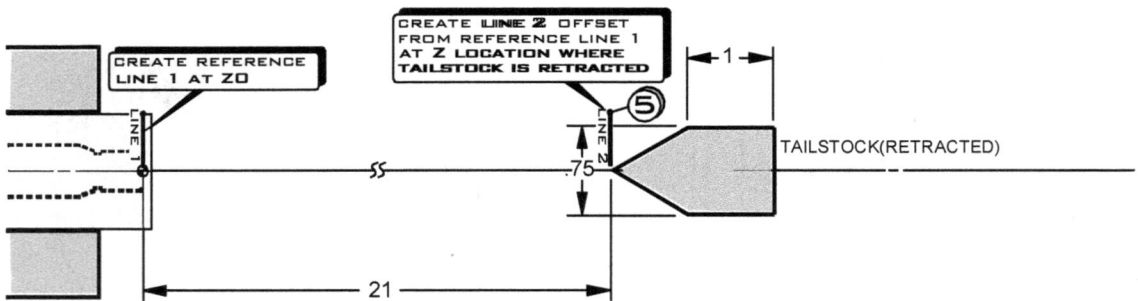

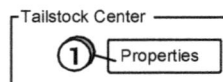

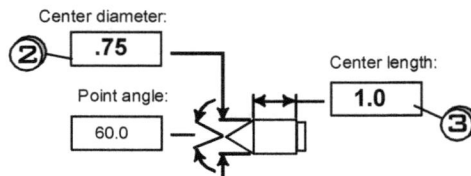

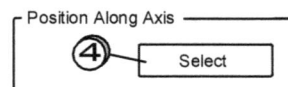

PROCESS PLAN 4P-19

No.	Operation	Tooling
1	**FACE END** BAR STOCK Material Removed .10	1/32 TNR, RH OD TURNING TOOL R1/32
2	**LATHE DRILL X .15 DEEP** Depth **-.15** ● Absolute ○ Incremental Material Removed NO CHECK □ Drill tip compensation	.25 DIA CENTER DRILL Ø.25
3	**LATHE STOCK ADVANCE** Level: 3 CREATE OFFSET LINE 2 AT Z LOCATION WHERE THE CHUCK GRIPS THE STOCK WHEN IT IS ADVANCED CREATE OFFSET LINE 1 AT Z LOCATION WHERE THE CHUCK GRIPS THE STOCK CHUCK LINE 2 .25 LINE 1 1.0	

PROCESS PLAN 4P-19(*continued*)

No.	Operation	Tooling

Toolpaths

Misc Ops → Stock Advance

Geometry — ☑ Transfer geometry ① Select

Original Position

Transferred Position

Stock Position — Original Position Z: -16.25 Select ④ — Transferred Position Z: -1 Select ⑥

Tool Positioning — Stock Advance Method — ⑧ ⦿ Push stock ☐ Use tool stop ○ Pull stock

PROCESS PLAN 4P-19(*continued*)

No.	Operation	Tooling
4	**LATHE TAILSTOCK(ADVANCE)** Toolpaths Misc Ops ➡ Tailstock Lathe tailstock Operation ① ● Advance ○ Retract **Tailstock Retract Position** TAILSTOCK **Tailstock Advance Position** TAILSTOCK	

PROCESS PLAN 4P-19(*continued*)

No.	Operation	Tooling
5	ROUGH TURN OD AND SEMIFINISH OD CONTOUR; LEAVE .002 FOR FINISH CUT.	1/32 TNR, RH OD TURNING TOOL

☑ Lead In/Out

Lead In | Lead Out
Adjust Contour
☑ Extend/shorten end of contour
Ammount .135   ● Extend   ○ Shorten
☐ Add Line

Tool parameters

T0303 R0.0315
OD ROUGH RIGHT

① User defined ▼
② Define

Home Position
D: 10   Select
Z: 40 ③   From Machine
④ ✓ ✗ ?

L R1/32

☑ Semi Finish → **Semi Finish Parameters**

Number of passes
1

Stepover
.005

Stock to leave in X:
.002

Stock to leave in Z:
0

✓ ✗ ?

.135

Material Removed

CHAIN 1 END PT
⑥

.002 LEFT AFTER SEMI-FINISH CUT

⑤

CHAIN 1 START PT

PROCESS PLAN 4P-19(*continued*)

No.	Operation	Tooling
6	**FINISH OD CONTOUR**	**1/64 TNR, RH OD FINISHING TOOL**

Operation 6:

☑ Lead In/Out

Lead In / **Lead Out**

┌ Adjust Contour ┐

☑ Extend/shorten end of contour

Ammount **.135**   ◉ Extend   ○ Shorten

☐ Add Line

Tool parameters

T0202 R0.0156
OD FINISH RIGHT

① User defined ▼

② Define

Home Position ✕

D: 10   Select

Z: 40 ③   From Machine

④ ✓   ✕   ?

←.135

CHAIN1 END PT   ⑥

Material Removed

.002

⑤   CHAIN1 START PT

R1/64

No.	Operation	Tooling
7	**ROUGH OD GROOVE IN A SINGLE PASS**	**R.01, W .125 OD GROOVING RIGHT**

🅧 Grooving Options ✕

┌ Groove Definition ┐

○ 1 Point
◉ 2 Points
○ 3 Lines
○ Chain

┌ Point Selection ┐

◉ Manual
○ Window

Rough step: Stock to leave in X:
Percent of tool width ▼   **0**

100   Stock to leave in Z:
0

Material Removed
①
②

1
5
.10
A

DETAIL A
.093
.5
.125
.125   R.01

GC-4125

PROCESS PLAN 4P-19(*continued*)

No.	Operation	Tooling
8	CUT 1-14 UNF OD THREAD	60°V,RH,OD THREADING TOOL

Tool parameters

Home Position

D: 10   Select

Z: 40 ③   From Machine

④ ✓ ✗ ?

Thread Form

Select from table

Thread Table

Thread Form | Unified - UN, UNC, UNF, UNEF ▼

Common diameter/lead combinations up to 4 inches

Basic major	Lead	Major diameter	Minor diameter	Comment ▼
1.6250	10.0000	1.6250	1.5167	1-5/8,10

Thread cut parameters

Thread shape parameters

NC code format: Canned ▼

Overcut
.06

End Position
15.25

StartPosition
16.25

16.25

15.25

.06

Material Removed

②

④

DETAIL A

.375

16 ER 14UN

← 1 →

6

A

SER 1000 M16

PROCESS PLAN 4P-19(*continued*)

No.	Operation	Tooling
9	Rough and Finish OD Groove ; leave .01 in X and Z for finishing.	R.01, W .125 OD GROOVING RIGHT

PROCESS PLAN 4P-19(*continued*)

No.	Operation	Tooling
10	ROUGH AND FINISH OD GROOVE ; LEAVE .01 IN X AND Z FOR FINISHING.	R.01, W .375 OD GROOVING RIGHT

Grooving Options

Groove Definition
- ◯ 1 Point
- ◉ 2 Points
- ◯ 3 Lines
- ◯ Chain

Tool parameters

Home Position

D: 10   Select

Z: 40 ③   From Machine

④ ✓   ✗   ?

Material Removed

① ②

DETAIL A

.275

.5   .375

.375   R.01

GC-4375

PROCESS PLAN 4P-19(*continued*)

No.	Operation	Tooling
1 1	LATHE TAILSTOCK(RETRACT)	

Toolpaths

Misc Ops ➡ Tailstock

Lathe tailstock

Operation

○ Advance ①—◉ Retract

Tailstock Advance Position

TAILSTOCK

Tailstock Retract Position

TAILSTOCK

PROCESS PLAN 4P-19(*continued*)

No.	Operation	Tooling
12	LATHE CUTOFF	.01 TNR, RH OD CUTOFF TOOL

Operation contents (No. 12):

Tool parameters

Cutoff: Select boundary point:

➤ Click ① the boundary point

Corner Geometry
- ○ None
- ○ Radius　　0.003
- ◉ Chamfer　　Parameters..
- ☑ Clearance Cut..

Home Position ☒
- D:　10　　　　Select
- Z:　40　③　　From Machine
- ④ ✓　　✖　　?

Clearance Cut ☒

Entry ammount　.1
X increment　.01
Z increment　.01

☐ Peck

✓　✖　?

Material Removed ①

Tooling (No. 12):

1.5

A—

DETAIL A

.125

.125　R.01

.5

GC-4125

4-20) Generate the 2D profile of the part shown in Figure 4p-23. Use PROCESS
PLAN 4P-20 as an aid in generating the part program to produce the part.

Material: 303 Stainless

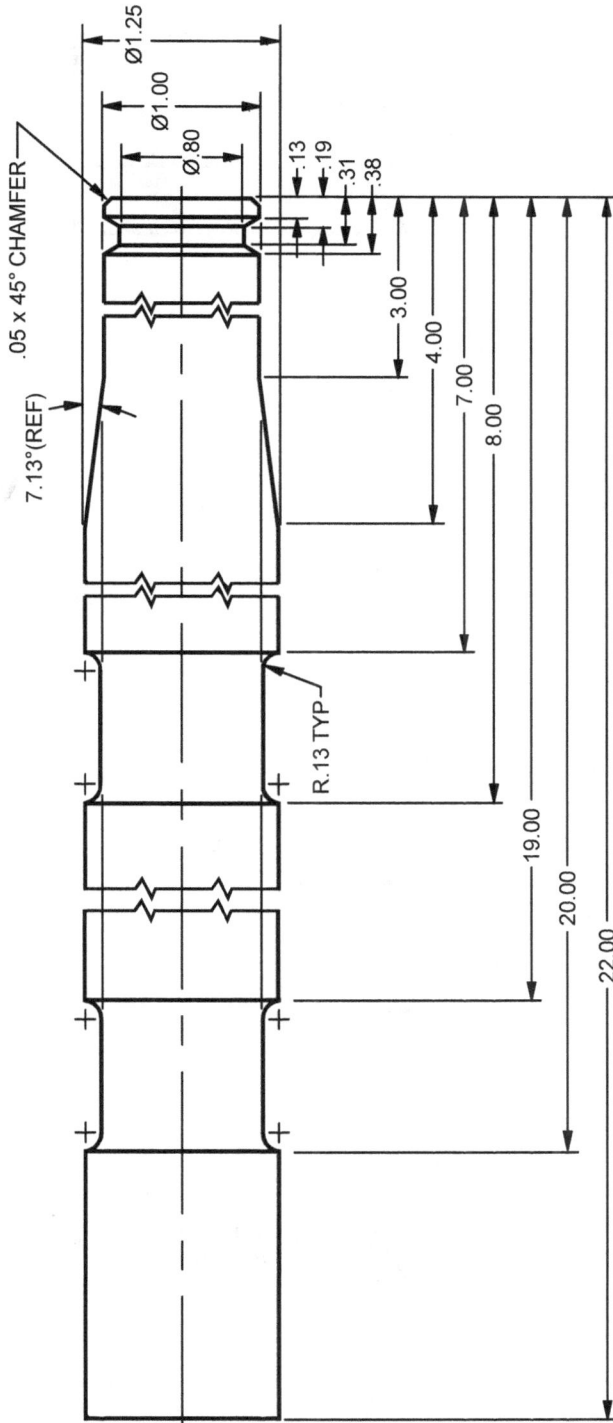

Figure 4p-23

2D PROFILE OF FINISHED
PART OPERATOR CREATES
IN CAD

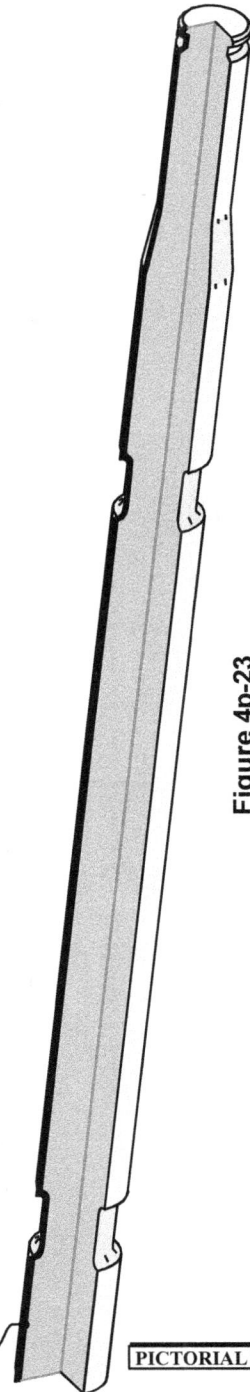

PICTORIAL

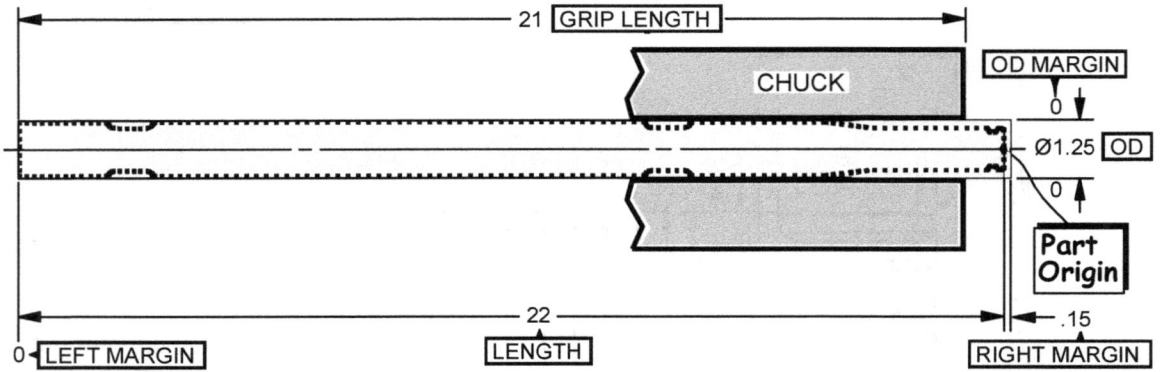

Tailstock Dimensions and   Retract Position

Level: 2

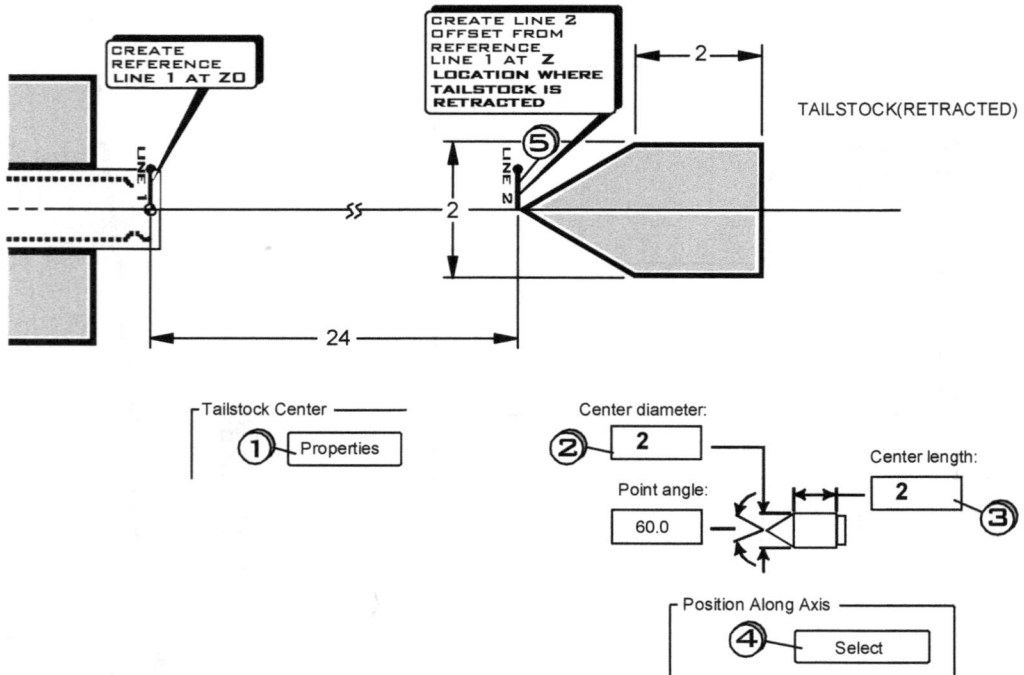

Draw the upper half of the steady rest

Steady rest boundary

2

2

Part boundary

Stock Setup

Steady Rest

Properties ①

Programmed Point

Select ⑤

Lathe Collision Avoidance Boundary(defined)

Select ②

Unselect

③

④

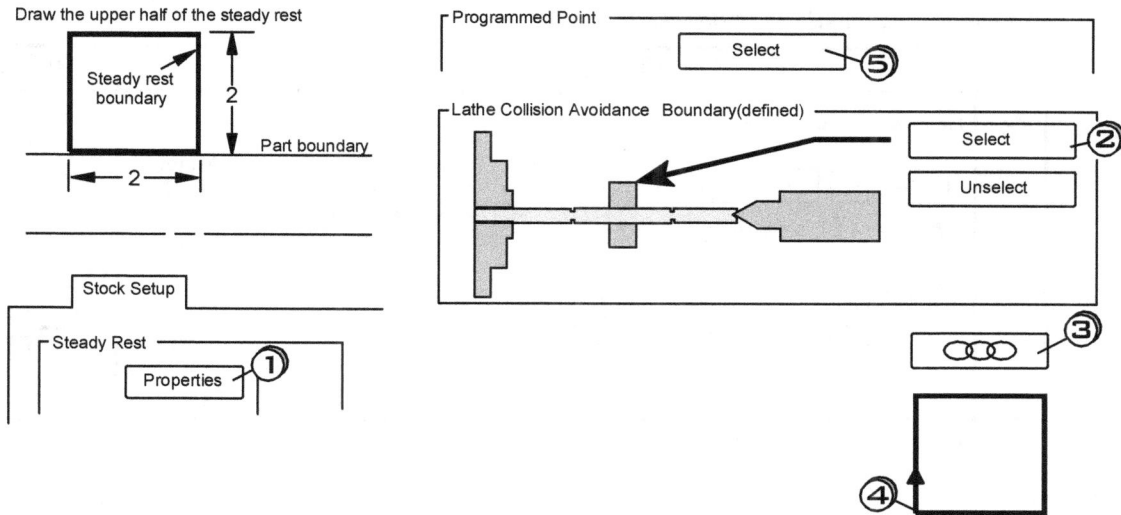

Steady Rest Dimensions and Retract Position

Level: 3

STEADY REST
(RETRACTED)

CREATE LINE 3 OFFSET FROM REFERENCE LINE 1 AT Z LOCATION WHERE STEADY REST **IS RETRACTED**

⑥

LINE 3

TAILSTOCK(RETRACTED)

SS

22

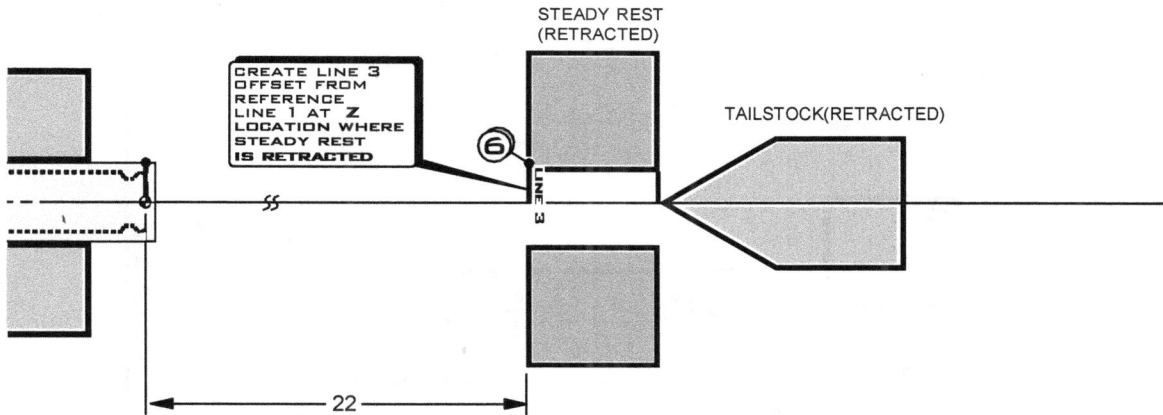

PROCESS PLAN 4P-20

No.	Operation	Tooling
1	FACE END	1/32 TNR, RH OD TURNING TOOL

Material Removed

.15

R1/32

PROCESS PLAN 4P-20(*continued*)

No.	Operation	Tooling
2	LATHE DRILL X .15 DEEP Depth **-.15** ⦿ Absolute ◯ Incremental [NO CHECK] ☐ Drill tip compensation Material Removed	.25 DIA CENTER DRILL ⌀.25
3	LATHE STOCK ADVANCE Level: 4 CREATE OFFSET LINE 5 AT Z LOCATION WHERE THE CHUCK GRIPS THE STOCK WHEN IT IS ADVANCED CHUCK CREATE OFFSET LINE 4 AT Z LOCATION WHERE THE CHUCK GRIPS THE STOCK LINE 5 LINE 4 ⊢1⊣ ⊢1⊣	

PROCESS PLAN 4P-20(*continued*)

No.	Operation	Tooling

Toolpaths

Misc Ops ➡ Stock Advance

Geometry
☑ Transfer geometry
① Select

CHUCK

Stock Position

Original Position
Z: -21
④ Select

Transferred Position
Z: -1
⑥ Select

Original Position

CHUCK

Transferred Position

Tool Positioning
Stock Advance Method
○ Push stock
⑧ ⦿ Pull stock

Stock clearance
.1 ⑨

Grip length
.25 ⑩

Tool X position 0.0

PROCESS PLAN 4P-20(*continued*)

No.	Operation	Tooling
4	LATHE STEADY REST(ADVANCE) Toolpaths Misc Ops ➡ Steady Rest Lathe steady rest Steady Rest Position Original Position — Z: 20 — Select Transferred Position — Z: 6.6479 — Select ① **Steady Rest Original(Retract) Position** CLICK MIDPOINT OF LINE ② **Steady Rest Transferred (Advance)Position**	
5	LATHE TAILSTOCK(ADVANCE) Toolpaths Misc Ops ➡ Tailstock Lathe tailstock Operation ① ⦿ Advance ◯ Retract **Tailstock Retract Position** **Tailstock Advance Position**	

PROCESS PLAN 4P-20(*continued*)

No.	Operation	Tooling
6	ROUGH TURN OD AND SEMIFINISH OD CONTOUR; LEAVE .002 FOR FINISH CUT.	1/32 TNR, RH OD TURNING TOOL

☑ Lead In/Out

Lead In / Lead Out

Adjust Contour

☑ Extend/shorten end of contour

Ammount [.135]   ⦿ Extend   ○ Shorten

☐ [Add Line]

Tool parameters

T0303 R0.0315
OD ROUGH RIGHT

① [User defined ▾]

② [Define]

Home Position ☒

D: [12] ④   [Select]

Z: [40] ③   [From Machine]

⑤ [✓] [✗] [?]

R 1/32

☑ Semi Finish →

Semi Finish Parameters ☒

Number of passes
[1]

Stepover
[.005]

Stock to leave in X:
[.002]

Stock to leave in Z:
[0]

[✓] [✗] [?]

CHAIN 1 END PT ⑥

.002 LEFT AFTER SEMI-FINISH CUT

Material Removed

CHAIN 1 START PT

⑤

PROCESS PLAN 4P-20(*continued*)

No.	Operation	Tooling
7	**FINISH OD CONTOUR** ☑ Lead In/Out Lead In / **Lead Out** ┌ Adjust Contour ─ ☑ Extend/shorten end of contour Ammount `.135` ◉ Extend ○ Shorten ☐ Add Line Tool parameters T0202 R0.0156 OD FINISH RIGHT ① User defined ▼ ② Define... **Home Position** ☒ D: `12` ④ Select Z: `40` ③ From Machine ⑤ ✓ ✗ ? CHAIN 1 END PT ⑥ .002 Material Removed CHAIN 1 START PT ⑤	1/64 TNR, RH OD FINISHING TOOL R1/64
8	**ROUGH AND FINISH OD GROOVE ; LEAVE .01 IN X AND Z FOR FINISHING.** ┌ Groove Definition ─ ○ 1 Point ○ 2 Points ○ 3 Lines ◉ Chain ○ Multiple chains **Home Position** ☒ D: `12` ④ Select Z: `40` ③ From Machine ⑤ ✓ ✗ ? ☑ Lead Out Exit Vector ☑ Use Exit Vector ┌ Fixed Direction ─ ○ None ◉ Tangent ○ Perpendicular Material Removed CHAIN 1 END PT ② ① CHAIN 1 START PT	R.01, W .125 OD GROOVING RIGHT 1 5 .10 A DETAIL A .093 .5 .125 .125 R.01 GC-4125

PROCESS PLAN 4P-20(*continued*)

No.	Operation	Tooling
9	ROUGH AND FINISH OD GROOVE AT Z-7 ; LEAVE .01 IN X AND Z FOR FINISHING.	R.01, W .25 OD GROOVING RIGHT
10	ROUGH AND FINISH OD GROOVE AT Z-19 ; LEAVE .01 IN X AND Z FOR FINISHING.	

Groove Definition (9): ○ 1 Point ○ 2 Points ○ 3 Lines ● Chain ○ Multiple chains

Home Position: D: 12 ④ Select Z: 40 ③ From Machine ⑤ ✓ ✗ ?

☑ Lead Out Exit Vector ☑ Use Exit Vector Fixed Direction ○ None ● Tangent ○ Perpendicular

CHAIN1 END PT ② Material Removed ① CHAIN1 START PT

DETAIL A: .144 .5 .250 .250 R.01 GC-4250

CHUCK

PROCESS PLAN 4P-20(*continued*)

No.	Operation	Tooling
11	**Lᴀᴛʜᴇ ᴛᴀɪʟsᴛᴏᴄᴋ(ʀᴇᴛʀᴀᴄᴛ)** TOOLPATHS Misc Ops ➡ Tailstock Lathe tailstock Operation ○ Advance　①◉ Retract **Tailstock Retract Position**	
12	**Lᴀᴛʜᴇ sᴛᴇᴀᴅʏ ʀᴇsᴛ(ʀᴇᴛʀᴀᴄᴛ)** TOOLPATHS Misc Ops ➡ Steady Rest Lathe steady rest Steady Rest Position Original Position — Z: 6.6479 — Select ① Transferred Position — Z: **20** ③ — Select **Steady Rest Original(Retract) Position** CLICK MIDPOINT OF LINE ②	
13	**Lᴀᴛʜᴇ ᴄʜᴜᴄᴋ(ᴜɴ-ᴄʟᴀᴍᴘ)** TOOLPATHS Misc Ops ➡ Chuck Lathe chuck Operation ○ Clamp　◉ Un-clamp ① CHUCK	

CHAPTER - 5

C-AXIS CAD OPERATIONS

5-1 Chapter Objectives

After completing this chapter you will be able to create geometry for C-axis:

1. Face Contour
2. Cross Contour
3. Contour
4. Face Drill
5. Cross Drill
6. Drill

5-2 An Example Of Generating C-axis Wireframe Geometry

EXAMPLE 5-1

Generate the wireframe geometry on levels 1 through 5 as illustrated in the **PICTORIALS** in Figure 5-3.

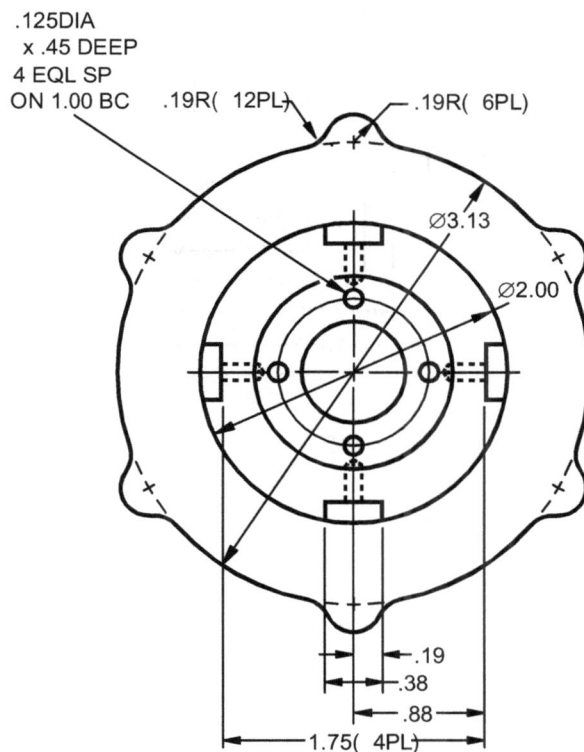

Figure 5-1

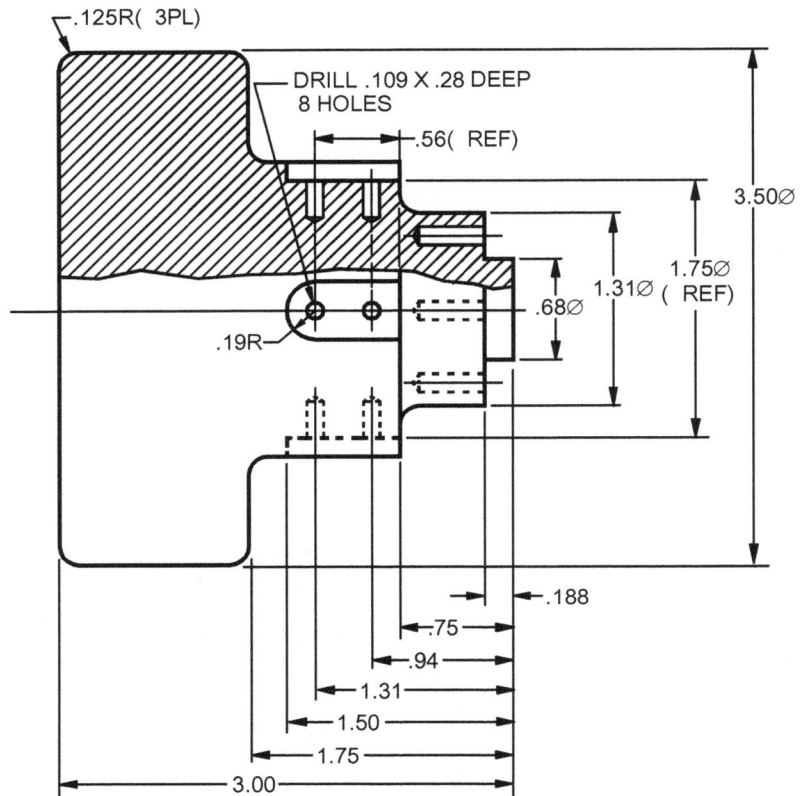

.125R(3PL)

DRILL .109 X .28 DEEP
8 HOLES

.56(REF)

3.50∅

1.75∅
(REF)

1.31∅

.68∅

.19R

.188

.75

.94

1.31

1.50

1.75

3.00

Figure 5-2

PICTORIALS

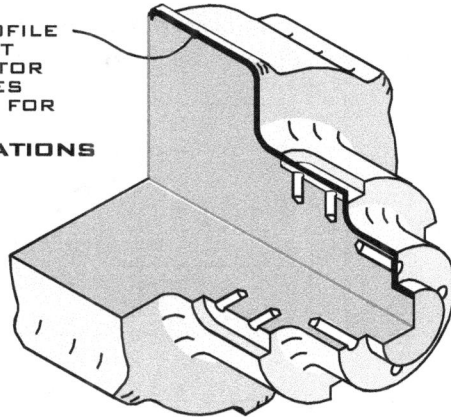

2D PROFILE
OF PART
OPERATOR
CREATES
IN CAD FOR
TURN
OPERATIONS

Level: 1

3D PROFILES OF PART OPERATOR CREATES
IN CAD FOR **C-AXIS MILL/TURN OPERATIONS** SHOWN **BOLD**

Level: 2

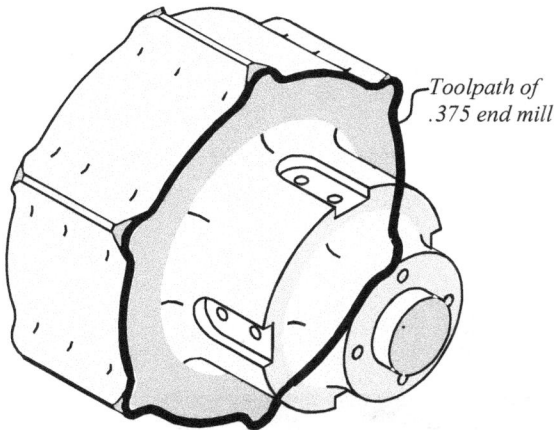

*Toolpath of
.375 end mill*

Level: 3

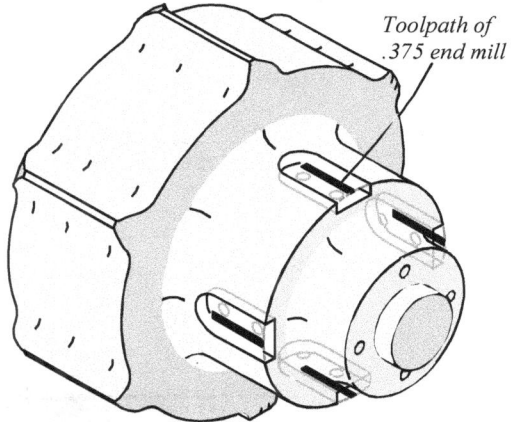

*Toolpath of
.375 end mill*

Level: 4

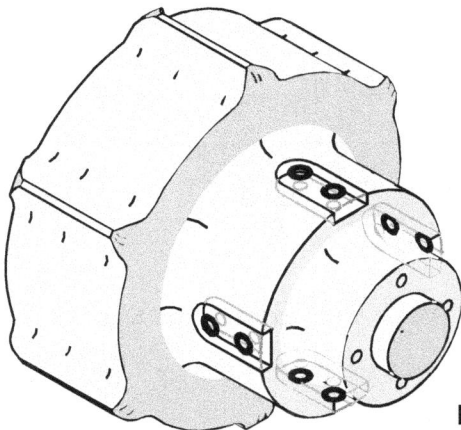

Level: 5

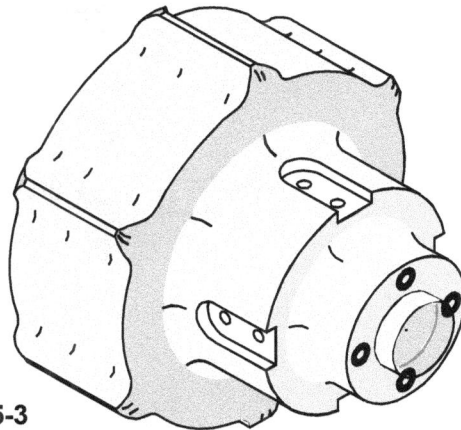

Figure 5-3

➤ Direct *Mastercam* to create a new file.

Start by assigning *Mastercam* to **diameter** programming mode with the working construction plane, **Cplane,** set to **+DZ.** Be sure the default **Gview** is set to top **T** .

```
           Lathe Radius   ▶ ②
           Lathe Diameter ▶   ▓+D+Z(WCS) ③
                         ①
```
WCS:TOP Tplane:TOP Cplane +D+Z   3D │Gview│WCS │Planes│ Z: 0.0 ▼ │ ⬚ │⬚│⬚ │ Level: 1 ▼ │Attributes ✳ ▼│── ▼│── ▼│ Groups │?
                                                                              ④

➤ Click ① the Planes button

➤ Click ② the Lathe Diameter arrow ▶

➤ Click ③   ▓+D+Z(WCS)

On ⬚ **Level: 1** ⬚, create the 2D profile illustrated in the   **PICTORIAL.**

➤ Click ④ the Level: button

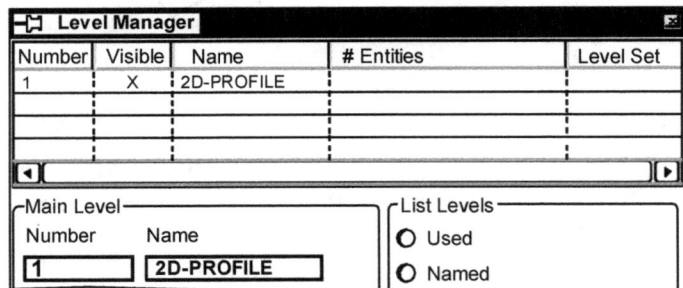

```
⊞  Level Manager                                          ▣
Number│ Visible│ Name      │ # Entities          │ Level Set
1     │   X    │ 2D-PROFILE │                    │
      │        │            │                    │
      │        │            │                    │
◄│                                                       │►
┌Main Level────────────┐  ┌List Levels────┐
 Number     Name          ○ Used
 ⬚1⬚       2D-PROFILE     ○ Named
```

➤ Press [Tab] ; enter the new level name   **2D-PROFILE**

➤ Draw the stock outline on level 4. Press [Esc] to cancel the function

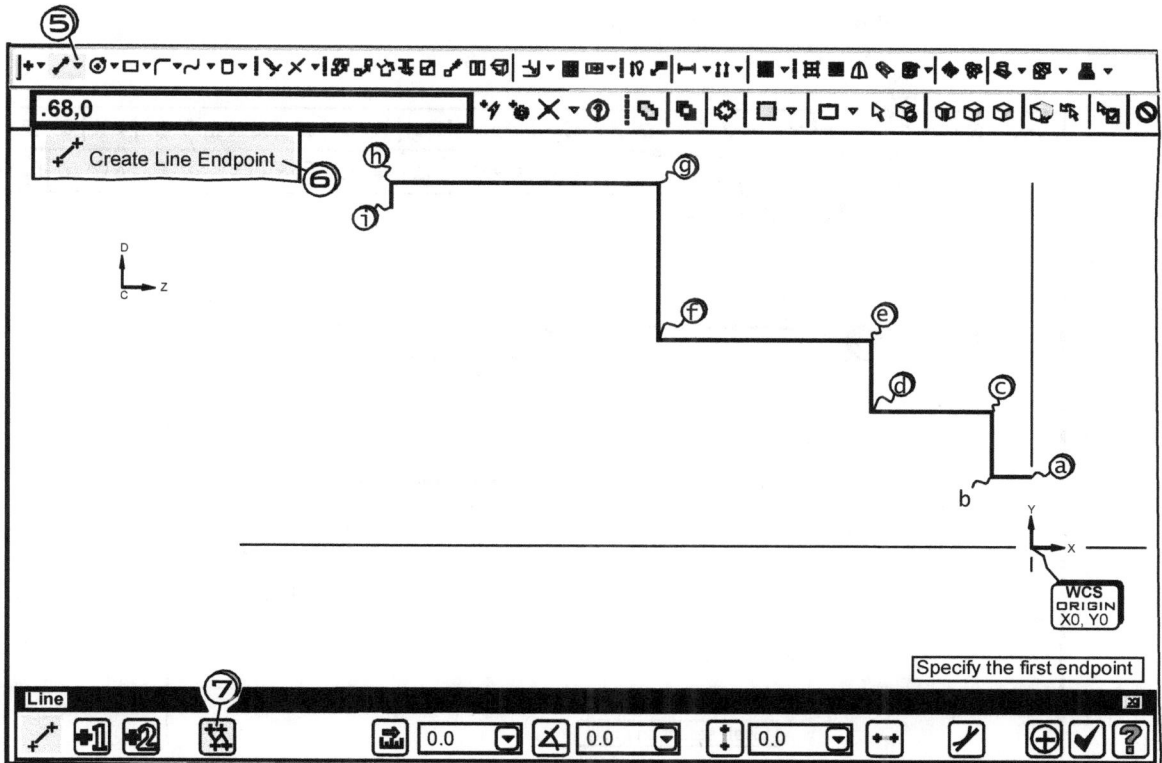

> Click ⑤ the line down arrow ▾

> Click ⑥ ✐ Create Line Endpoint

Specify the first endpoint

> Click ⑦ the multi line button 🔀
> or tap the [M] key

> Enter coords [.68,0] [Enter] ◄----ⓐ

> Enter coords [.68,-.188] [Enter] ◄----ⓑ

> Enter coords [1.31,-.188] [Enter] ◄----ⓒ

> Enter coords [1.31,-.75] [Enter] ◄----ⓓ

> Enter coords [2,-.75] [Enter] ◄----ⓔ

> Enter coords [2,-1.75] [Enter] ◄----ⓕ

> Enter coords [3.5,-1.75] [Enter] ◄----ⓖ

> Enter coords [3.5,- 3] [Enter] ◄----ⓗ

> Enter coords [3.5-.25,-3] [Enter] ◄----ⓘ

> Press [Esc] to cancel the operation

Create the R.25, R.125 and R.063 fillets

Fillet: select an entity

▶ Click ⑧ the down arrow ▾

▶ Click ⑨ ⌐ Fillet Entities

▶ Tap the R key for *Radius*; enter **.125**

Fillet: select an entity

▶ Click ⑩ the first line entity

Fillet: select another entity

▶ Click ⑪ the second line entity

Fillet: select an entity

▶ Click ⑫ the first line entity

Fillet: select another entity

▶ Click ⑬ the second line entity

Fillet: select an entity

▶ Click ⑭ the first line entity

Fillet: select another entity

▶ Click ⑮ the second line entity

Fillet: select an entity

▶ Click ⑯ the first line entity

Fillet: select another entity

▶ Click ⑰ the second line entity

▶ Tap Esc for function cancel

On Level: 2 ,create the C-axis 3D profile at **Z-1.75** as illustrated in the **PICTORIAL.**

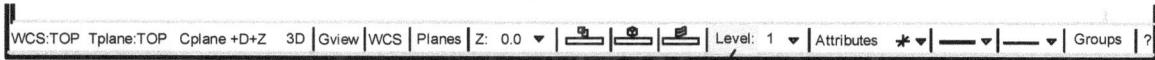

WCS:TOP Tplane:TOP Cplane +D+Z   3D │Gview│WCS │ Planes │ Z: 0.0 ▼│ ▭ │ ⊕ │ ▭ │ Level: 1 ▼ │Attributes ✱ ▼│ ── ▼│ ── ▼│ Groups │?

⑱

➤ Click ⑱ the   Level: button

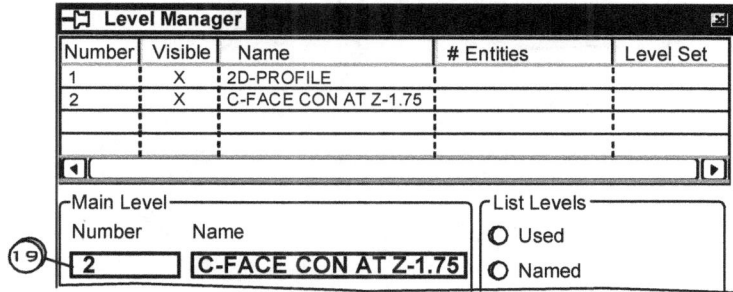

Level Manager   ☒

Number	Visible	Name	# Entities	Level Set
1	X	2D-PROFILE		
2	X	C-FACE CON AT Z-1.75		

◄ │ ►

┌Main Level──────────────────────┐   ┌List Levels──────┐
   Number   Name   ○ Used
⑲ │ 2 │ │ **C-FACE CON AT Z-1.75** │   ○ Named

➤ Click ⑲ the number box; enter **2**

➤ Press Tab ; enter the new level name **C-FACE CON AT Z-1.75**

Change the view to Isometric. Assign the working **Cplane** as **Right** and the working Z depth as **Z-1.75**

➤ *Right* Click; Click ⑳ ▧ Isometric(WCS)   Alt+7

➤ Click ㉑ Planes

➤ Click ㉒ ▧ Right(WCS)

➤ Click ㉓ ;enter **-1.75**

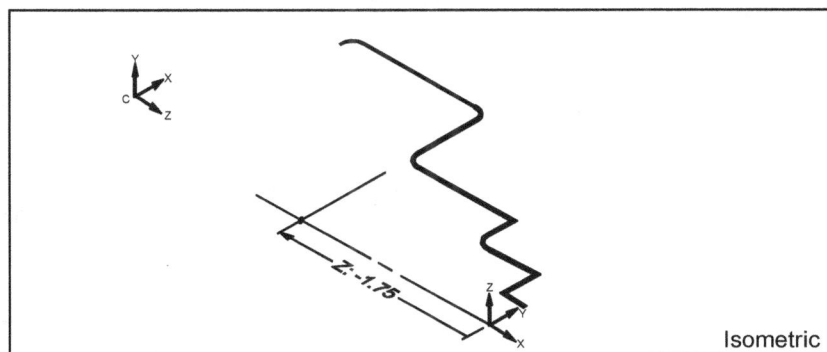

㉑   ㉓

▧ Isometric(WCS)   Alt+7   ⑳

WCS:TOP Tplane:TOP Cplane +D+Z   3D │Gview│WCS │ Planes │ Z: -1.75 ▼│ ▭ │ ⊕ │ ▭ │ Level: 1 ▼ │Attributes ✱ ▼│ ── ▼│ ── ▼│ Groups │?

▧ Right(WCS) ㉒

Isometric

Generate a template composed of the R1.565 and R.19 arcs.

➤ Click ㉔ the circle down arrow ▾

➤ Click ㉕ 🏹 Create Arc Polar

Enter the center point

➤ Enter 0,0   Enter↵

➤ Tap the R key for *Radius*

➤ Enter **3.13/2** Tab

➤ Tab into the *Start Angle* box

Sketch the initial angle

➤ Enter **330**

➤ Tab into the *Final Angle* box

Sketch the final angle

➤ Enter **30**   Enter↵

Enter the center point

➤ Enter 3.13/2,0   Enter↵

➤ Tap the R key for *Radius*

➤ Enter **.19** Tab

➤ Tab into the *Start Angle* box

Sketch the initial angle

➤ Enter **-90**

➤ Tab into the *Final Angle* box

Sketch the final angle

➤ Enter **90**   Enter↵

➤ Press Esc to cancel the function

Complete the template shape by breaking the R1.565 arc in two pieces and adding the R.19 fillets.

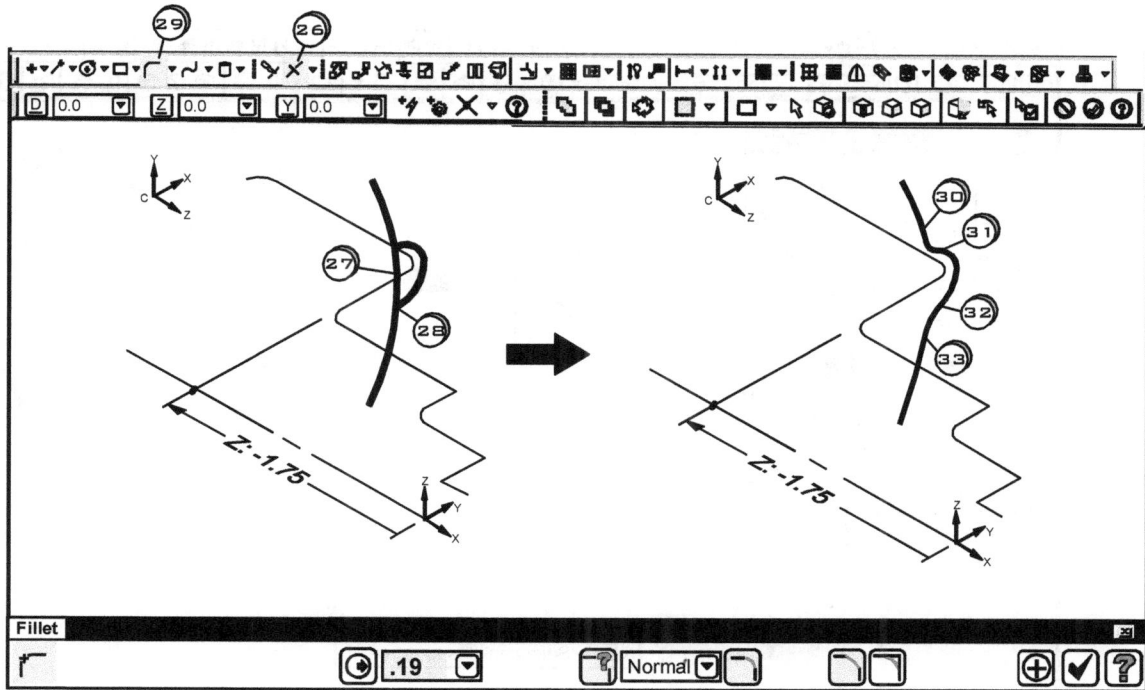

Fillet

➤ Click ㉖ Break two pieces

| Select an entity to break |

➤ Click ㉗

| Indicate the break position |

➤ Click ㉘

➤ Tap the Esc key

➤ Click ㉙ Fillet

➤ Tap the R key for *Radius*; enter **.19**

➤ Click ㉚

➤ Click ㉛

➤ Click ㉜

➤ Click ㉝

➤ Tap Esc for function cancel

Use the **XFORM**, **Rotate**, **Copy** command to make *5 copies* to complete the C contour.

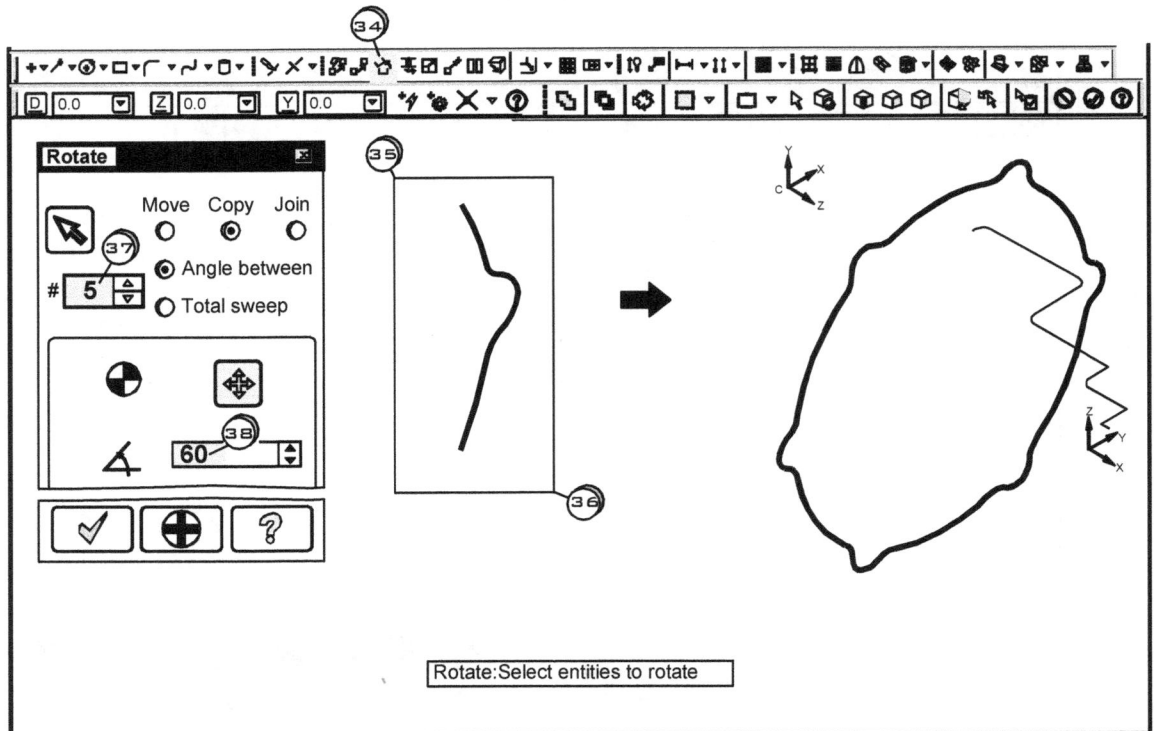

Rotate:Select entities to rotate

➤ Click ③④ the Xform Rotate icon ✿

Rotate:Select entities to rotate

➤ Click ③⑤ ③⑥ the window corners [Enter]

➤ Click ③⑦ in the # copies box ; enter [5]

➤ Click ③⑧ in the angle box ∠ ;enter the copy angle [60] [Enter]

➤ Press [Esc] for function cancel

On Level: 3 ,create the .56 long C-axis cross contour lines at   Z-.75. Refer to the
PICTORIAL.

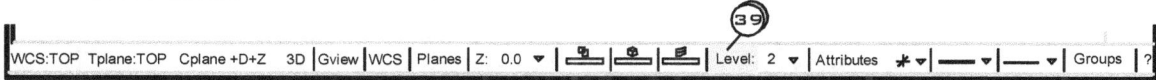

WCS:TOP Tplane:TOP Cplane +D+Z   3D │Gview│WCS │Planes │ Z: 0.0 ▼│ ▢ │ ▢ │ ▢ │ Level: 2 ▼│Attributes ✱ ▼│── ▼│── ▼│Groups │ ?

⟩⟩⟩ Click ③⑨ the Level: button

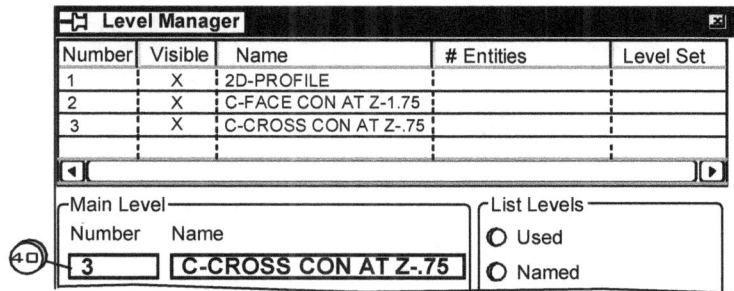

┌───┐
│ ▛▟ **Level Manager**   ▨ │
├──────┬─────────┬──────────────────┬───────────┬──────────┤
│Number│ Visible │ Name   │ # Entities│ Level Set│
├──────┼─────────┼──────────────────┼───────────┼──────────┤
│ 1   │ X   │ 2D-PROFILE   │   │   │
│ 2   │ X   │ C-FACE CON AT Z-1.75 │   │   │
│ 3   │ X   │ C-CROSS CON AT Z-.75 │   │   │
└──────┴─────────┴──────────────────┴───────────┴──────────┘

┌─Main Level──────────────────────────┐ ┌─List Levels──────┐
│ Number   Name   │ │ ○ Used   │
│ ┌───┐ ┌────────────────────────┐   │ │ ○ Named   │
│ │ 3 │ │**C-CROSS CON AT Z-.75**│   │ │   │
│ └───┘ └────────────────────────┘   │ └──────────────────┘
└─────────────────────────────────────┘

⟩⟩⟩ Click ④⑩ the number box; enter **3**

⟩⟩⟩ Press Tab ; enter the new level name **C-CROSS CON AT Z-.75**

Assign the working **Cplane** as **Top** and the working Z depth as **Z-.75**

⟩⟩⟩ Click ④① Planes

⟩⟩⟩ Click ④② 🔲 Top (WCS)

⟩⟩⟩ Click ④③ ;enter **-.75**

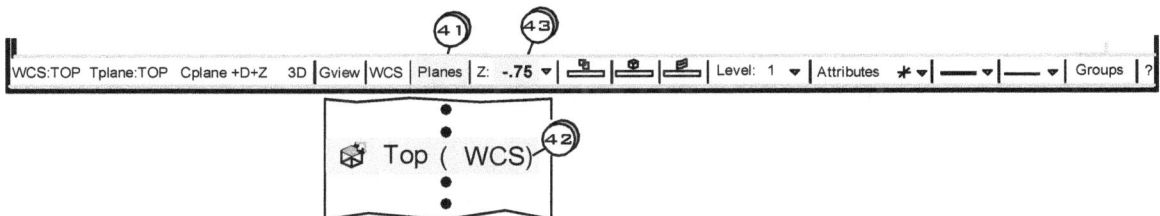

WCS:TOP Tplane:TOP Cplane +D+Z   3D │Gview│WCS │Planes │ Z: -.75 ▼│ ▢ │ ▢ │ ▢ │ Level: 1 ▼│Attributes ✱ ▼│── ▼│── ▼│Groups │ ?

🔲 Top (WCS) ④②

Create a 3D polar line of length .56in.

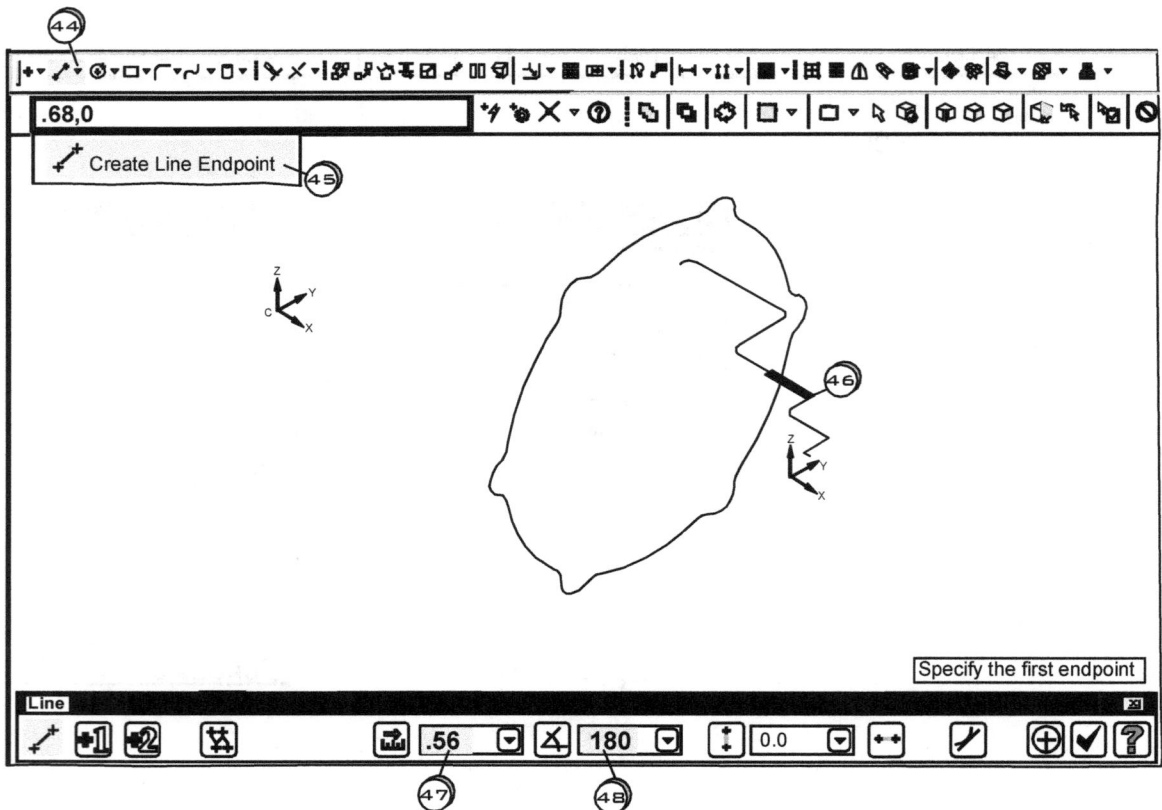

➤ Click ④④ the line down arrow ▾

➤ Click ④⑤ ✦ Create Line Endpoint

     Specify the first endpoint

➤ Click ④⑥ near the line's **_endpoint_**

     Specify the second endpoint

➤ Click ④⑦ in the length box ; enter **.56**

➤ Click ④⑧ in the angle box ; enter **180**

➤ Press [Esc] to cancel the operation

Select **Cplane** Right . Use the **XFORM**, **Rotate**, **Copy** command to make *3 copies* of the .56in 3D line

➤ Click ㊽   Planes

➤ Click ㊿   ☐ Right(WCS)

㊾

WCS:TOP Tplane:TOP Cplane +D+Z   3D │Gview│WCS │ Planes │ Z: -.75 ▼│ ▱ │ ▱ │ ▱ │ Level: 1 ▼│ Attributes ✱ ▼│ ── ▼│ ── ▼│ Groups │ ?│

•
•
☐ Right(WCS) 50
•
•

51

│ + ▼ ✏ ▼ ⊙ ▼ ▢ ▼ ⌐ ▼ ◿ ▼ ▢ ▼ │ ❯ ✕ ▼ │ ▨ ◨ ▨ | ▦ ▨ ◢ Ⅲ ◎ │ ⌐ ▼ ▮ ▣ ▼ │ ♪ ◿ ▮ ┝ ▼ ¦¦ ▼ │ ▮ ▼ │ ▦ ▮ ▲ ◈ ❀ ▼ │ ◆ ❀ ◢ ▼ ◈ ▼ ▲ ▼ │

│▯│0.0 ▼│ Z │0.0 ▼│ Y │0.0 ▼│ ⚡ ✶ ❖ ✕ ▼ ◎ │ ◨ │ ▨ │ ◈ │ ▢ ▼ │ ▢ ▼ ▷ ◈ │ ◈ ◈ ◈ │ ◈ ❀ │ ◈ │ ⊘ ⊘ ⊘ │

Rotate   ⊠

|➤| Move Copy Join
                   ○   ◉   ○
53   ◉ Angle between
|3| ▲▼   ○ Total sweep

⊕   ⊕

◿ |90| ▼

| ✓ | ⊕ | ? |

Rotate:Select entities to rotate

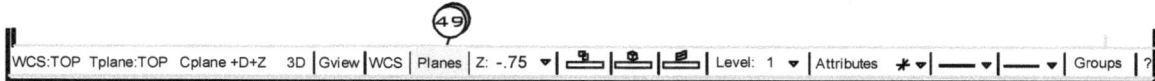

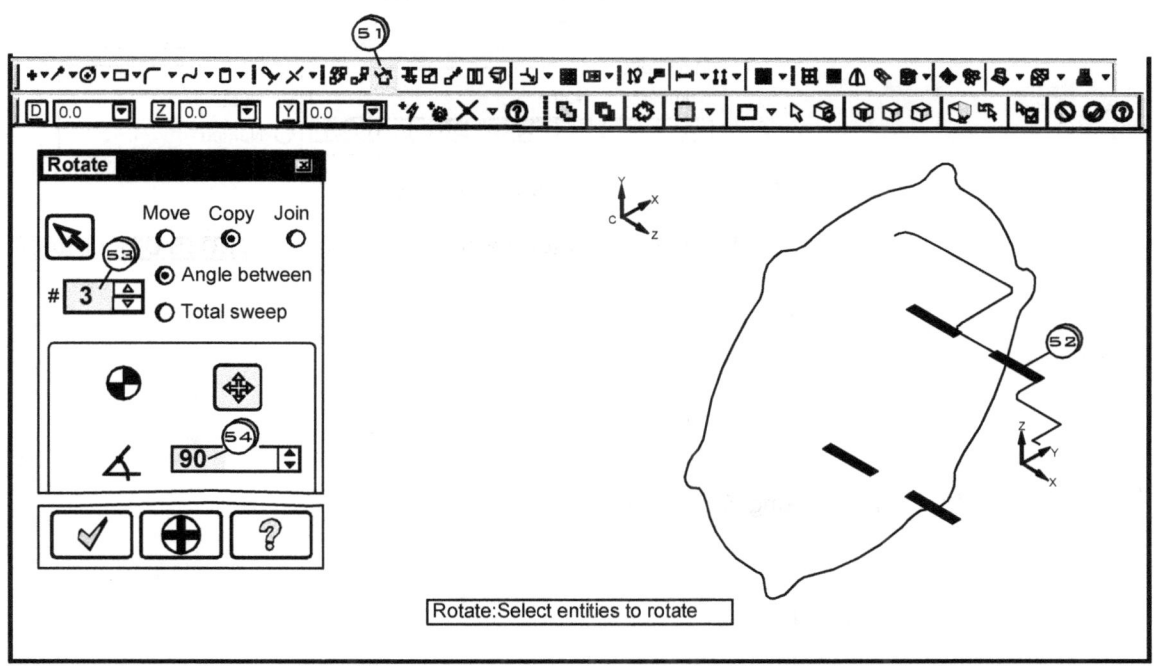

➤ Click 51   the Xform Rotate icon ☆

Rotate:Select entities to rotate

➤ Click 52 the line entity   [Enter ←]

➤ Click 53 in the # copies box ; enter   **3**

➤ Click 54 in the angle box ◿ ;enter the copy angle [90 ▼] [Enter ←]

➤ Press [Esc] for function cancel

On ⎡Level: 4⎤ , generate the .109 Dia cross drill circles. See the **PICTORIAL.**

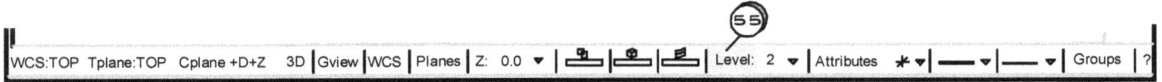

(55)

WCS:TOP Tplane:TOP Cplane +D+Z 3D │Gview│WCS │ Planes │ Z: 0.0 ▼ │ ⬜ ⬜ ⬜ │ Level: 2 ▼ │ Attributes ✱ ▼│ ── ▼│ ── ▼│ Groups │ ?

➤ Click (55) the Level: button

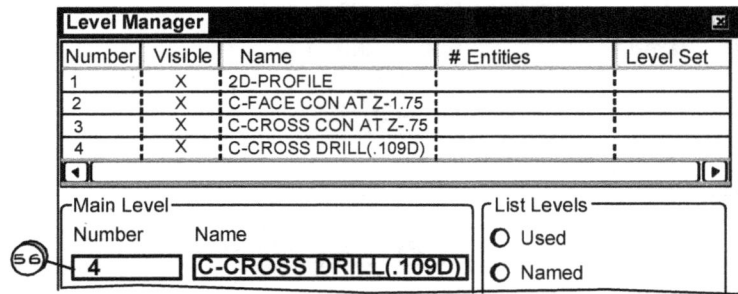

Level Manager   ▣

Number	Visible	Name	# Entities	Level Set
1	X	2D-PROFILE		
2	X	C-FACE CON AT Z-1.75		
3	X	C-CROSS CON AT Z-.75		
4	X	C-CROSS DRILL(.109D)		

◀ ▶

┌Main Level──────────────────┐ ┌List Levels─┐
Number   Name   ○ Used
(56) ⎡4⎤   ⎡C-CROSS DRILL(.109D)⎤   ○ Named

➤ Click (56) the number box; enter **4**

➤ Press ⎡Tab⎤ ; enter the new level name ⎡C-CROSS DRILL(.109D)⎤

Assign the working **Cplane** as **Top**

➤ Click (57) Planes

➤ Click (58) ⬛ Top (WCS)

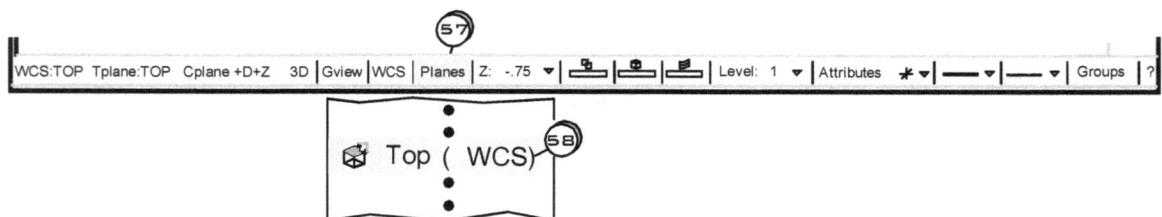

(57)

WCS:TOP Tplane:TOP Cplane +D+Z 3D │Gview│WCS │ Planes │ Z: -.75 ▼ │ ⬜ ⬜ ⬜ │ Level: 1 ▼ │ Attributes ✱ ▼│ ── ▼│ ── ▼│ Groups │ ?

•
⬛ Top (WCS) (58)
•
•

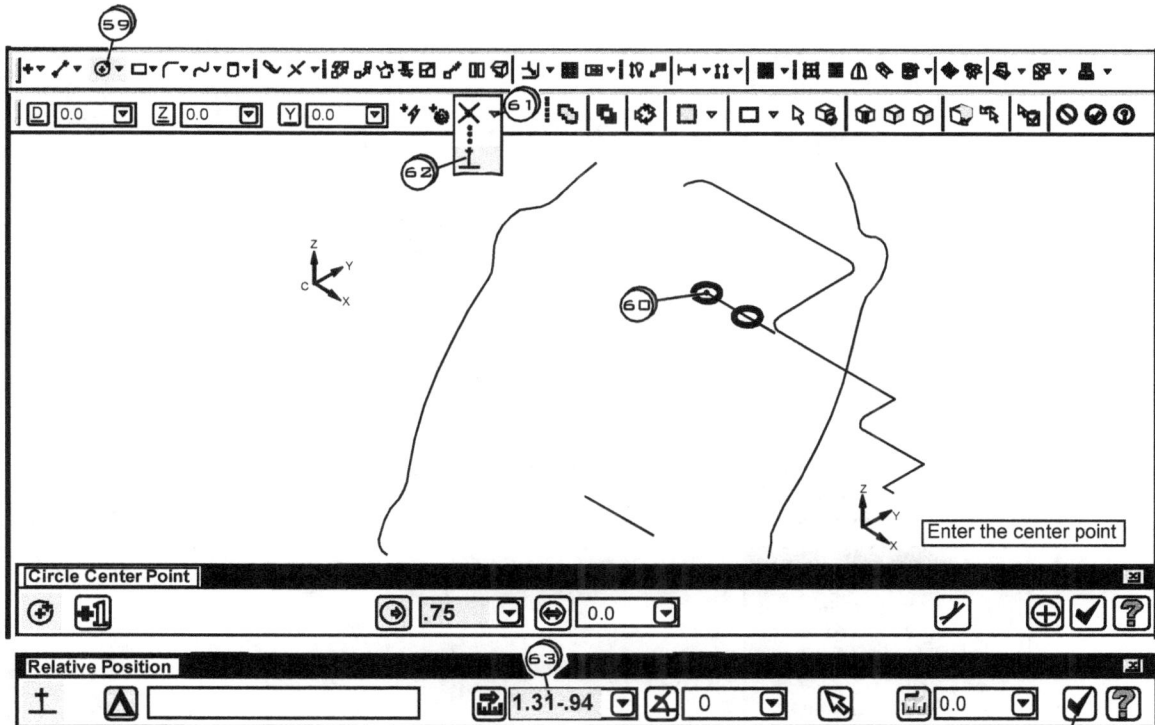

Enter the center point

Circle Center Point

Relative Position

➣ Click ⑤⑨ Circle Center Point

Enter the center point

➣ Click ⑥⓪ the end of the line

➣ Tap the ☐D☐ key for *Diameter; enter* **.109** Enter

➣ Click ⑥① the down arrow ▾

➣ Click ⑥② ⊥ Relative

Enter a known point or change to Along mode

➣ Click ⑥⓪ the end of the line

➣ Click ⑥③ in the *radius* box ☐ ; Enter the relative radius 1.31-.94 ▾

➣ Click ⑥④ the OK button ☑

➣ Tap the ☐D☐ key for *Diameter; enter* **.109** Enter

Select **Cplane** Right . Use the **XFORM**, **Rotate**, **Copy** command to make *3 copies* of the .109 dia circles.

➤ Click (65)   Planes

➤ Click (66)   ▣ Right(WCS)

(65)

WCS:TOP Tplane:TOP Cplane +D+Z   3D | Gview | WCS | Planes | Z: -.75 ▾ | ⬚ | ⬚ | ⬚ | Level: 1 ▾ | Attributes ✳▾ | ──▾ | ──▾ | Groups | ?

•
•
▣ Right(WCS)   (66)
•
•

(67)

| + ▾ ✎ ▾ ◎ ▾ □ ▾ ⌐ ˅ ⌐ ▾ ○ ▾ | ⋎ ✕ ▾ | ⧉ ⌂ ✿ ⁊ ⊟ ♪ Ⅲ ⊡ | ⤸ ▾ ▪ ▥ ▾ | ⋔ ⤶ | ⊢ ▾ ⊪ ▾ | ▉ ▾ ⊞ ▉ ⚑ ❧ ⬥ ▾ | ◆ ❀ ⬥ ▾ | ⬥ ▾ ▲ ▾ |
| D ▯ 0.0 ▯ Z ▯ 0.0 ▯ Y ▯ 0.0 ▯ | ✦ ✹ ✕ ▾ ⌀ | ◫ | ◱ | ◉ | □ ▾ | □ ▾ ◣ ⊗ | ⬒ ⬓ ⬔ | ◨ ◱ | ◈ ◉ ⑦ |

┌──────────────────┐
│ Rotate   ⛶│
├──────────────────┤
│   Move Copy Join │
│ ▐◥   ○   ◉   ○ │
│   (70)   │
│   ◉ Angle between   │
│ # ▯3▯ ▲▾   │
│   ○ Total sweep   │
│ ┌──────────────────────┐ │
│ │ ◕   ┌───┐   │ │
│ │   │ ✛ │   │ │
│ │   (71)└───┘   │ │
│ │ ∠ ▯90▯ ▾   │ │
│ └──────────────────────┘ │
│ ┌────┬────┬────┐   │
│ │ ✓ │ ⊕ │ ? │   │
│ └────┴────┴────┘   │
└──────────────────┘

(69) (68)

┌─────────────────────────────┐
│ Rotate:Select entities to rotate │
└─────────────────────────────┘

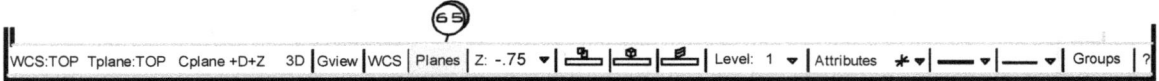

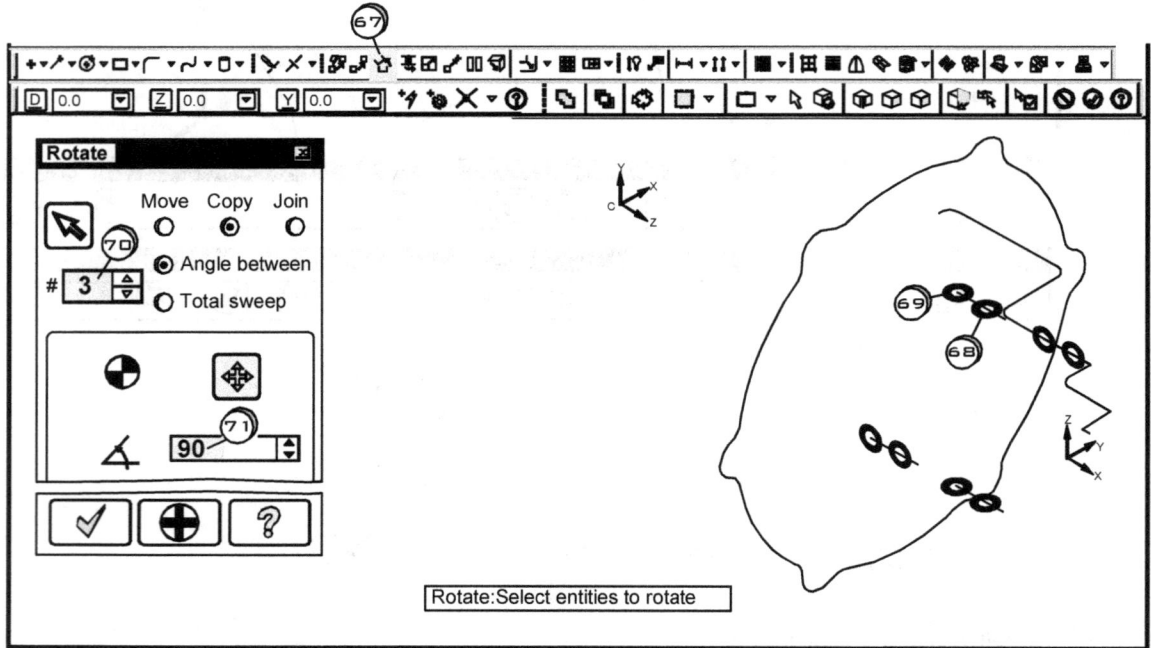

➤ Click (67) the Xform Rotate icon ✿

┌─────────────────────────────┐
│ Rotate:Select entities to rotate │
└─────────────────────────────┘

➤ Click (68) (69) the circles   │ Enter ⏎ │

➤ Click (70) in the # copies box ; enter   **3**

➤ Click (71) in the angle box ∠ ;enter the copy angle ▯90▯ ▾ │ Enter ⏎ │

➤ Press │Esc│ for function cancel

On ⌐Level: 5⌐, create the .125 Dia face drill circles. Refer to the **PICTORIAL.**

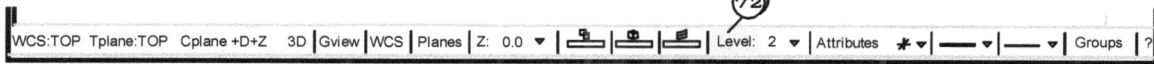

⟨72⟩

WCS:TOP Tplane:TOP Cplane +D+Z   3D │Gview│WCS │ Planes │ Z: 0.0 ▼│ 🔲 │ 🔲 │ 🔲 │ Level: 2 ▼ │ Attributes ✳ ▼│ ── ▼│ ── ▼│ Groups │?

➤ Click ⟨72⟩ the   Level: button

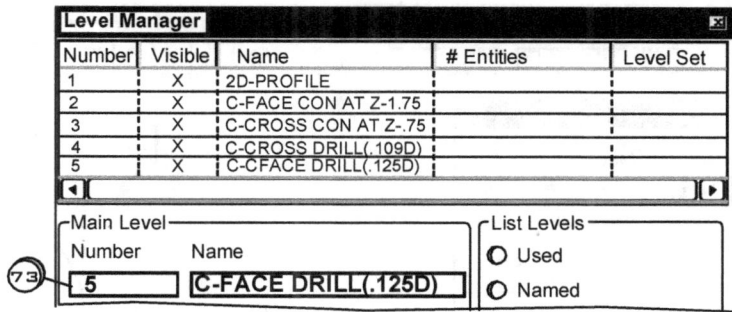

Number	Visible	Name	# Entities	Level Set
1	X	2D-PROFILE		
2	X	C-FACE CON AT Z-1.75		
3	X	C-CROSS CON AT Z-.75		
4	X	C-CROSS DRILL(.109D)		
5	X	C-CFACE DRILL(.125D)		

Level Manager

┌Main Level─────────────────┐   ┌List Levels─┐
Number   Name   ○ Used
⟨73⟩ │ 5 │ │ C-FACE DRILL(.125D) │   ○ Named

➤ Click ⟨73⟩ the number box; enter **5**

➤ Press ⌐Tab⌐ ; enter the new level name ⌐C-FACE DRILL(.125D)⌐

➤ Click ⟨74⟩ enter the working Z depth as **0**

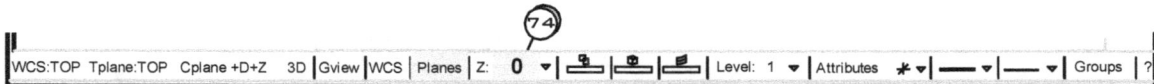

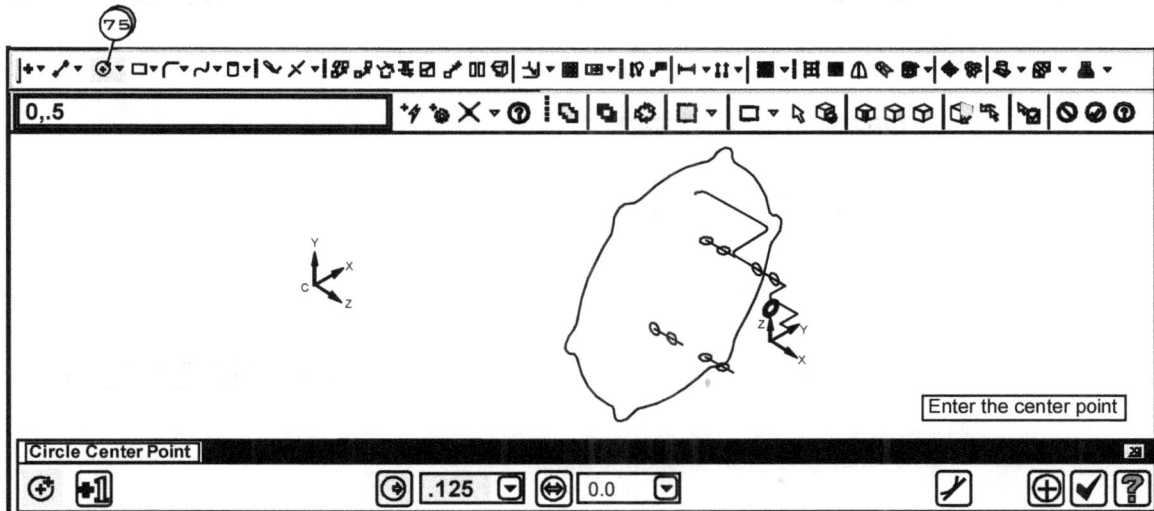

⟨74⟩

WCS:TOP Tplane:TOP Cplane +D+Z   3D │Gview│WCS │ Planes │ Z: **0** ▼│ 🔲 │ 🔲 │ 🔲 │ Level: 1 ▼ │ Attributes ✳ ▼│ ── ▼│ ── ▼│ Groups │?

⟨75⟩

│+▼ ✦▼ ✦▼ ⊙▼ □▼ ⌐▼ ⌐▼ □▼│ ✎ ✗▼│🔧🔧🔧🔧🔧 ✎ 🔧🔧│ ⌐▼ ■ ⌐▼│🔧🔧 ┤ 🔧▼│ ■▼│▦ ■ ⚠ ✎ 🔧▼│◆🔧│🔧▼ 🔧▼ ♣▼

│ 0,.5 │   ✦✦✗▼⊙│🔧🔧│🔧│🔧│□▼│□▼ ▶🔧│🔧🔧🔧│🔧🔧│🔧│⊗⊘⊙

                                                    ⌐ Enter the center point ⌐

Circle Center Point

⊕ ⊞1   ⊙ │.125│▼ ⊕ │0.0│▼   ✎   ⊕ ✔ ?

➤ Click ⟨75⟩ Circle Center Point

⌐ Enter the center point ⌐

➤ Enter the xy coordinates of the circle center **0,.5**

➤ Tap the ⌐D⌐ key for *Diameter; enter* **.125**   ⌐Enter⌐

Complete the face drill circle geometry by using **XFORM**, **Rotate**, **Copy** to make *3 copies.*

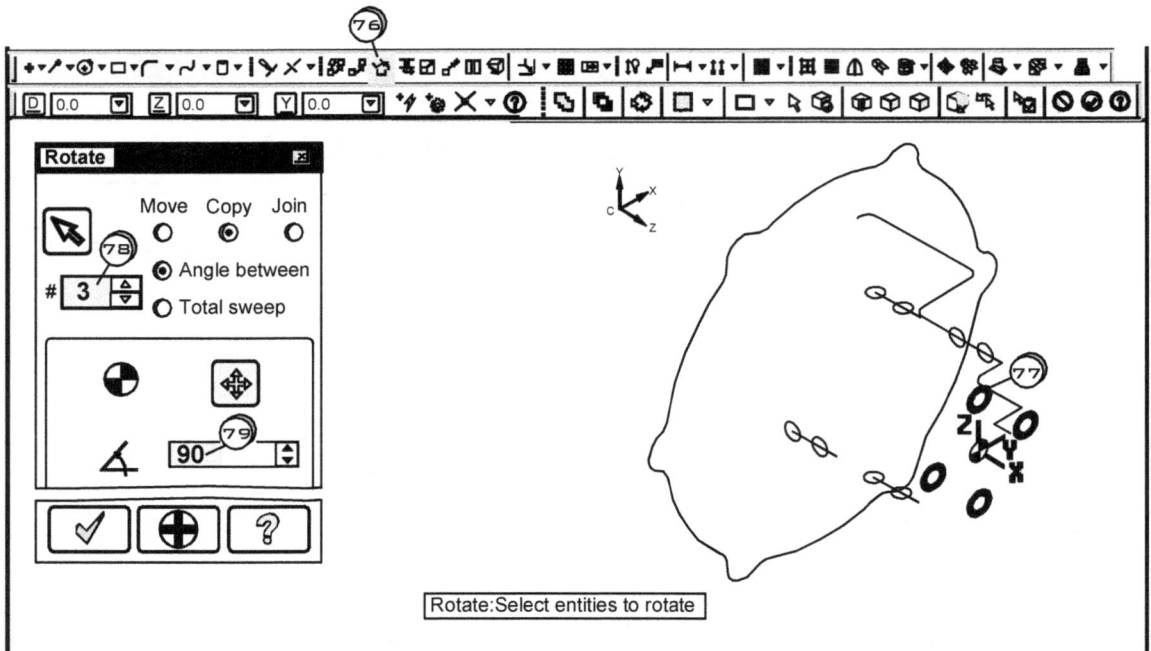

Rotate:Select entities to rotate

⟫ Click ⑦⑥ the Xform Rotate icon ☼

Rotate:Select entities to rotate

⟫ Click ⑦⑦ the circle   [Enter ◄┘]

⟫ Click ⑦⑧ in the # copies box ; enter **3**

⟫ Click ⑦⑨ in the angle box ∠ ;enter the copy angle [90 ▼] [Enter ◄┘]

⟫ Press [Esc] for function cancel

EXERCISES

In each case use the **PICTORIALS** as an aid in creating geometry on the levels indicated.

YOUR INITIALS

5-1) File Name: **EX5-1JV**

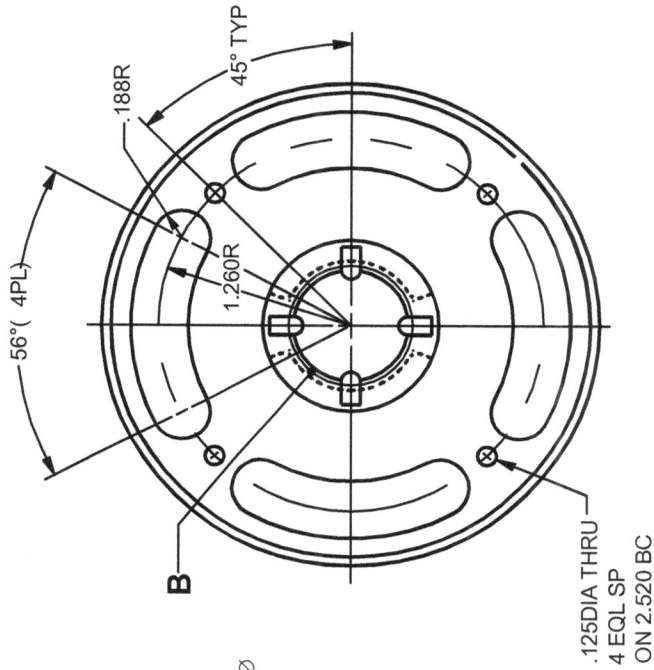

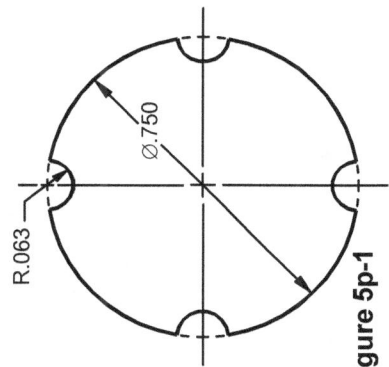

DETAIL B

R.063

Ø.750

Figure 5p-1

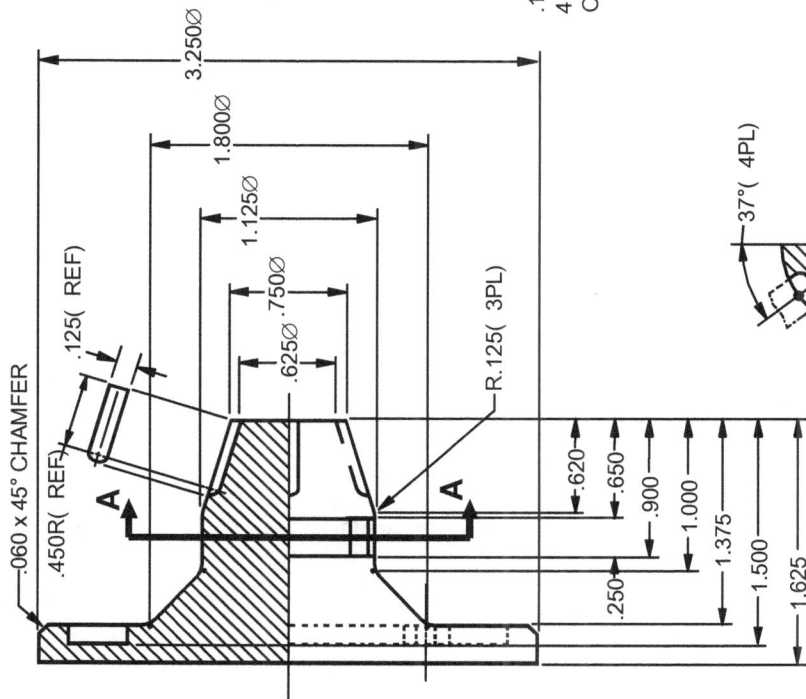

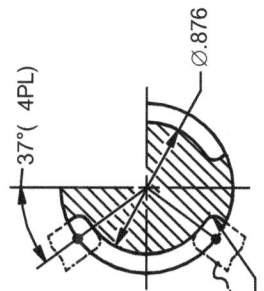

SECTION A-A

Ø.876

37°(4PL)

.250 BULL END MILL

.063R TYP

PICTORIALS

2D PROFILE
OF PART
OPERATOR
CREATES
IN CAD FOR
**TURN
OPERATIONS**

Level: 1

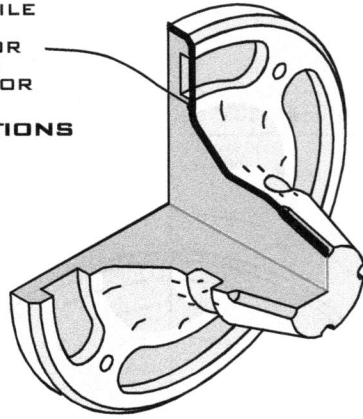

3D PROFILES OF PART OPERATOR CREATES
IN CAD FOR **C-AXIS MILL/TURN OPERATIONS** SHOWN **BOLD**

Level: 2

Level: 3

*Toolpath of
.375 end mill*

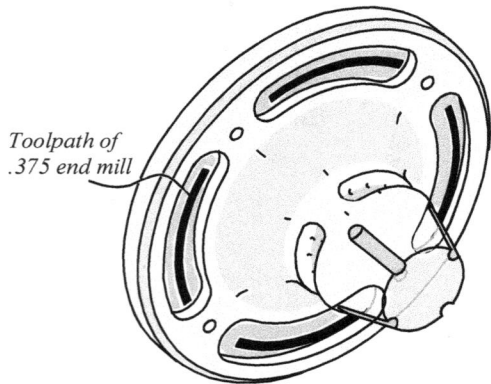

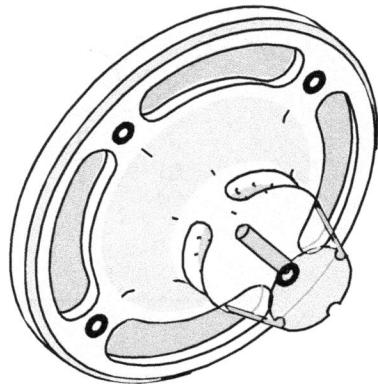

Level: 4

Level: 5

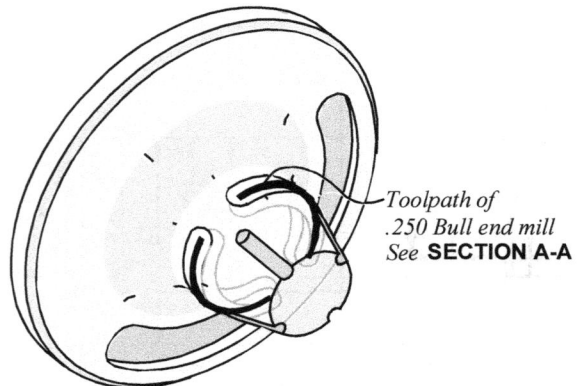

*Toolpath of
.250 Bull end mill
See* **SECTION A-A**

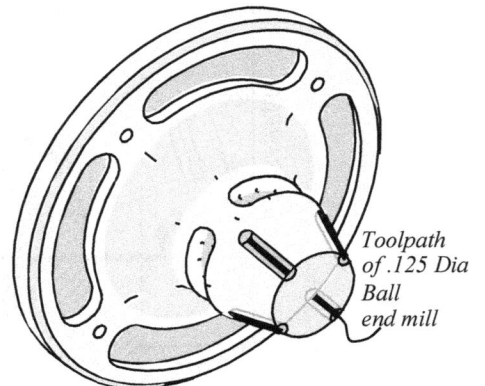

*Toolpath
of .125 Dia
Ball
end mill*

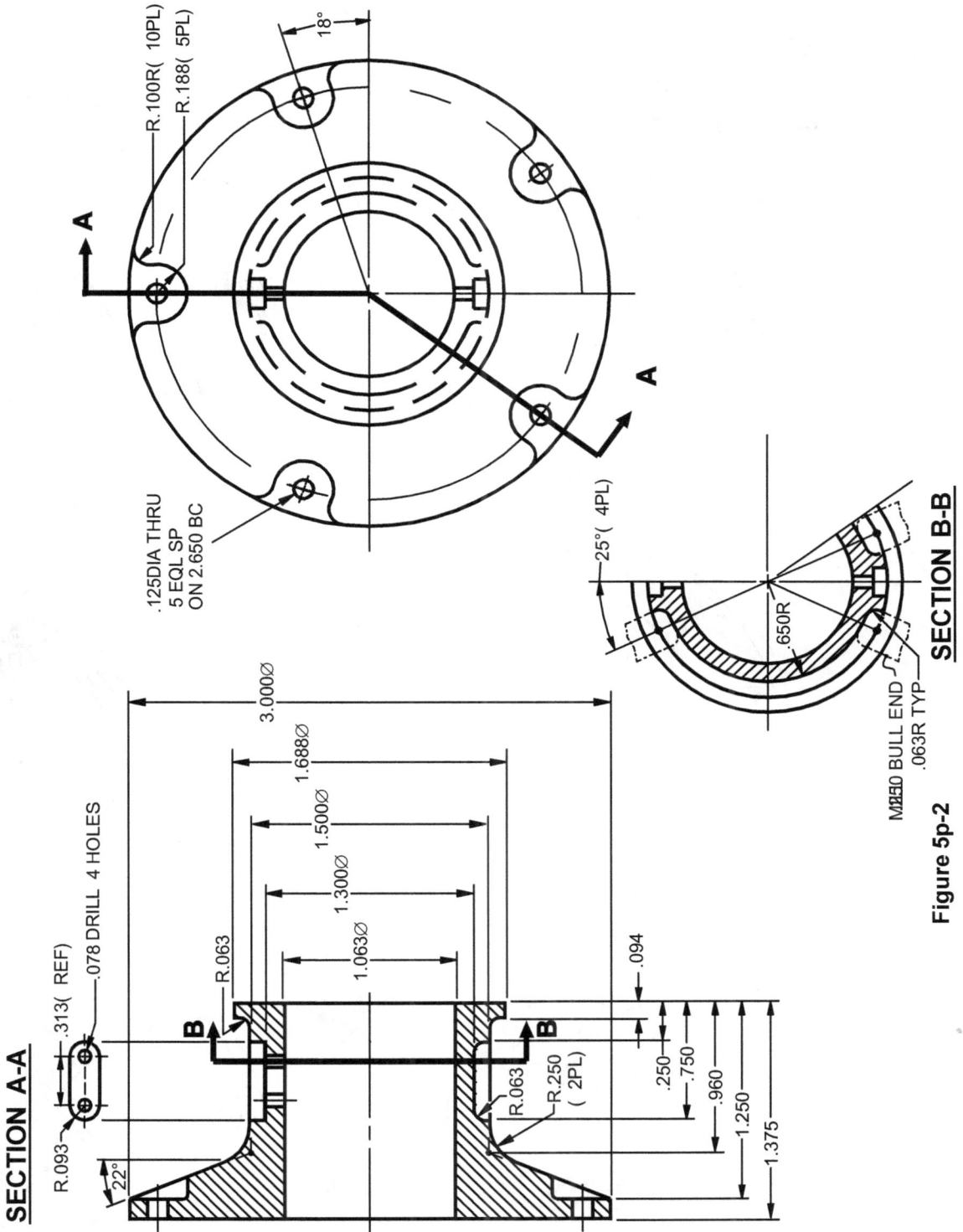

5-2) File Name: **EX5-2JV**

YOUR INITIALS

18°

R.100R(10PL)
R.188(5PL)

A

A

.125 DIA THRU
5 EQL SP
ON 2.650 BC

25°(4PL)

650R

M350 BULL END
.063R TYP

SECTION B-B

Figure 5p-2

3.000⌀
1.688⌀
1.500⌀
1.300⌀
1.063⌀

R.063

.313(REF)
.078 DRILL 4 HOLES

B

B

R.063
R.250
(2PL)

.094
.250
.750
.960
1.250
1.375

R.093
22°

SECTION A-A

PICTORIALS

Level: 1

2D PROFILE
OF PART
OPERATOR
CREATES
IN CAD FOR
**TURN
OPERATIONS**

3D PROFILES OF PART OPERATOR CREATES
IN CAD FOR **C-AXIS MILL/TURN OPERATIONS**
SHOWN **BOLD**

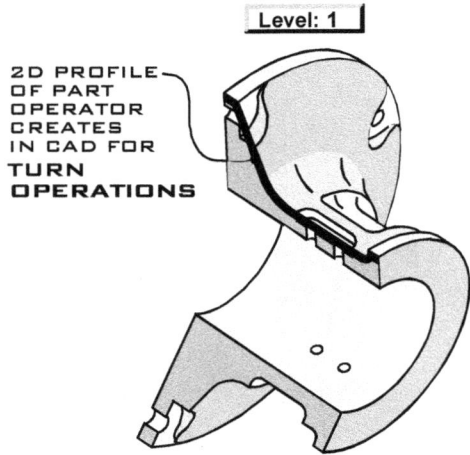

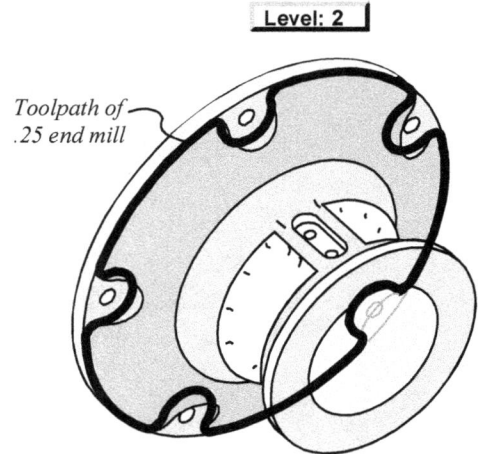

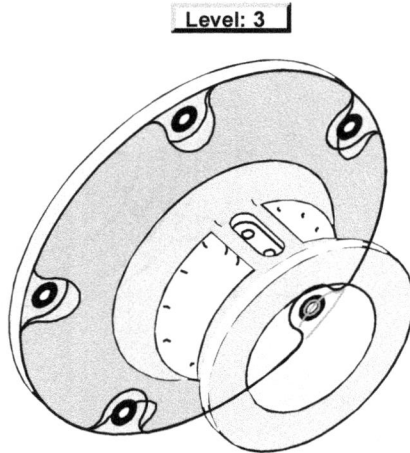

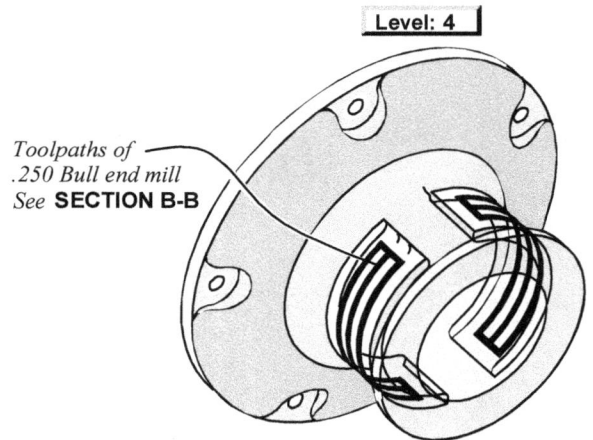

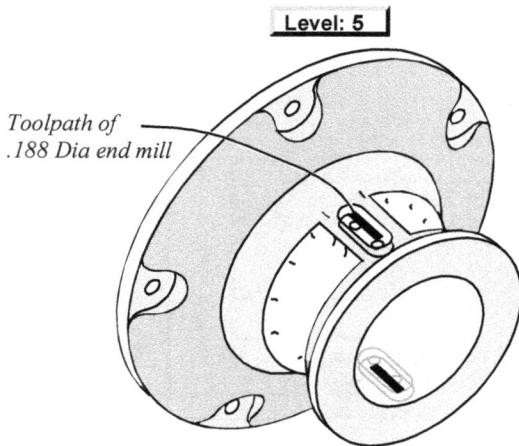

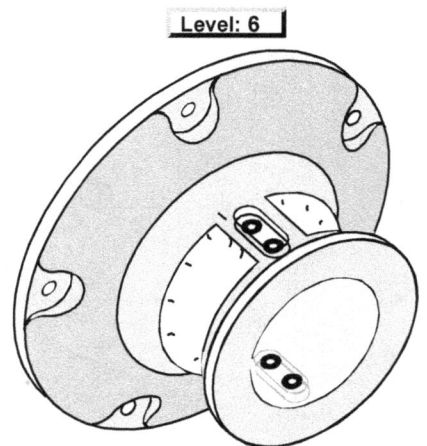

Level: 2

*Toolpath of
.25 end mill*

Level: 3

Level: 4

*Toolpaths of
.250 Bull end mill
See* **SECTION B-B**

Level: 5

*Toolpath of
.188 Dia end mill*

Level: 6

YOUR INITIALS

5-3) | File Name: **EX5-3JV** |

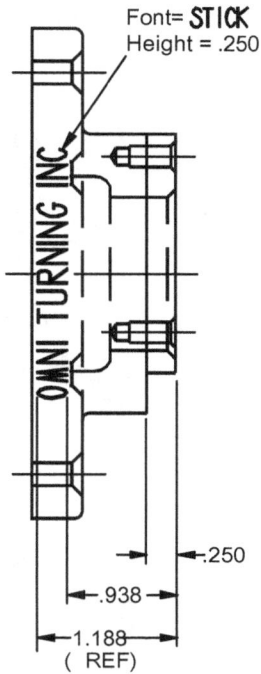

Font= **STICK**
Height = .250

OMNI TURNING INC.

.250
.938
1.188
(REF)

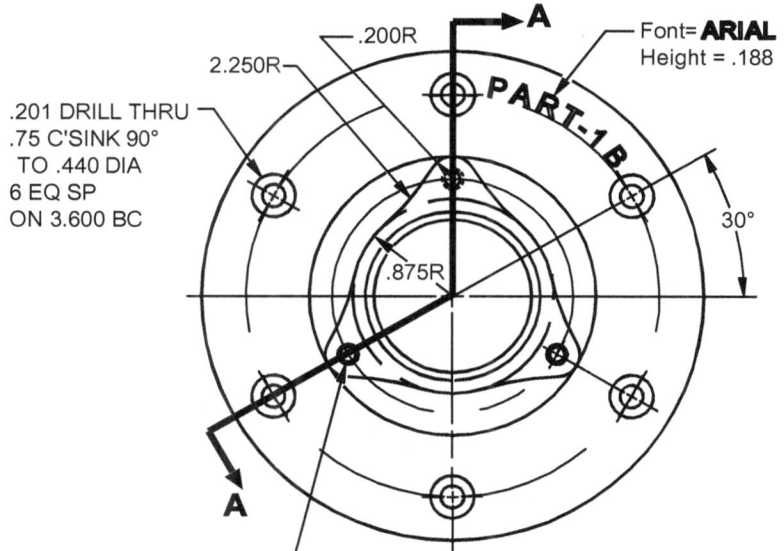

.200R

2.250R

.201 DRILL THRU
.75 C'SINK 90°
TO .440 DIA
6 EQ SP
ON 3.600 BC

.875R

Font= **ARIAL**
Height = .188

PART-1B

30°

#25(.150) DRILL X.600 DEEP
.190-24UNC-2B X.400 DEEP
C'SINK TO .240 DIA 3 EQ SP
ON 2.100 BC

SECTION A-A

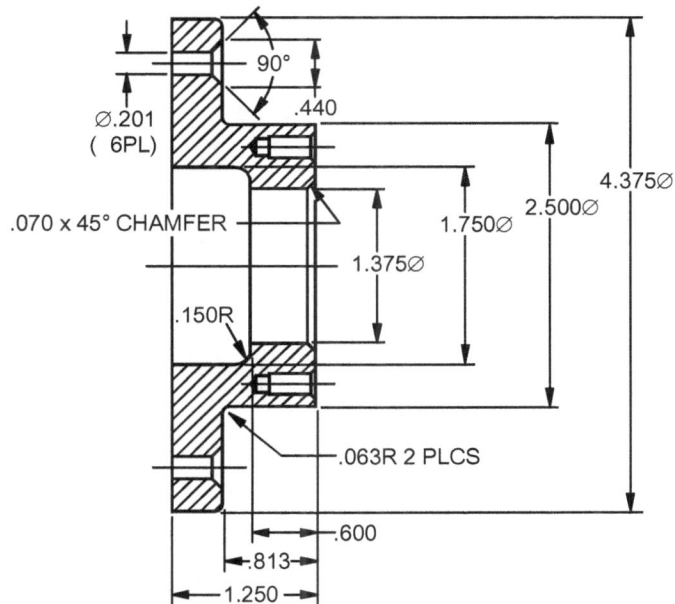

90°

.440

⌀.201
(6PL)

.070 x 45° CHAMFER

1.375⌀

1.750⌀

2.500⌀

4.375⌀

.150R

.063R 2 PLCS

.600
.813
1.250

Figure 5p-3

PICTORIALS

Level: 1

2D PROFILE
OF PART
OPERATOR
CREATES
IN CAD FOR
**TURN
OPERATIONS**

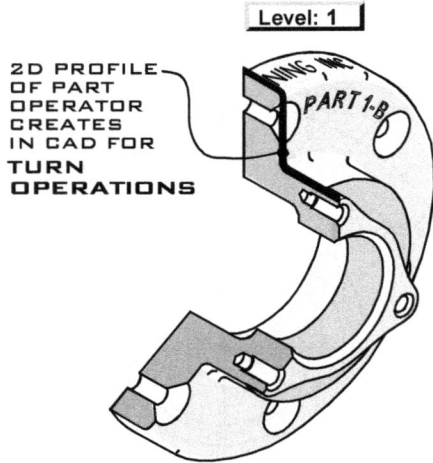

3D PROFILES OF PART OPERATOR CREATES
IN CAD FOR **C-AXIS MILL/TURN OPERATIONS**
SHOWN **BOLD**

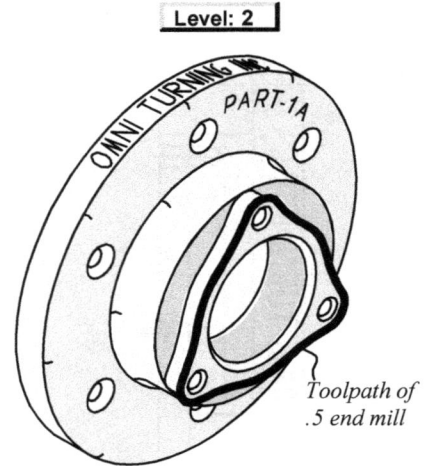

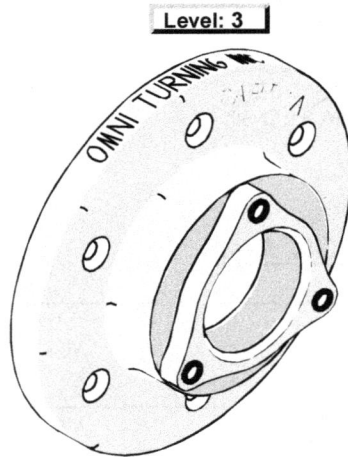

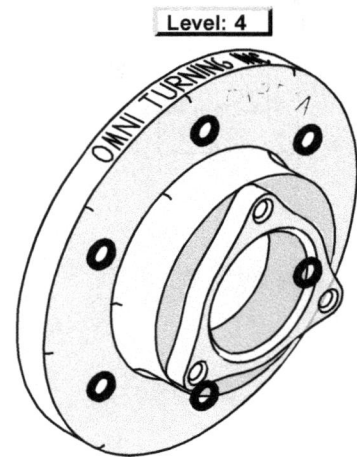

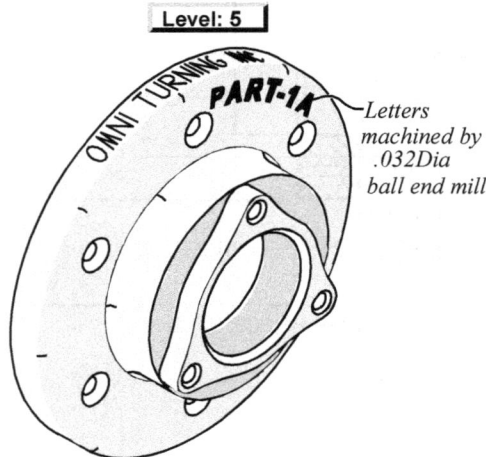

PART 1-B

Level: 2

PART-1A

*Toolpath of
.5 end mill*

Level: 3

Level: 4

Level: 5

PART-1A

*Letters
machined by
.032Dia
ball end mill*

*Letters
machined by
.063Dia
ball end mill*

Level: 6

OMNI TURNING

YOUR INITIALS

5-4) File Name: **EX5-4JV**

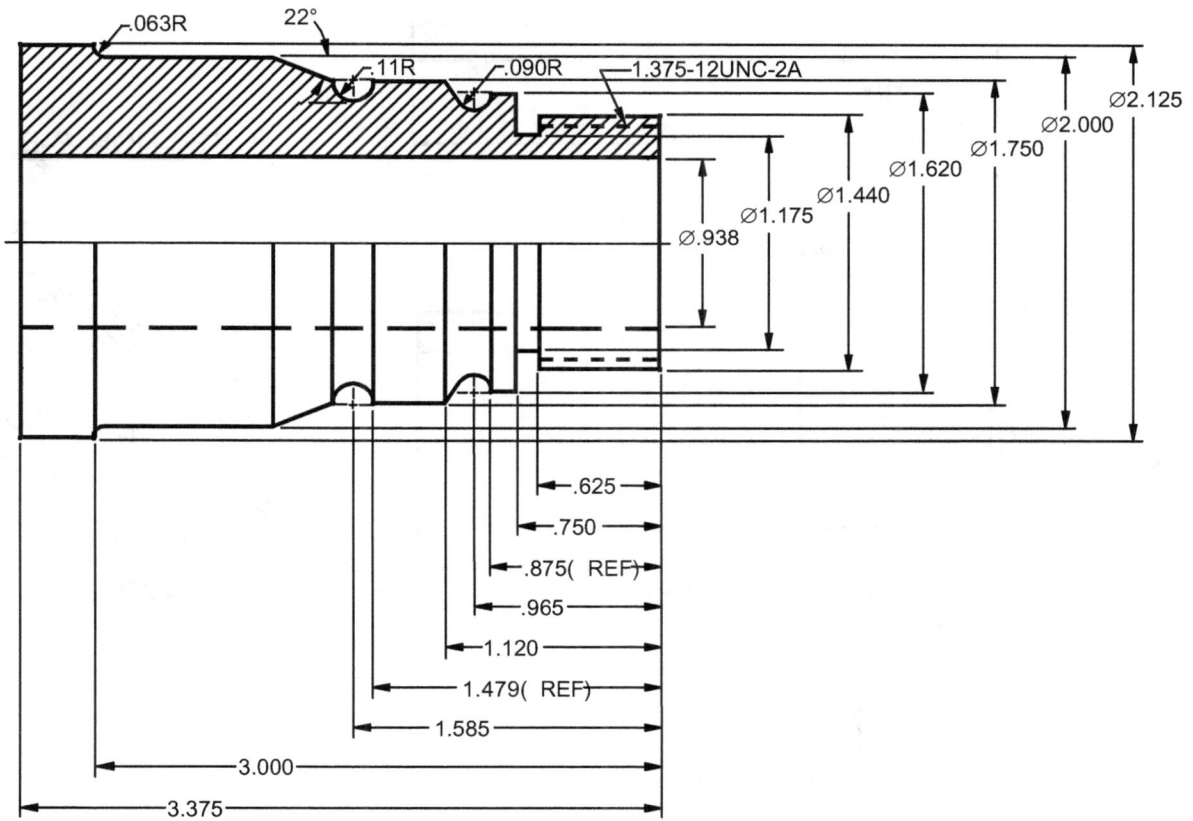

Figure 5p-4

EXERCISE 5-5 CONT

TRUE SHAPE OF
UNROLLED TOOH

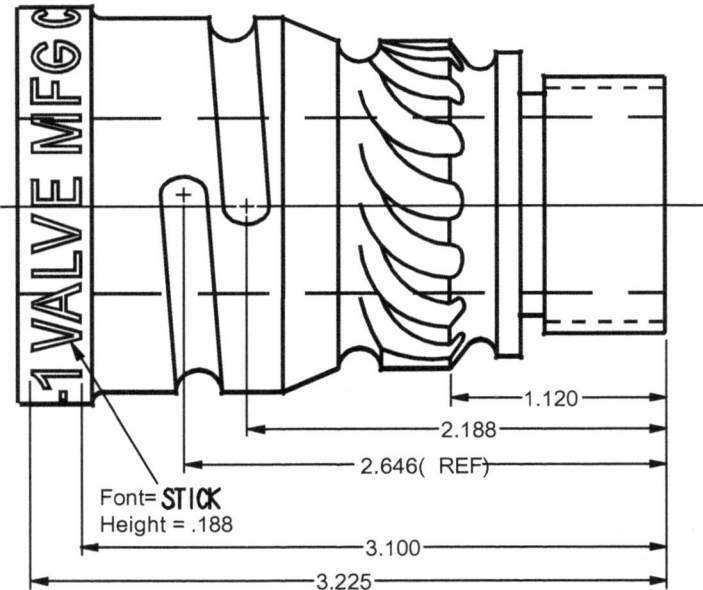

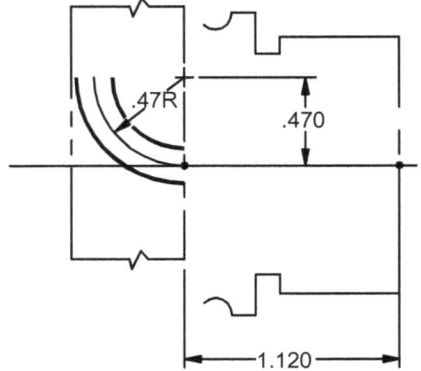

.47R

.470

1.120

1.120

2.188

2.646(REF)

Font= STICK
Height = .188

3.100

3.225

.125(REF)

.125R

B

B

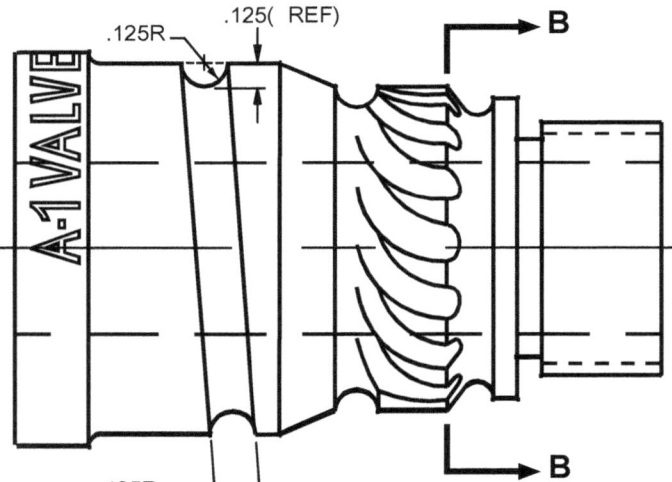

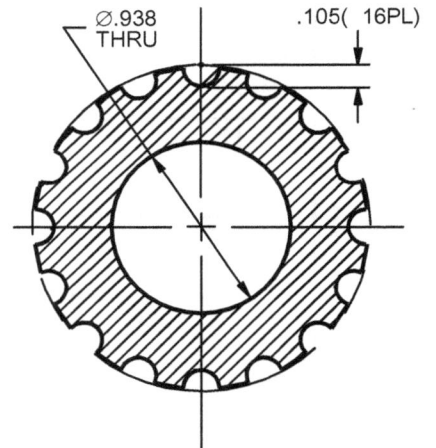

SECTION B-B

Ø.938
THRU

.105(16PL)

A-1VALVE

.125R

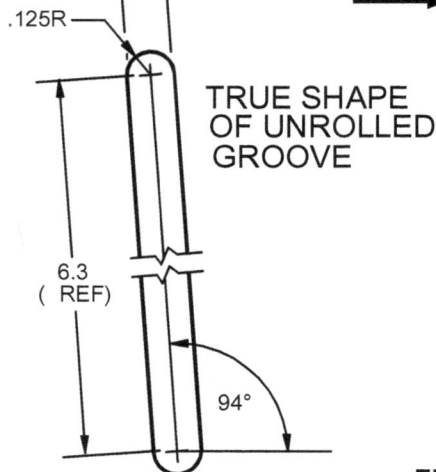

TRUE SHAPE
OF UNROLLED
GROOVE

6.3
(REF)

94°

Figure 5p-5

PICTORIALS

2D PROFILE
OF PART
OPERATOR
CREATES
IN CAD FOR
**TURN
OPERATIONS**

Level: 1

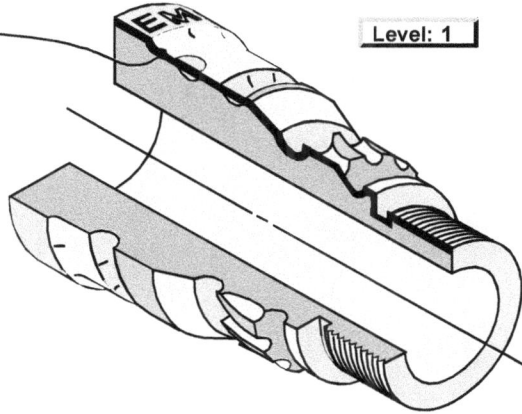

2D PROFILES OF PART OPERATOR CREATES
IN CAD FOR **C-AXIS MILL/TURN OPERATIONS** SHOWN **BOLD**

Level: 2

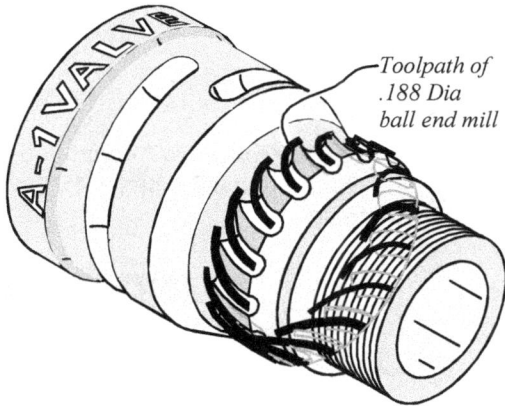

*Toolpath of
.188 Dia
ball end mill*

Level: 3

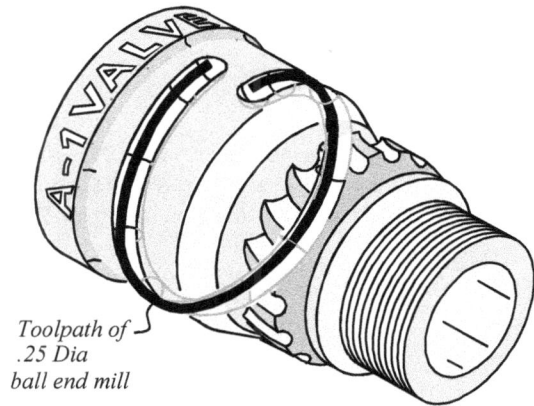

*Toolpath of
.25 Dia
ball end mill*

Level: 4

*Letters
machined by
.032Dia
ball end mill*

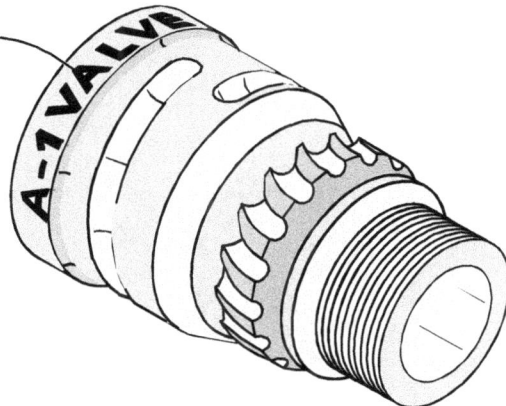

a) Create the 2D profile for turning on **Level: 1** use **Cplane: DZ+**

origin(0,0)

b) Generate the .47R arc for C axis contouring on **Level: 2** at **Z = -1.12** use **Cplane: Top**

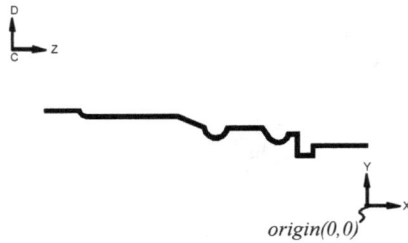

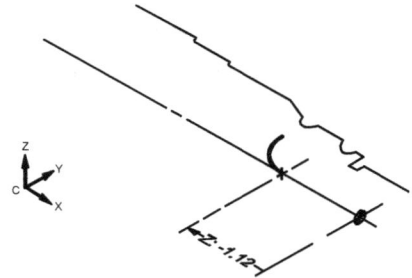

Z = -1.12

d) Select **Cplane: Right** . Use the **Xform, Rotate, Copy** command to make **15** copies of the rolled arc at a rotation angle of **22.5°** .

Xform:
Rotate

c) Use the **Xform, Roll** command to roll the .47R arc around the part's **1.75** in rotary diameter.

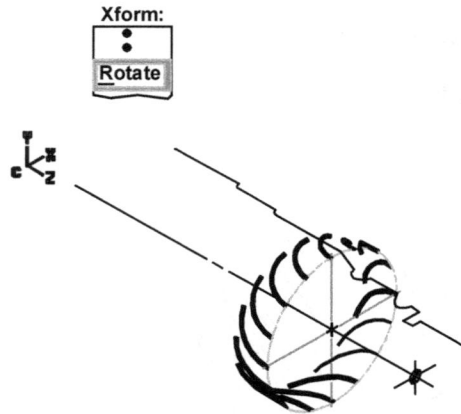

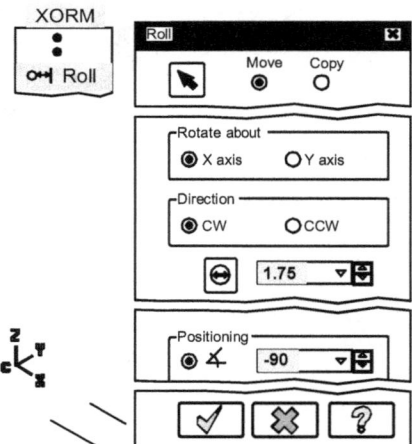

XORM
Roll

Roll

Move   Copy

Rotate about
X axis   Y axis

Direction
CW   CCW

1.75

Positioning
-90

ø1.75

e) Hide **Level: 2**. Select **Level: 3** and **Cplane: Top.** Create a polar line at **Z=-2.188** having an angle of **94°** and length **6.3**.

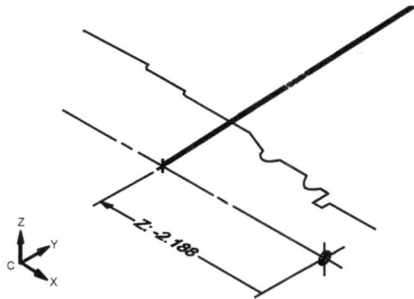

f) Use the *Xform, Roll* command to roll the 6.3in line around the part's **2**in rotary diameter.

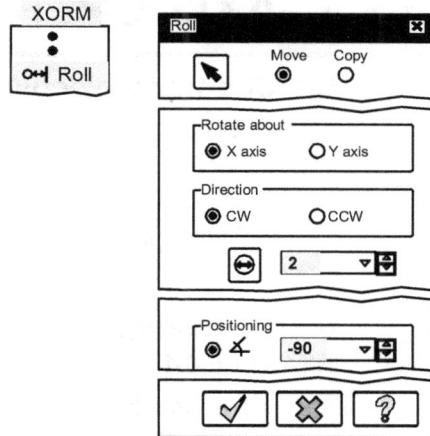

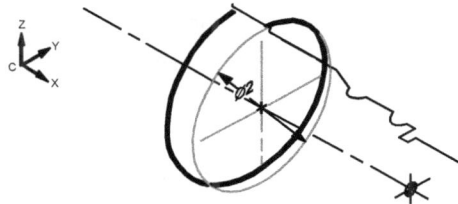

h) Use the *Xform, Roll* command to roll the letters around the part's **2.125**in rotary diameter.

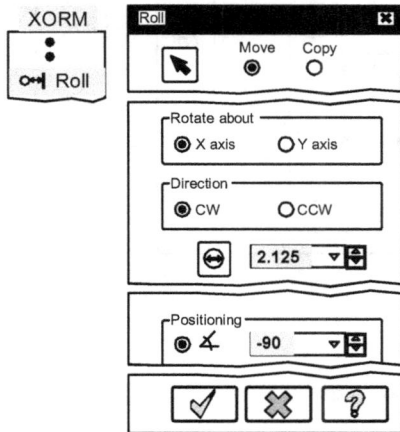

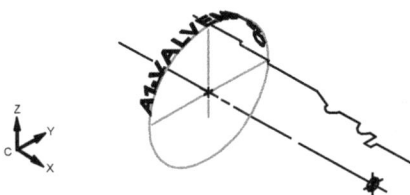

g) Hide **Level: 3**. Select **Level: 4**. Insert Letters having **Drafting Globals** as shown below:

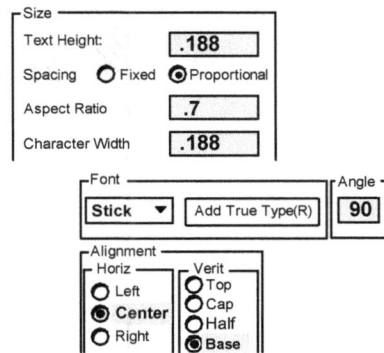

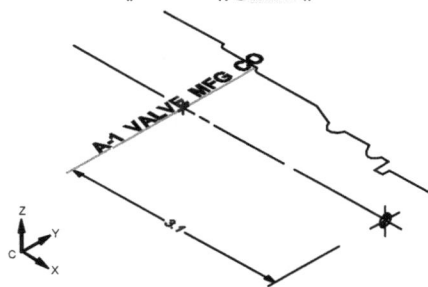

CHAPTER - 6

C-AXIS OPERATIONS

6-1 Chapter Objectives

After completing this chapter you will be able to generate toolpaths for C-axis:

1. Face Contour
2. Cross Contour
3. Face Drill
4. Cross Drill

6-2 An Example of C-axis Machining

EXAMPLE 6-1

Open the file EX5-1 in the folder ☐ CHAPTER 5

Using PROCESS PLAN 6-1 direct Mastercam to execute C-axis machining on the part shown in Figure 6-1.

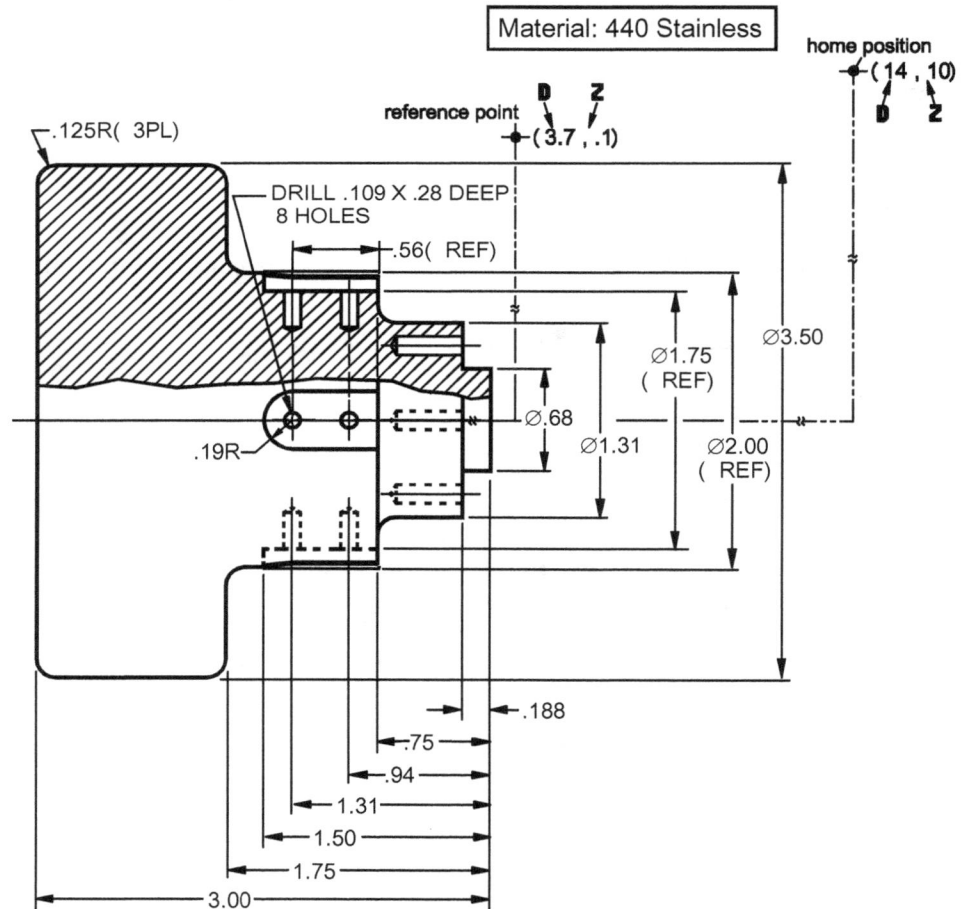

Figure 6-1

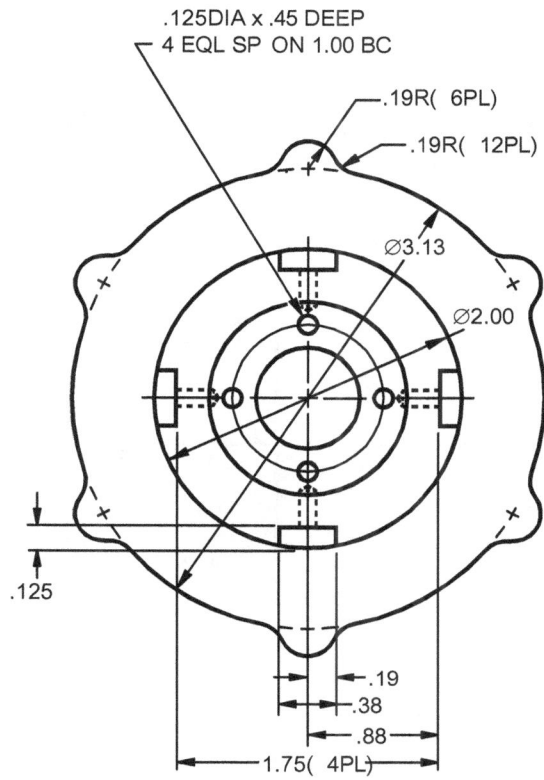

.125 DIA x .45 DEEP
4 EQL SP ON 1.00 BC

.19R(6PL)

.19R(12PL)

Ø3.13

Ø2.00

.125

.19

.38

.88

1.75(4PL)

Figure 6-2

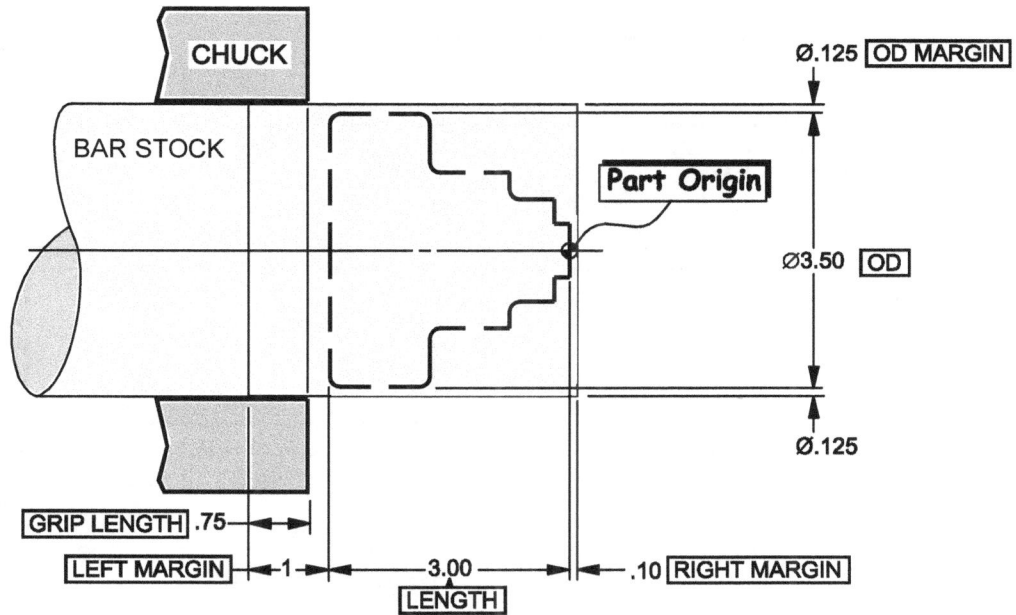

CHUCK

BAR STOCK

Ø.125 OD MARGIN

Part Origin

Ø3.50 OD

Ø.125

GRIP LENGTH .75

LEFT MARGIN 1

3.00

LENGTH

.10 RIGHT MARGIN

PROCESS PLAN 6-1

No.	Operation	Tooling
1	Face End to .10in	1/32TNR - RH - ROUGH - OD - TURNING TOOL R1/32
2	Rough Turn OD; leave .01 in X and Z for finishing	
3	Finish OD Contour	1/64TNR - RH - OD - FINISHING TOOL R1/64
4	Rough and finish OD groove; leave .01in X and Z for finishing	R.01, W.25 OD GROOVING RIGHT DETAIL A .093 .5 .125 .125 R.01 KENNAMETAL GC-4125 1 5 .10 A
5	C-Axis Face Contour X 1.3 Deep	.375 DIA END MILL
6	C-Axis Cross Contour X .125 Deep	.375
7	C-Axis Cross Drill X .405 Deep	.1094(7/64) DRILL .109
8	C-Axis Face Drill X .45 Deep	.125(1/8) DRILL .125
9	Lathe Cutoff	.016TNR - RH - OD - CUTOFF TOOL DETAIL A R.016 1.8 A

B) IDENTIFY THE STOCK TO BE MACHINED

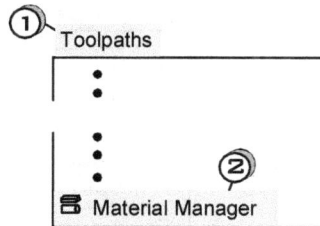

① Toolpaths

```
•
•
•
•
⊟ Material Manager  ②
```

➤ Click ① Toolpaths

➤ Click ② ⊟ Material Manager

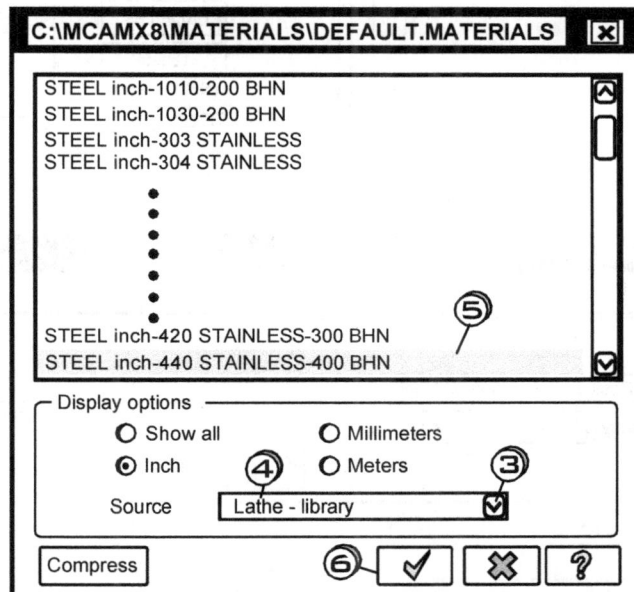

C:\MCAMX8\MATERIALS\DEFAULT.MATERIALS ☒

```
STEEL inch-1010-200 BHN
STEEL inch-1030-200 BHN
STEEL inch-303 STAINLESS
STEEL inch-304 STAINLESS
        •
        •
        •
        •
        •
        •
STEEL inch-420 STAINLESS-300 BHN        ⑤
STEEL inch-440 STAINLESS-400 BHN
```

Display options

○ Show all   ○ Millimeters
⊙ Inch   ④   ○ Meters   ③
Source   [Lathe - library   ☑]

[Compress]   ⑥ [✓] [✗] [?]

➤ Click ③ the down arrow ☑

➤ Click ④ Lathe - library

➤ Click ⑤ STEEL inch-440 STAINLESS-400 BHN

➤ Click ⑥ the OK button [✓]

B) SPECIFY THE SIZE OF THE STOCK BOUNDARY

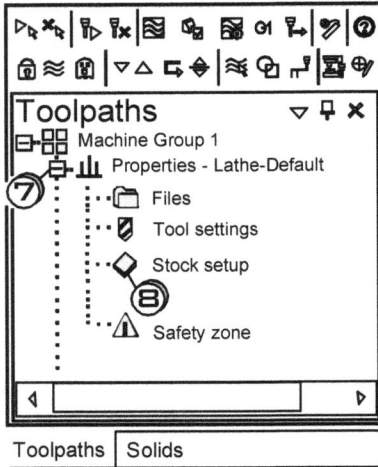

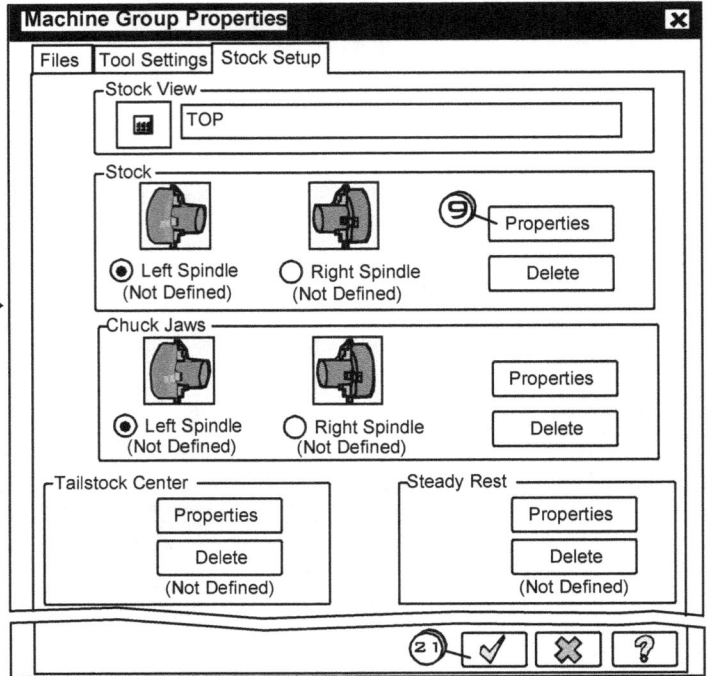

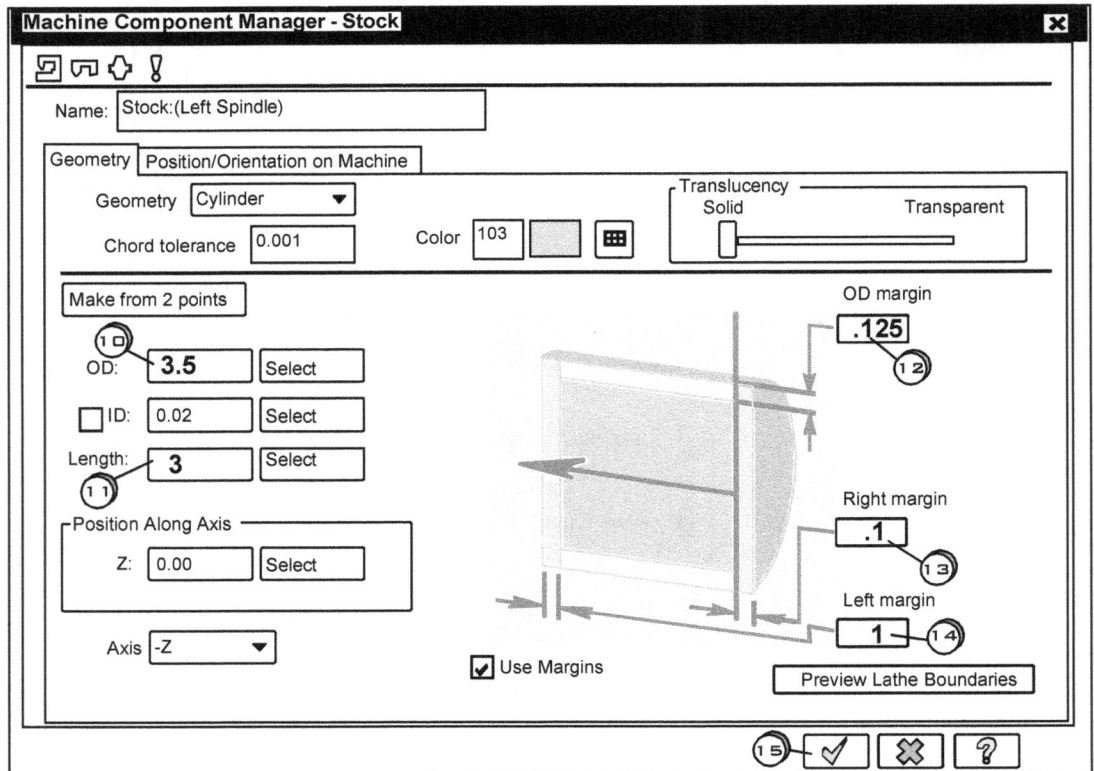

➤ Click ⑦ the minus sign ⊟

➤ Click ⑧ ◇ Stock setup

➤ Click ⑨ the [Properties] button

➤ Click ⑩ in the OD box; enter **3.5**

➤ Click ⑪ in the Length box; enter **3**

➤ Click ⑫ in the OD margin box; enter **.125**

➤ Click ⑬ in the Right margin box; enter **.1**

➤ Click ⑭ in the Left margin box; enter **1**

➤ Click ⑮ the OK button ☑

C) SPECIFY THE CHUCK BOUNDARY

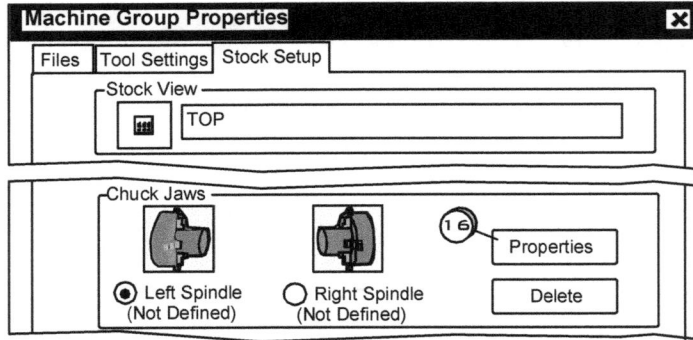

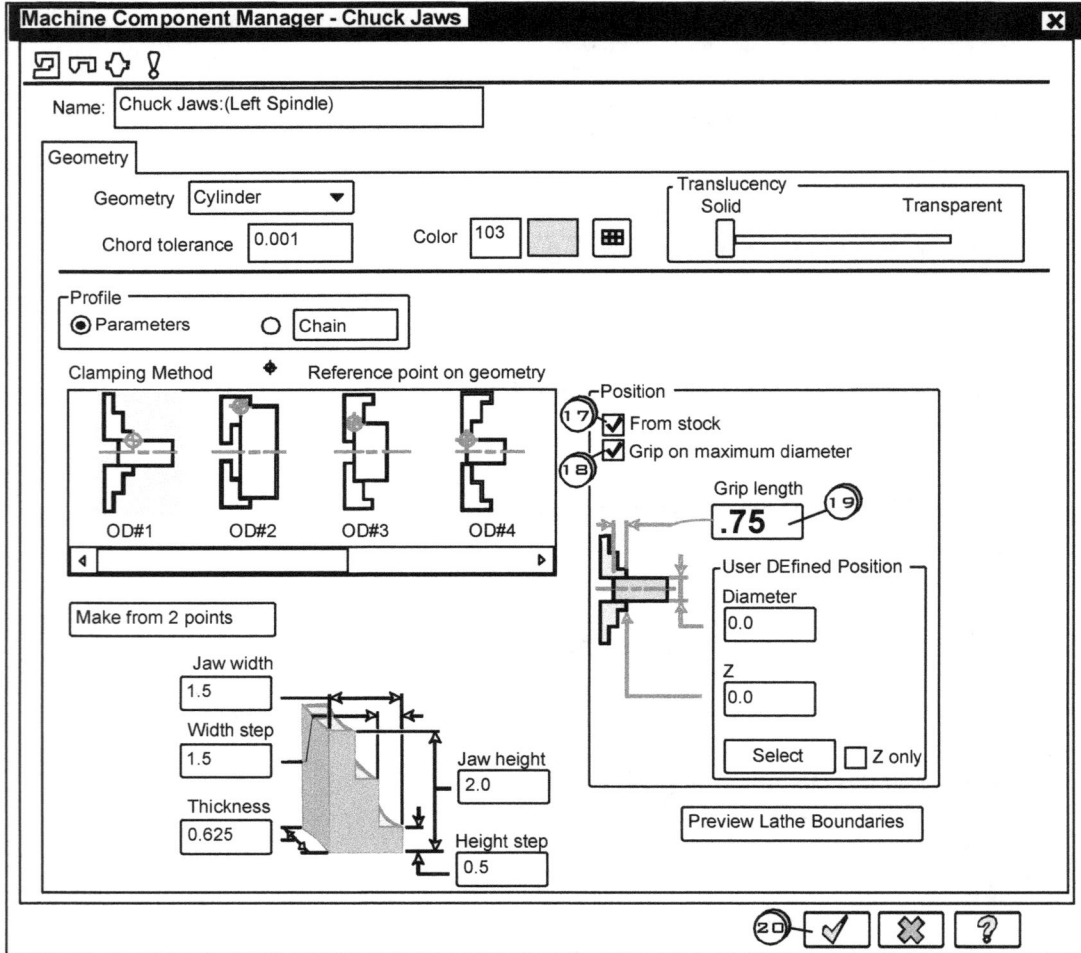

Machine Group Properties ☒

Files | Tool Settings | Stock Setup

Stock View

⊞ | TOP

Chuck Jaws

⦿ Left Spindle
(Not Defined)

○ Right Spindle
(Not Defined)

(16) | Properties

Delete

Machine Component Manager - Chuck Jaws ☒

Name: Chuck Jaws:(Left Spindle)

Geometry

Geometry | Cylinder ▼

Chord tolerance | 0.001

Color | 103 | ⊞

Translucency
Solid ▭────── Transparent

Profile
⦿ Parameters   ○ Chain

Clamping Method   ✦   Reference point on geometry

OD#1   OD#2   OD#3   OD#4

◄ ▬▬▬▬▬▬▬ ►

Make from 2 points

Jaw width
1.5

Width step
1.5

Thickness
0.625

Jaw height
2.0

Height step
0.5

Position
(17) ☑ From stock
(18) ☑ Grip on maximum diameter

Grip length (19)
.75

User DEfined Position
Diameter
0.0

Z
0.0

Select   ☐ Z only

Preview Lathe Boundaries

(20) ✓ | ☒ | ?

➤ Click (16) the [Properties] button

➤ Click (17) the check *on* ☑ From stock

➤ Click (18) the check *on* ☑ Grip on maximum diameter

➤ Click (19) enter Grip length **.75**

➤ Click (20) the OK button ✓

➤ Click (21) the OK button ✓

D) OPERATION#1- FACE THE END
- ◆ OBTAIN THE NEEDED R1/32 OD ROUGH RIGHT TOOL
- ◆ SPECIFY THE REFERENCE POINT FOR THE TOOL
- ◆ SPECIFY HOME(TOOL CHANGE) POSITION FOR THE TOOL

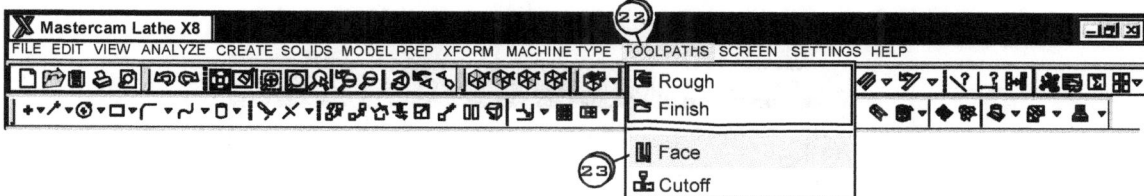

➤ Click ㉒ TOOLPATHS ➤ Click ㉓ ▥ Face

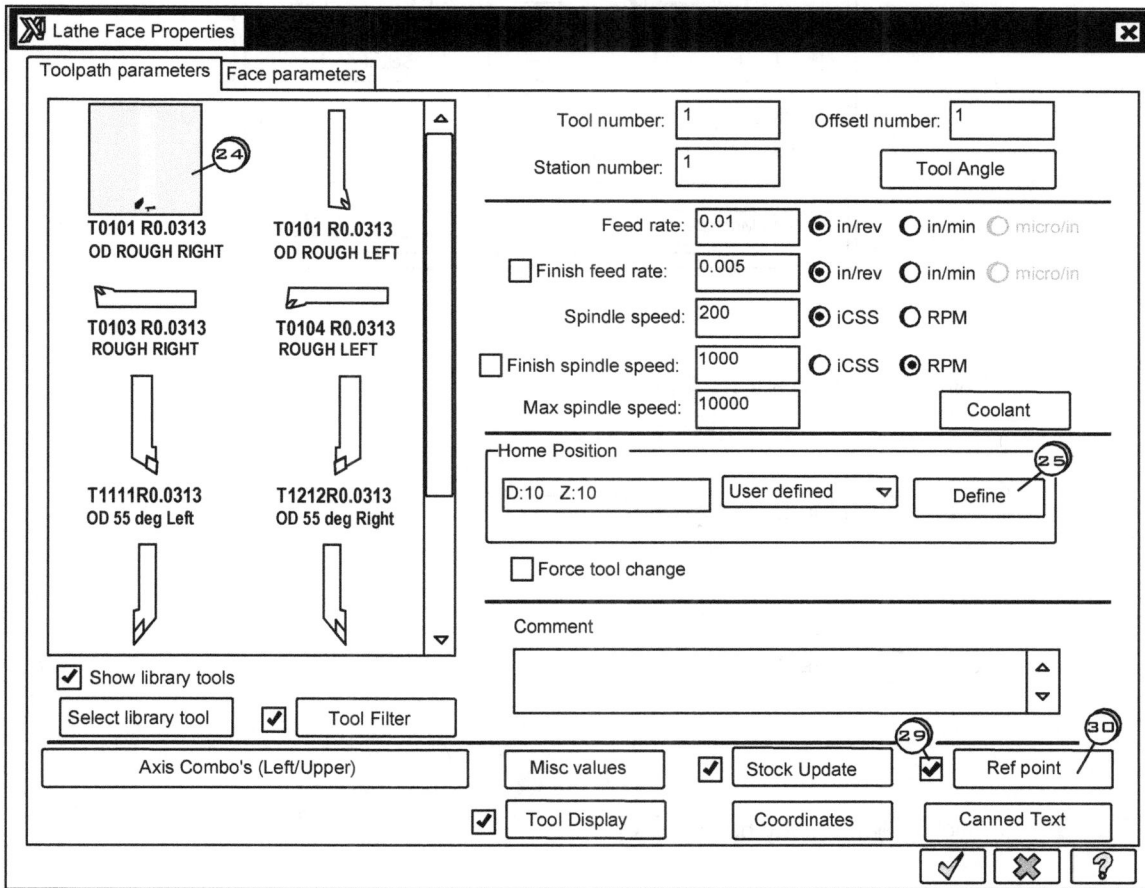

➤ Click ㉔ the OD ROUGH RIGHT tool
➤ Click ㉕ the [Define] button
➤ Click ㉖ in the D box; enter [14]
➤ Click ㉗ in the Z box; enter [10]
➤ Click ㉘ the OK button [✓]

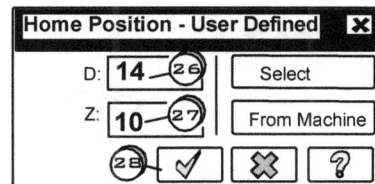

➤ Click ㉙ the check on ☑ | Ref point |

➤ Click ㉚ | Ref point |

Reference Points

☑ Approach ㉛

D: **3.7** ㉜ ☑

Z: **.1** ㉝ ☑

| Select |

⦿ Absolute
◯ Incremental

| From Machine |

☑ Retract

D: **3.7** ☑

Z: **.1** ☑

| Select |

⦿ Absolute
◯ Incremental

㉞ → ←

㉟ ✓ ✗ ?

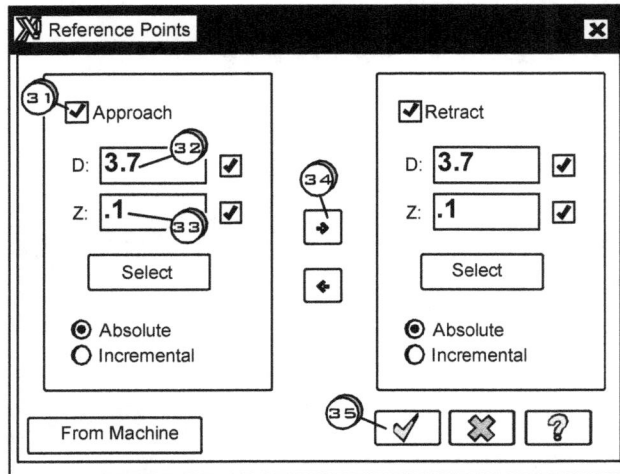

➤ Click ㉛ the check on ☑ Approach

➤ Click ㉜ in the D box; enter **3.7**

➤ Click ㉝ in the Z box; enter **.1**

➤ Click ㉞ the copy button →

➤ Click ㉟ the OK button ✓

◆ ENTER THE FACE MACHINING PARAMETERS

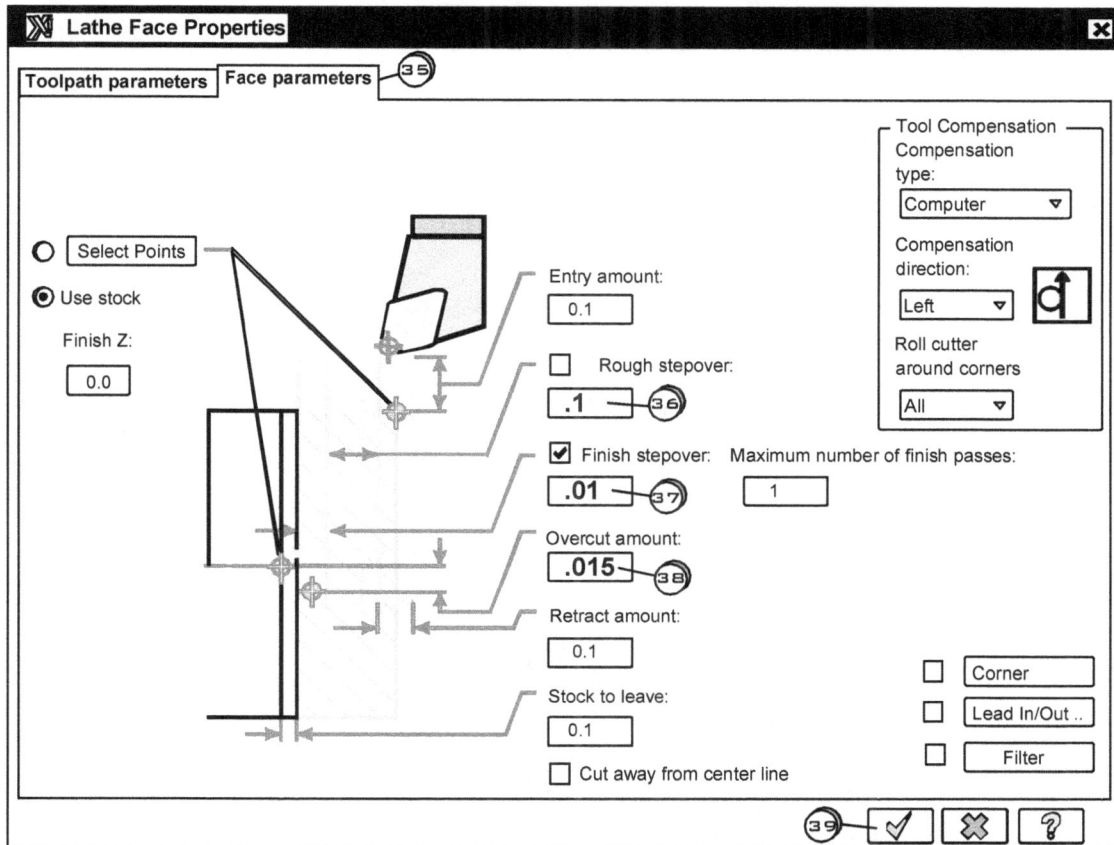

➤ Click ㉟ the ⌐Face parameters⌐ tab

➤ Click ㊱ in the Rough stepover: box; enter .1

➤ Click ㊲ in the Finish stepover: box; enter .01

➤ Click ㊳ in the Overcut amount: box; enter .015

➤ Click ㊴ the OK button ✓

E) OPERATION#2- ROUGH TURN THE OD,
LEAVE .01 IN X AND Z FOR FINISHING

CHAIN THE REQUIRED OD GEOMETRY

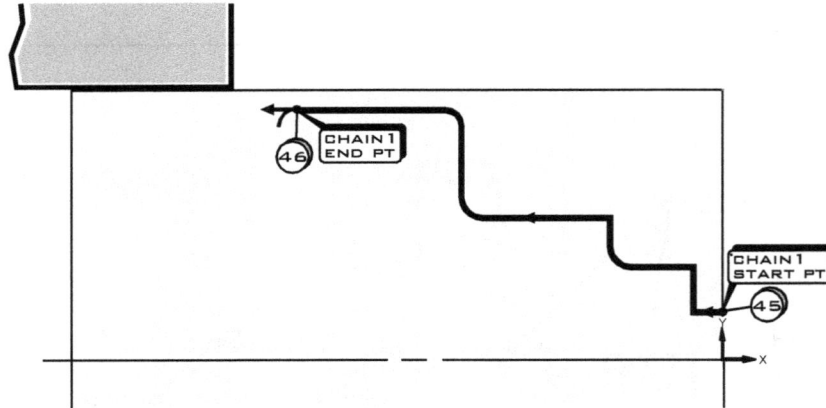

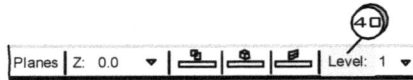

➤ Click ㊵ the Level: button

Level Manager

Number	Visible	Name
1	X	2D-PROFILE
2		C-FACE CON AT Z-1.75
3		C-CROSS CON AT Z-.75
4		C-CROSS DRILL(.109D)
5		C-CFACE DRILL(.125D)

TOOLPATHS
Rough
Finish

➤ Click ㊶ Level 1 as the current level

➤ Click the visibility (X) off for all the other levels

➤ Click ㊷ TOOLPATHS

➤ Click ㊸ Rough

➤ Click ㊹ the chain button

➤ Click ㊺ the start point ㊻ the end point

➤ Click ㊼ the OK button

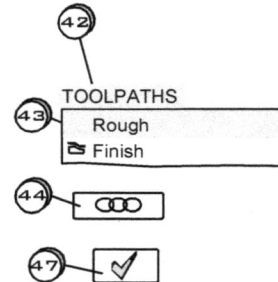

◆ ACCEPT THE DEFAULT T0101 R0.0313 OD ROUGH RIGHT TOOL

◆ ENTER THE ROUGH OD MACHINING PARAMETERS

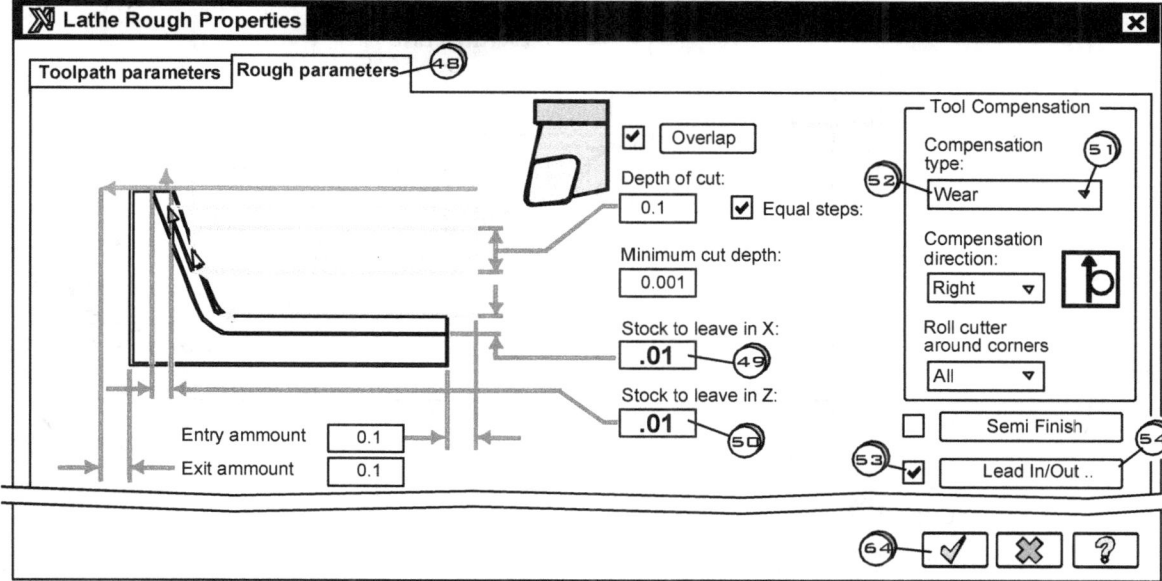

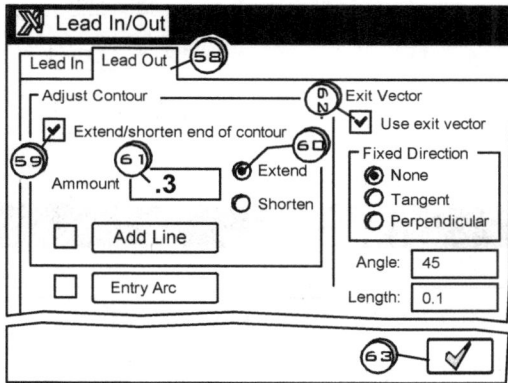

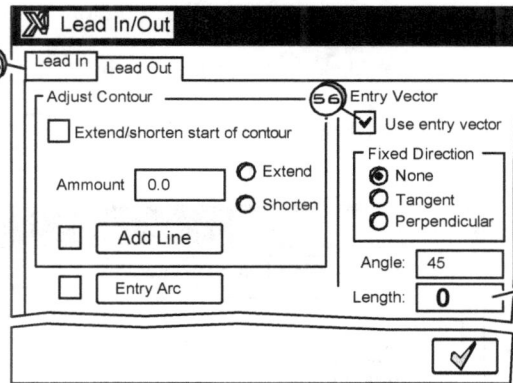

≫ Click ④⑧ the ⌐Rough parameters tab

≫ Click ④⑨ in the Stock to leave in X: box; enter .01

≫ Click ⑤⓪ in the Stock to leave in Z: box; enter .01

≫ Click ⑤① the Compensation type down button ▽

≫ Click ⑤② Wear

≫ Click ⑤③ check on ☑ for Lead In/Out

≫ Click ⑤④ the ⌐Lead In/Out .. button

≫ Click ⑤⑤ the ⌐Lead In tab

≫ Click ⑤⑥ check on ☑ for Use entry vector

≫ Click ⑤⑦ in the Length box; enter 0
  lead in will be from the reference point (2.7,.1)

≫ Click ⑤⑧ the ⌐Lead Out tab

≫ Click ⑤⑨ check on ☑ for Extend/shorten

≫ Click ⑥⓪ the Extend radio button ◉

≫ Click ⑥① in the Ammount: box; enter .3

≫ Click ⑥② check on ☑ for Use exit vector
  lead out will be a polar line at a 45° angle from the
  horizontal of length 0.1

≫ Click ⑥③ the OK button ✓

≫ Click ⑥④ the OK button ✓

F) OPERATION#3- FINISH THE OD CONTOUR

- ◆ OBTAIN THE NEEDED R1/64 OD FINISH RIGHT TOOL
- ◆ SPECIFY HOME(TOOL CHANGE) POSITION FOR THE TOOL

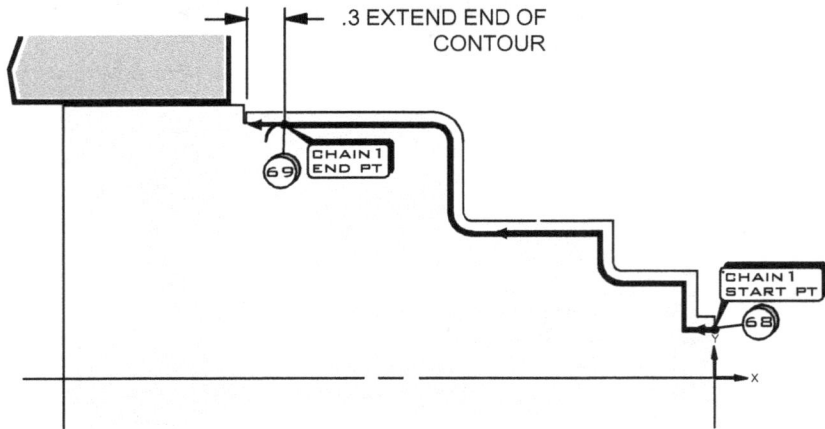

.3 EXTEND END OF CONTOUR

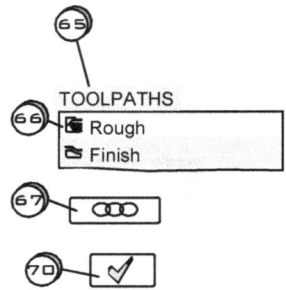

➢ Click ⟨65⟩ TOOLPATHS

➢ Click ⟨66⟩ ☞ Finish

➢ Click ⟨67⟩ the chain button ⟨OOO⟩

➢ Click ⟨68⟩ the start point ⟨69⟩ the end point

➢ Click ⟨70⟩ the OK button ✓

Lathe Finish Properties ✕

Toolpath parameters | Finish parameters

T0103 R0.0313
ROUGH RIGHT

T0104 R0.0313
ROUGH LEFT

T1111R0.0313
OD 55 deg Left

T1212R0.0313
OD 55 deg Right

⟨71⟩

T2121R0.0156
OD FINISH RIGHT

T2222R0.0156
OD FINISH LEFT

Tool number:	1	Offsetl number:	1
Station number:	1	Tool Angle

Feed rate: 0.01 ● in/rev ○ in/min ○ micro/in

☐ Finish feed rate: 0.005 ● in/rev ○ in/min ○ micro/in

Spindle speed: 200 ● iCSS ○ RPM

☐ Finish spindle speed: 1000 ○ iCSS ● RPM

Max spindle speed: 10000 [Coolant]

┌ Home Position ─────────
D:10 Z:10 | User defined ▽ | Define

☐ Force tool change

➢ Click ⟨71⟩ the OD FINISH RIGHT tool

♦ ENTER THE FINISH OD MACHINING PARAMETERS

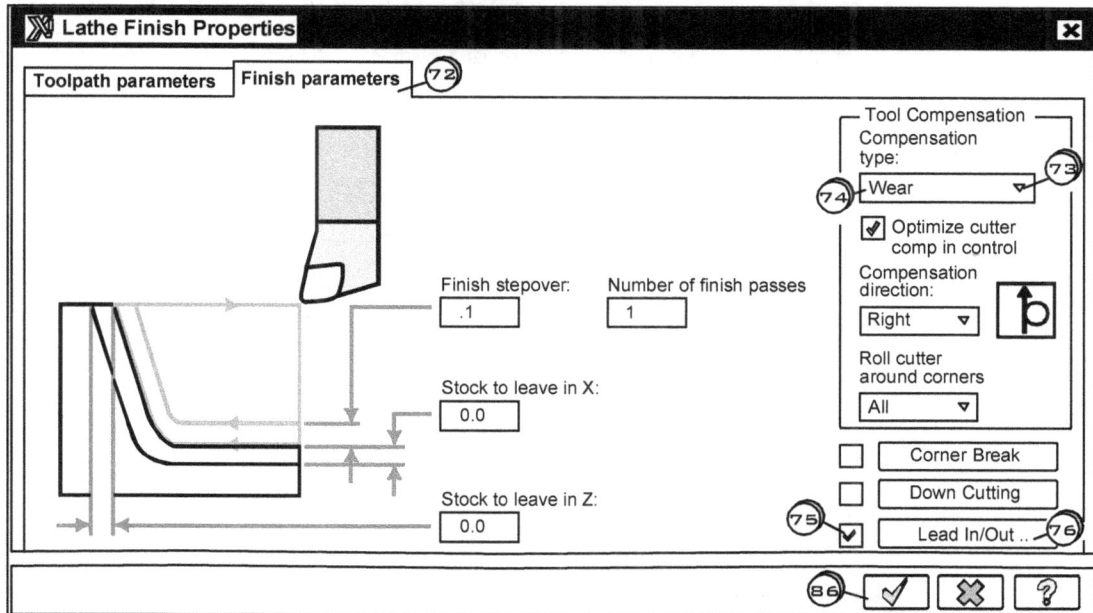

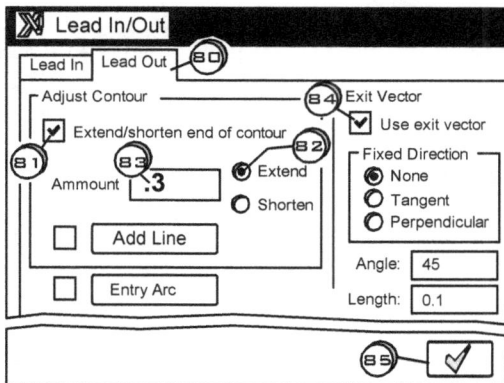

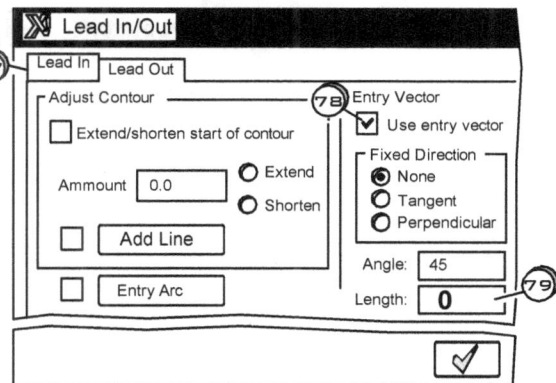

▷ Click ⑦② the ⌐Finish parameters tab

▷ Click ⑦③ the Compensation type down button ▽

▷ Click ⑦④ Wear

▷ Click ⑦⑤ check on ✔ for Lead In/Out

▷ Click ⑦⑥ the ⌐Lead In/Out ..⌐ button

▷ Click ⑦⑦ the ⌐Lead In tab

▷ Click ⑦⑧ check on ✔ for Use entry vector

▷ Click ⑦⑨ in the Length box; enter ⌐0⌐

▷ Click ⑧⓪ the ⌐Lead Out tab

▷ Click ⑧① check on ✔ for Extend/shorten

▷ Click ⑧② the Extend radio button ◉

▷ Click ⑧③ in the Ammount: box; enter ⌐.3⌐

▷ Click ⑧④ check on ✔ for Use exit vector
   lead out will be a polar line at a 45° angle from the
   horizontal of length 0.1

▷ Click ⑧⑤ the OK button ✔

▷ Click ⑧⑥ the OK button ✔

E) OPERATION#4 [Level: 1] **- ROUGH AND FINISH OD GROOVE;**
LEAVE .01 IN X AND Z FOR FINISHING
- ◆ **CHAIN THE GROOVE BOUNDARY**
- ◆ **OBTAIN THE NEEDED R.01 W.25 OD GROOVING RIGHT TOOL**

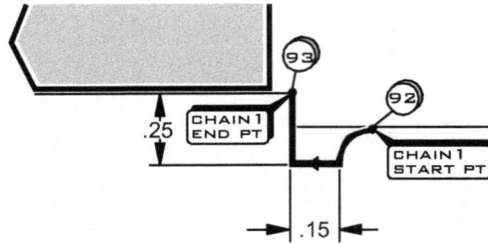

➤ Click ⓼⓻   TOOLPATHS

➤ Click ⓼⓼   ▣ Groove

➤ Click ⓼⓽ the Chain radio button *on* ⊙ Chain

➤ Click ⓽⓪   the OK button   ☑

➤ Click ⓽⓵   the chain button   ⟨∞⟩

➤ Click ⓽⓶   the start point ⓽⓷ the end point

➤ Click ⓽⓸   the OK button   ☑

Lathe Groove Properties   ✕

| Toolpath parameters | Groove shape parameters | Groove rough parameters | Groove finish parameters |

Tool number: 1   Offsetl number: 1

Station number: 1   Tool Angle

T4343 R0.01 W0.375   T4444 R0.01 W0.125
OD GROOVE CEN   OD GROOVE RIG

Feed rate: 0.01   ⊙ in/rev ○ in/min ○ micro/in

☐ Finish feed rate: 0.005   ⊙ in/rev ○ in/min ○ micro/in

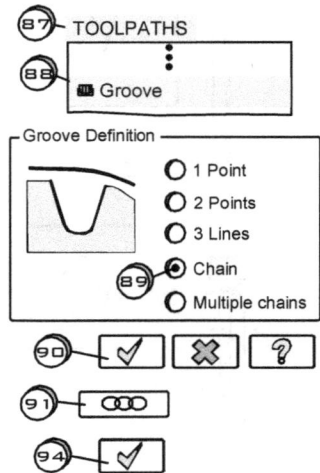

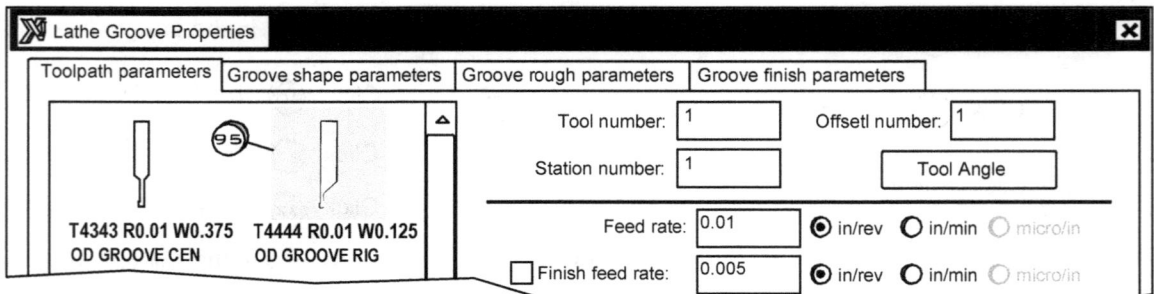

➤ Click ⓽⓹   the **T4444 R0.01 W0.125** tool
          OD GROOVE RIGHT

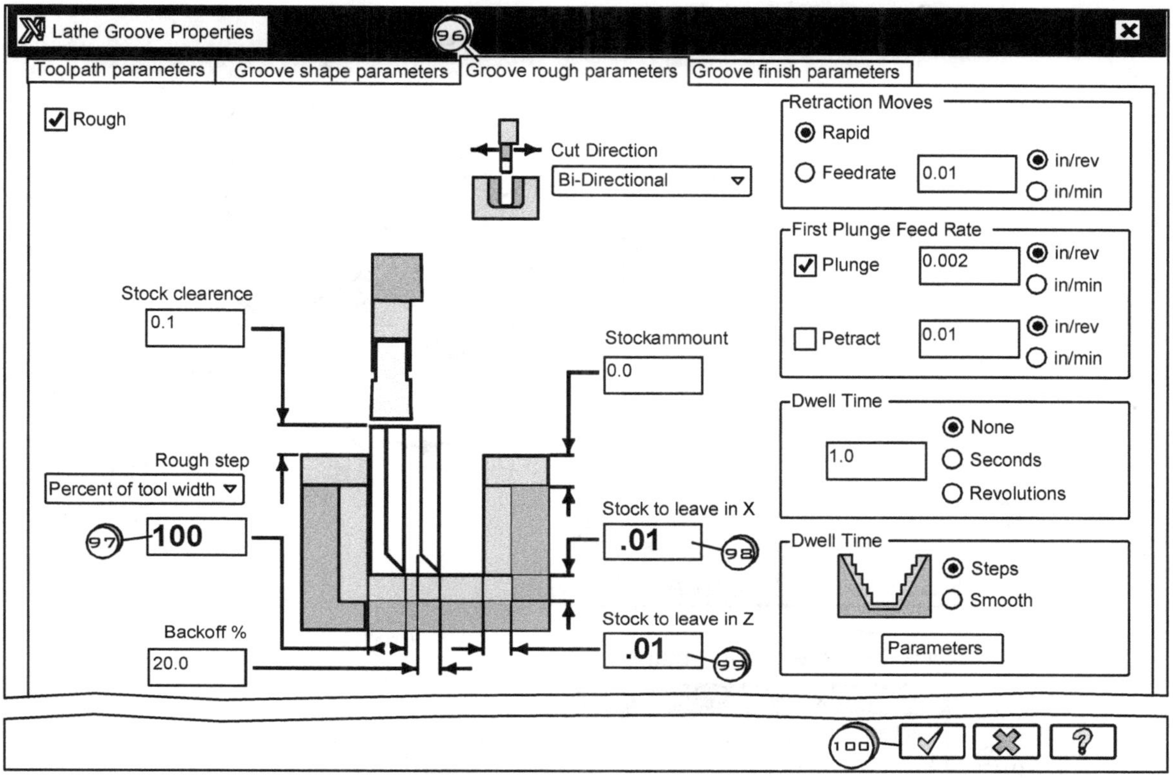
Lathe Groove Properties
(96)
Toolpath parameters
Groove shape parameters
Groove rough parameters
Groove finish parameters
Rough
Cut Direction
Bi-Directional
Retraction Moves
Rapid
Feedrate
0.01
in/rev
in/min
First Plunge Feed Rate
Plunge
0.002
in/rev
in/min
Petract
0.01
in/rev
in/min
Stock clearence
0.1
Stockammount
0.0
Dwell Time
None
1.0
Seconds
Revolutions
Rough step
Percent of tool width
(97)
100
Stock to leave in X
.01
(98)
Dwell Time
Steps
Smooth
Backoff %
20.0
Stock to leave in Z
.01
(99)
Parameters
(100)

- Click (96) Groove rough parameters tab
- Click (97) in the Percent of tool width box; enter 100
- Click (98) in the Stock to leave in X box; enter .01
- Click (99) in the Stock to leave in Z box; enter .01
- Click (100) the OK button

F) OPERATION#5 [Level: 2] - C-AXIS FACE CONTOUR X 1.3 DEEP

◆ CHAIN THE CONTOUR BOUNDARY

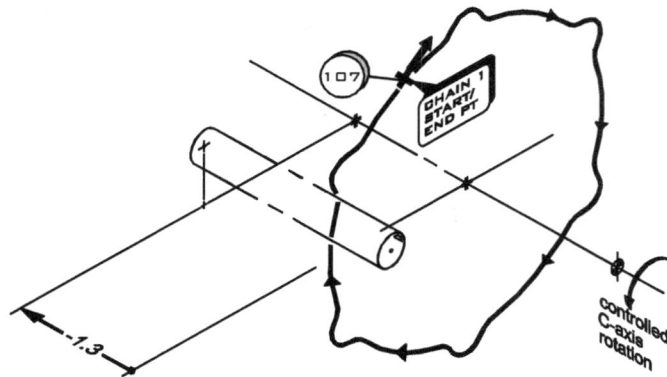

> Click (101) the Level: button

> Click (102) Level 2 as the current level

> Click the visibility (X) off for all the other levels

> Click (103) TOOLPATHS

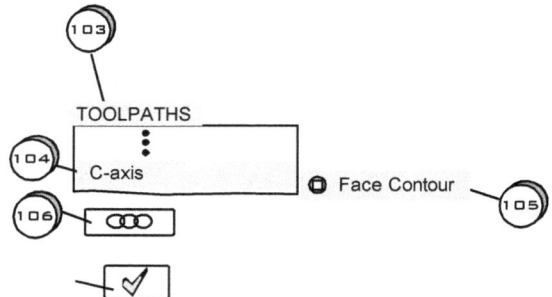

> Click (104) C-axis

> Click (105) ● Face Contour

> Click (106) the chain button [ⓒⓞⓞ]

> Click (107) the start/end point

> Click (108) the OK button [✓]

◆ OPEN THE TOOL PAGE AND SPECIFY A 3/8 FLAT ENDMILL

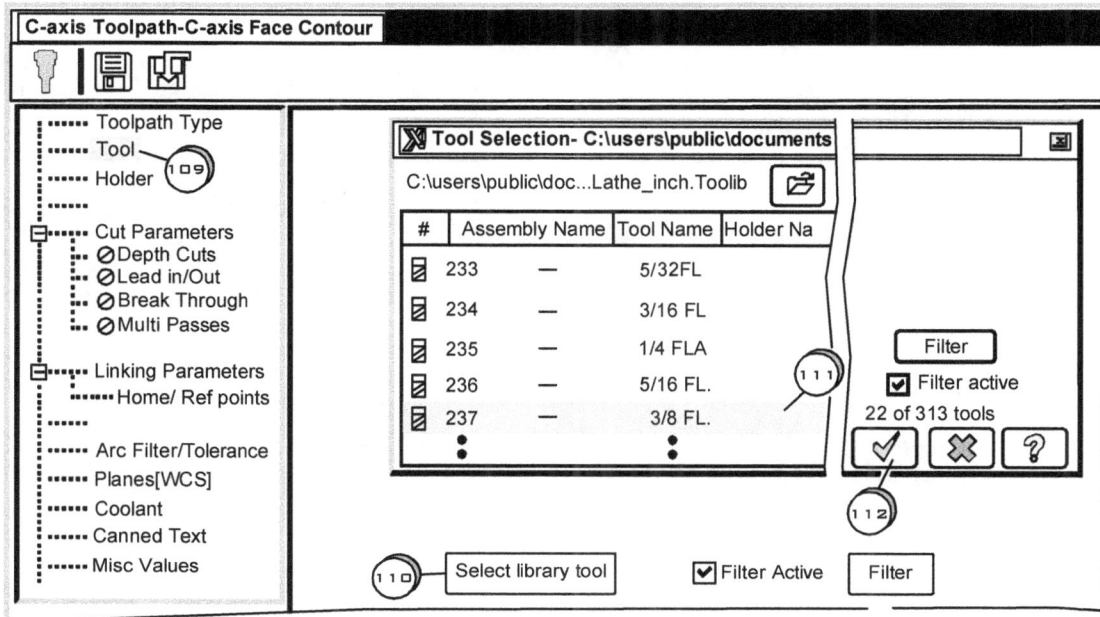

```
┌─────────────────────────────────────────────────────────────────────────┐
│ C-axis Toolpath-C-axis Face Contour                                       │
├─────────────────────────────────────────────────────────────────────────┤
│   ▽    ▣   ▨                                                              │
├─────────────────────────────────────────────────────────────────────────┤
│  ····· Toolpath Type      ┌──────────────────────────────────┐  ┌──────┐ │
│  ····· Tool                │ ▣ Tool Selection- C:\users\public\documents│ │
│  ····· Holder  (109)       │ C:\users\public\doc...Lathe_inch.Toolib  📂 │ │
│  ·····                     │ ┌──┬──────────────┬───────────┬────────┐   │ │
│  ┌─··· Cut Parameters      │ │ #│ Assembly Name│ Tool Name │Holder Na│  │ │
│  │··  ⊘Depth Cuts          │ ├──┼──────────────┼───────────┼────────┤   │ │
│  │··  ⊘Lead in/Out         │ │233│     —       │  5/32FL   │         │  │ │
│  │··  ⊘Break Through       │ │234│     —       │  3/16 FL  │         │  │ │
│  │··  ⊘Multi Passes        │ │235│     —       │  1/4 FLA  │         │ ┌─┐│ │
│  │                         │ │236│     —       │  5/16 FL. │    (111)│Filter│
│  ┌─··· Linking Parameters  │ │237│     —       │  3/8 FL.  │         │☑Filter active│
│  │··· Home/ Ref points     │ │  •          •                       │22 of 313 tools│
│  │                         │ │  •          •                       │✓  ✗  ?│ │
│  ····· Arc Filter/Tolerance│ └──────────────────────────────────┘        │ │
│  ····· Planes[WCS]         │                                      (112)   │ │
│  ····· Coolant             │                                             │ │
│  ····· Canned Text    (110)┌──────────────────┐      ☑ Filter Active  Filter│
│  ····· Misc Values         │ Select library tool│                         │ │
│                            └──────────────────┘                         │ │
└─────────────────────────────────────────────────────────────────────────┘
```

➤ Click (109) the Tool page

➤ Click (110) the [Select library tool] button

➤ Click (111) the 3/8 Endmill Flat

➤ Click (112) the OK button [✓] to bring the tool from the library into the part file
for the operation.

♦ **OPEN THE LINKING PARAMETERS PAGE TOOL PAGE AND SPECIFY A CLEARENCE OF 1 AND A DEPTH OF CUT -1.3**

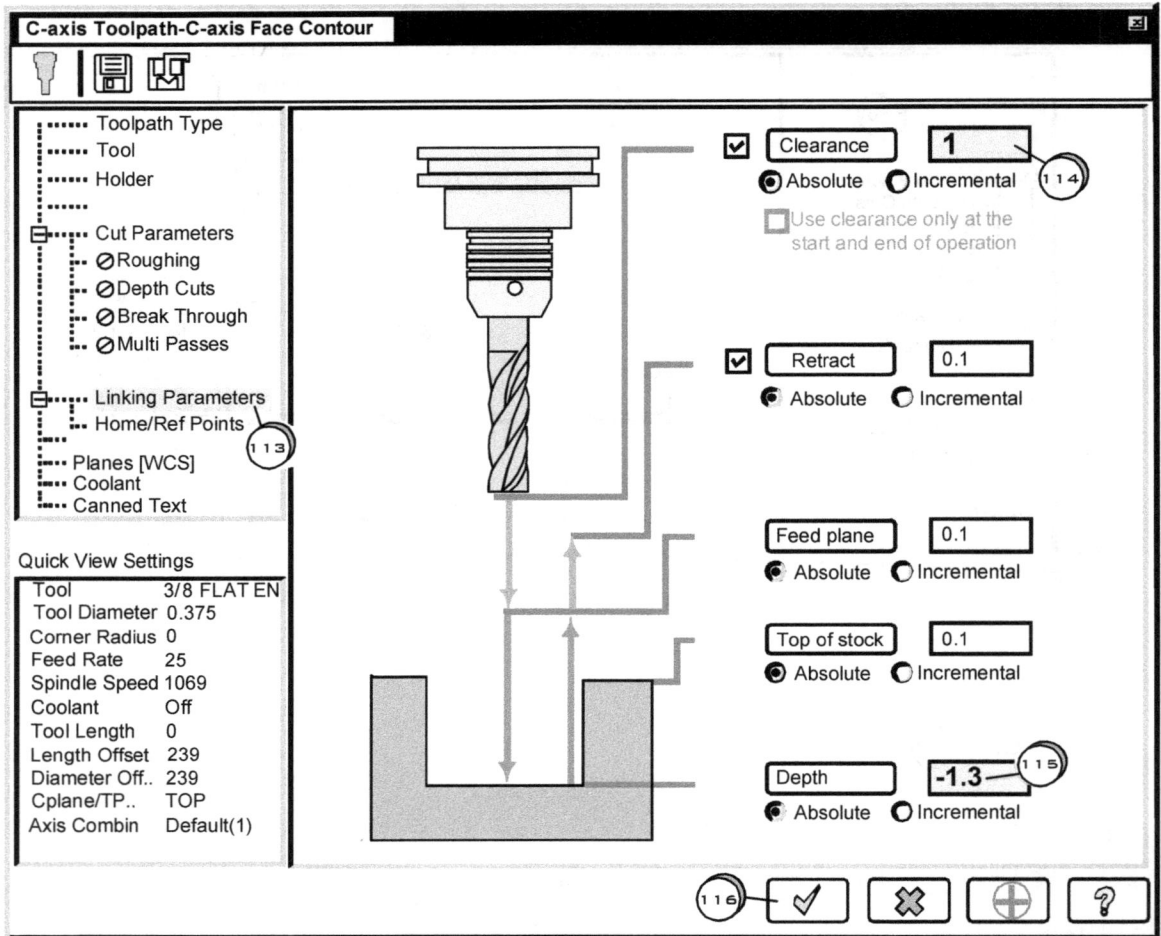

➤ Click (113) the   Linking Parameters   page.

➤ Click (114) ; enter   **1**   in the clearance box

➤ Click (115) in the Depth box and enter   **-1.3**

➤ Click (116) the OK button [✓] to *create* the operation

G) OPERATION#6 [Level: 3] - C-AXIS CROSS CONTOUR X .125 DEEP

♦ CHAIN THE CONTOUR BOUNDARY

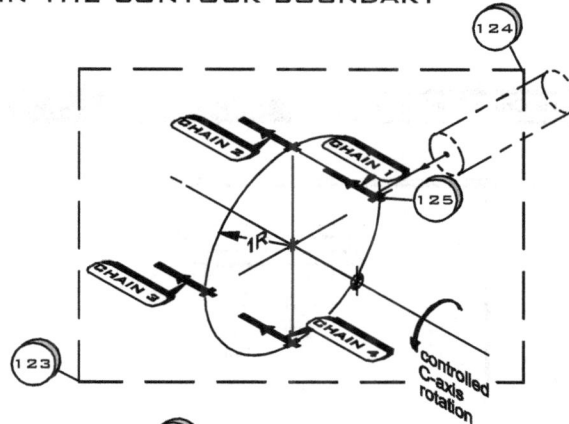

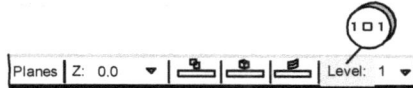

| Planes | Z: 0.0 ▼ | | | | Level: 1 ▼ |

➤ Click (117) the Level: button

Level Manager

Number	Visible	Name
1		2D-PROFILE
2		C-FACE CON AT Z-1.75
3	X	C-CROSS CON AT Z-.75
4		C-CROSS DRILL(.109D)
5		C-CFACE DRILL(.125D)

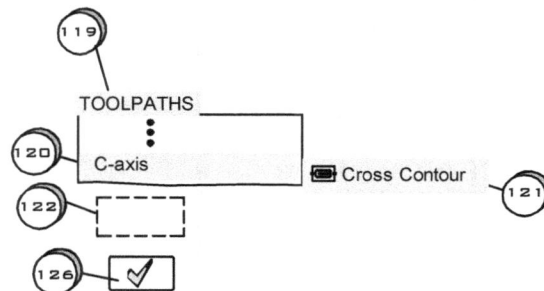

TOOLPATHS
⋮
C-axis   ▣ Cross Contour

➤ Click (118) Level 3 as the current level

➤ Click the visibility (X) off for all the other levels

➤ Click (119)   TOOLPATHS

➤ Click (120)   C-axis

➤ Click (121)   ▣ Cross Contour

➤ Click (122) the window button   [- - -]

➤ Click (123) (124)   the window corners

sketch approximate start point

➤ Click (125) the start point

➤ Click (126) the OK button   ☑

♦ SPECIFY CONTOUR MACHINING PARAMETERS USING THE
CURRENTLY SELECTED .375 DIA END MILL TOOL

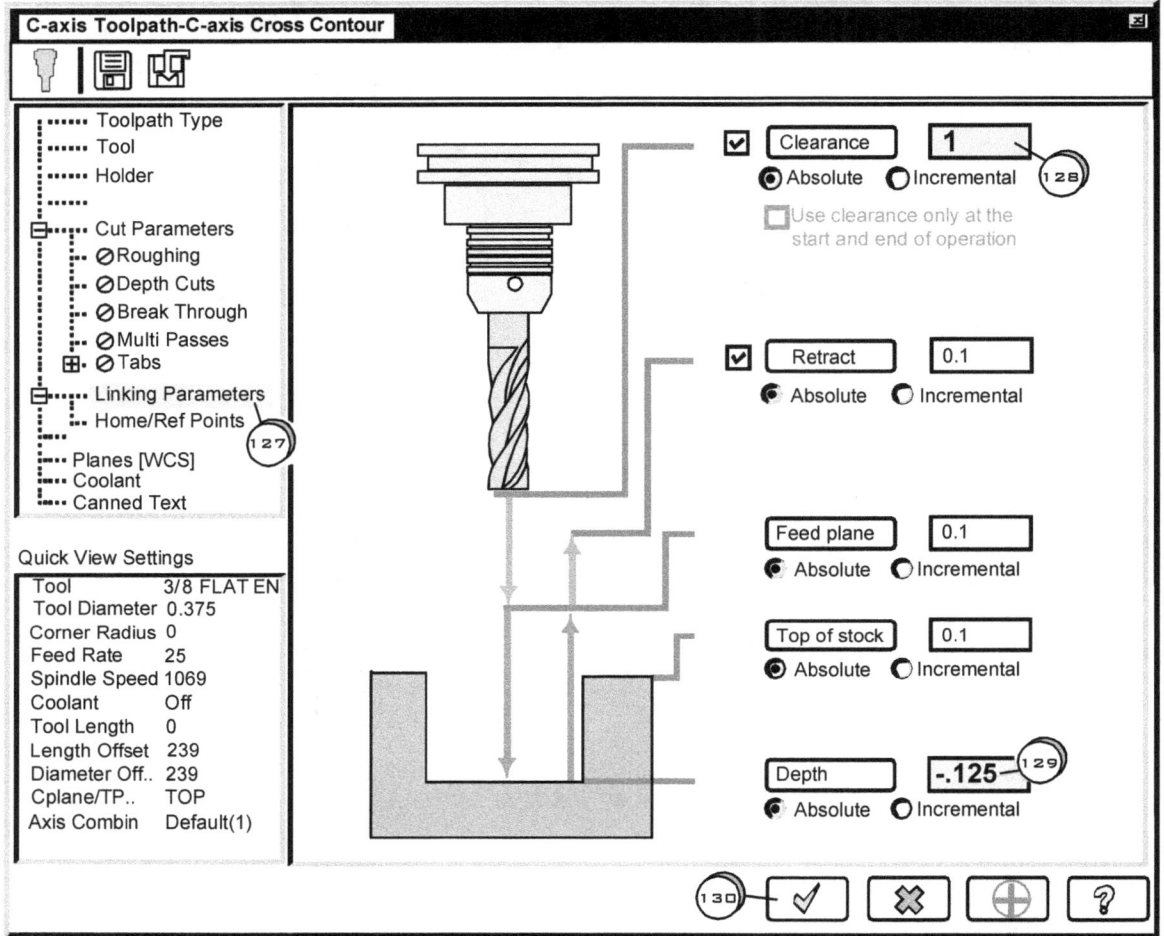

C-axis Toolpath-C-axis Cross Contour

- ..... Toolpath Type
- ..... Tool
- ..... Holder
- ....
- ⊞ ..... Cut Parameters
  - ⊘ Roughing
  - ⊘ Depth Cuts
  - ⊘ Break Through
  - ⊘ Multi Passes
  - ⊞ ⊘ Tabs
- ⊞ ..... Linking Parameters
  - ..... Home/Ref Points (127)
  - ... Planes [WCS]
  - ... Coolant
  - ... Canned Text

Quick View Settings

Tool	3/8 FLAT EN
Tool Diameter	0.375
Corner Radius	0
Feed Rate	25
Spindle Speed	1069
Coolant	Off
Tool Length	0
Length Offset	239
Diameter Off..	239
Cplane/TP..	TOP
Axis Combin	Default(1)

☑ Clearance **1** (128)
  ⦿ Absolute ○ Incremental
  ☐ Use clearance only at the start and end of operation

☑ Retract 0.1
  ⦿ Absolute ○ Incremental

Feed plane 0.1
  ⦿ Absolute ○ Incremental

Top of stock 0.1
  ⦿ Absolute ○ Incremental

Depth **-.125** (129)
  ⦿ Absolute ○ Incremental

(130) ✓ ✗ ⊕ ?

➤ Click (127) the Linking Parameters page.

➤ Click (128) ; enter **1** in the clearance box

➤ Click (129) in the Depth box and enter **-.125**

➤ Click (130) the OK button ✓ to *create* the operation

H) OPERATION#7 [Level: 4] - C-AXIS CROSS DRILL X .405 DEEP

♦ CREATE THE .1094(7/64) DRILL TOOLPATH

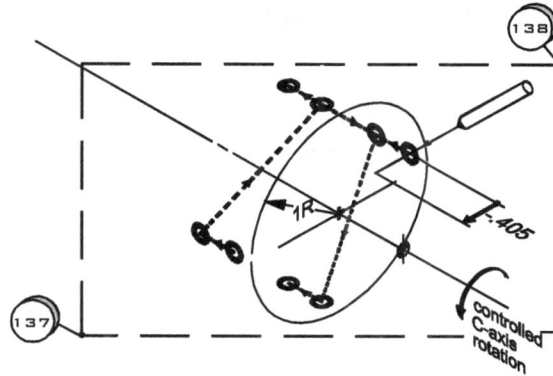

Planes | Z: 0.0 ▼ | 🔲 🔲 🔲 | Level: 1 ▼

➤ Click (131) the Level: button

Level Manager

Number	Visible	Name
1		2D-PROFILE
2		C-FACE CON AT Z-1.75
3		C-CROSS CON AT Z-.75
4	X	C-CROSS DRILL(.109D)
5		C-CFACE DRILL(.125D)

TOOLPATHS
⋮
C-axis   🔲Cross Drill

Window Points

☑

➤ Click (132) Level 4 as the current level

➤ Click the visibility (X) off for all the other levels

➤ Click (133) TOOLPATHS

➤ Click (134) C-axis

➤ Click (135) 🔲Cross Drill

➤ Click (136) the Window Points button [Window Points]

➤ Click (137) (138) the window corners

➤ Click (139) the OK button [☑]

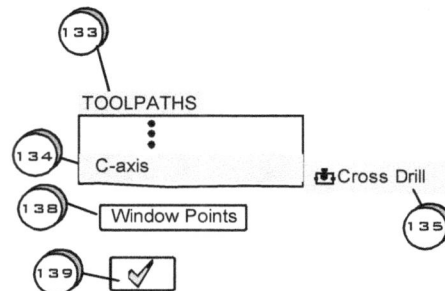

◆ OPEN THE TOOL PAGE AND SPECIFY A 7/64 DRILL

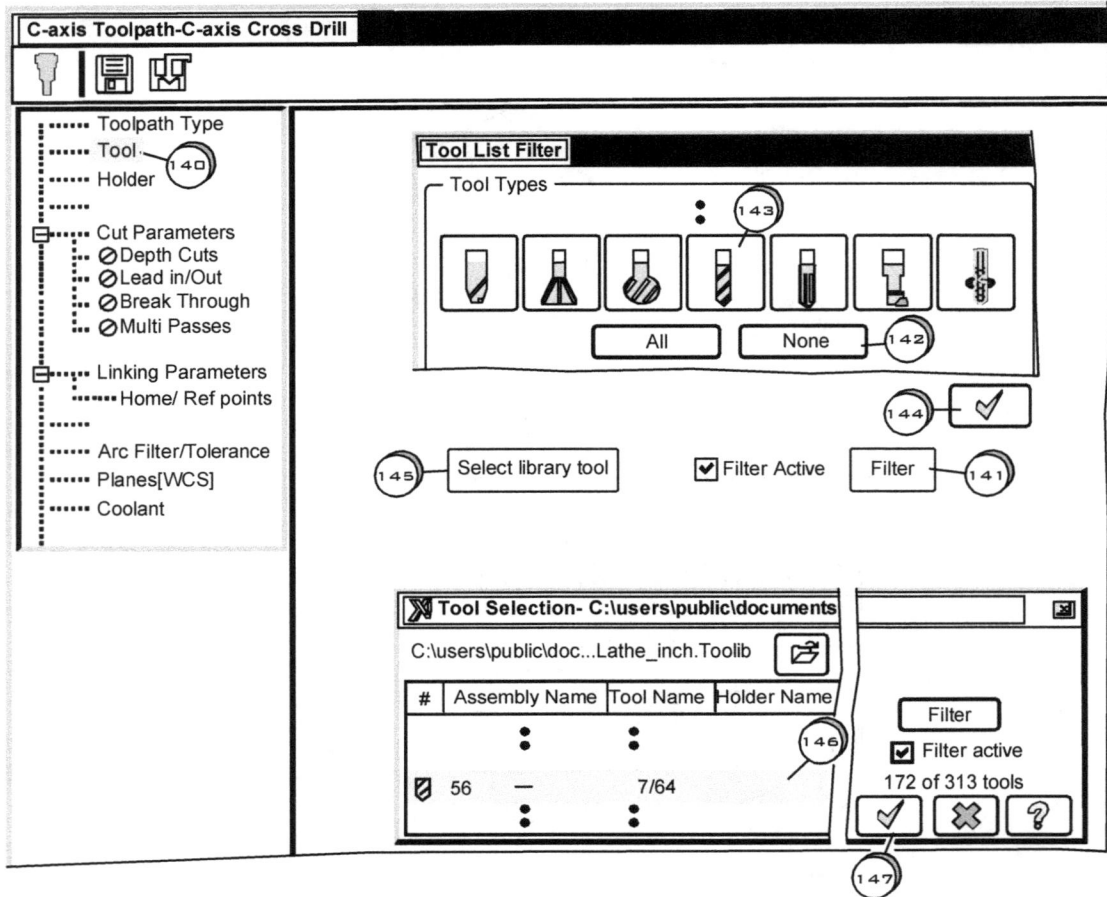

C-axis Toolpath-C-axis Cross Drill

- Toolpath Type
- Tool — (140)
- Holder
- Cut Parameters
  - ⊘ Depth Cuts
  - ⊘ Lead in/Out
  - ⊘ Break Through
  - ⊘ Multi Passes
- Linking Parameters
  - Home/ Ref points
- Arc Filter/Tolerance
- Planes[WCS]
- Coolant

Tool List Filter

Tool Types (143)

All   None (142)

(144) ✓

(145) Select library tool   ☑ Filter Active   Filter (141)

Tool Selection- C:\users\public\documents ☒

C:\users\public\doc...Lathe_inch.Toolib

#	Assembly Name	Tool Name	Holder Name
	⋮	⋮	
56	—	7/64	(146)
	⋮	⋮	

Filter

☑ Filter active

172 of 313 tools

✓ ✗ ？

(147)

▶ Click (140) the Tool page

▶ Click (141) the Filter button

▶ Click (142) the None button

▶ Click (143) drill button

▶ Click (144) the OK button ✓

▶ Click (145) the Select library tool button

▶ Click (146) the 7/64 Drill

▶ Click (147) the OK button ✓

to bring the tool from the library into the part file for the operation.

♦ SPECIFY CROSS DRILL MACHINING PARAMETERS USING THE
CURRENTLY SELECTED 7/64 DRILL TOOL

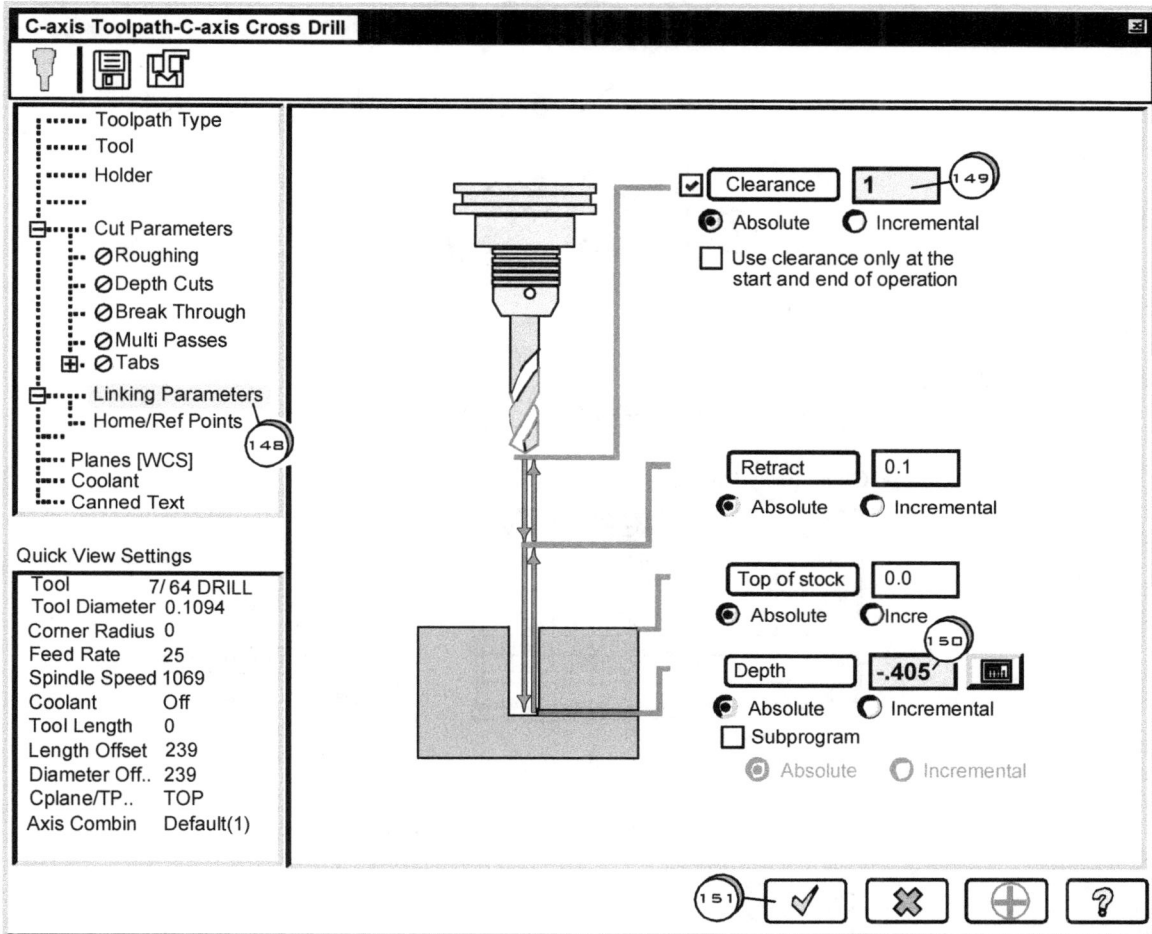

C-axis Toolpath-C-axis Cross Drill

- Toolpath Type
- Tool
- Holder
- Cut Parameters
  - ⊘Roughing
  - ⊘Depth Cuts
  - ⊘Break Through
  - ⊘Multi Passes
  - ⊘Tabs
- Linking Parameters
  - Home/Ref Points (148)
- Planes [WCS]
- Coolant
- Canned Text

Quick View Settings

Tool	7/64 DRILL
Tool Diameter	0.1094
Corner Radius	0
Feed Rate	25
Spindle Speed	1069
Coolant	Off
Tool Length	0
Length Offset	239
Diameter Off..	239
Cplane/TP..	TOP
Axis Combin	Default(1)

☑ Clearance 1 (149)
● Absolute ○ Incremental
☐ Use clearance only at the start and end of operation

Retract 0.1
● Absolute ○ Incremental

Top of stock 0.0
● Absolute ○ Incre (150)

Depth -.405
● Absolute ○ Incremental
☐ Subprogram
  ○ Absolute ○ Incremental

(151) ✓ ✗ ⊕ ?

➣ Click (148) the Linking Parameters page.

➣ Click (149) ; enter **1** in the clearance box

➣ Click (150) in the Depth box and enter **-.405**

➣ Click (151) the OK button ✓ to *create* the operation

I) OPERATION#8 ⬚Level: 5⬚ - C-AXIS FACE DRILL X .45 DEEP

♦ CREATE THE .125(1/8) DRILL TOOLPATH

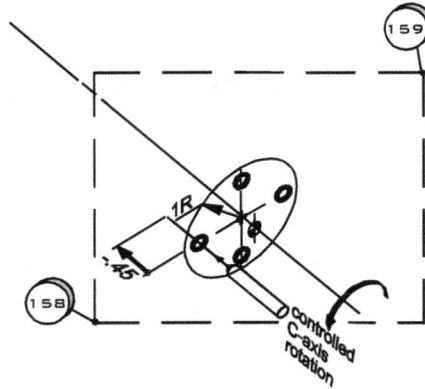

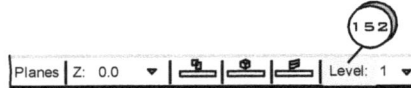

Planes | Z: 0.0 ▼ | 🔲 🔲 🔲 | Level: 1 ▼

▶ Click (152) the Level: button

Level Manager

Number	Visible	Name
1		2D-PROFILE
2		C-FACE CON AT Z-1.75
3		C-CROSS CON AT Z-.75
4		C-CROSS DRILL(.109D)
5	X	C-CFACE DRILL(.125D)

TOOLPATHS
⋮
C-axis

○Face Drill

Window Points

✓

▶ Click (153) Level 5 as the current level

▶ Click the visibility (X) off for all the other levels

▶ Click (154)   TOOLPATHS

▶ Click (155)   C-axis

▶ Click (156)   ○Face Drill

▶ Click (157) the Window Points button   ⬚Window Points⬚

▶ Click (158) (159) the window corners

▶ Click (160) the OK button   ⬚✓⬚

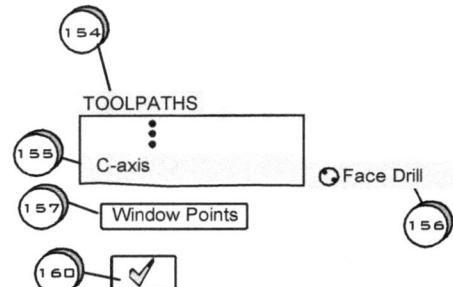

♦ OPEN THE TOOL PAGE AND SPECIFY A .125(1/8) DRILL

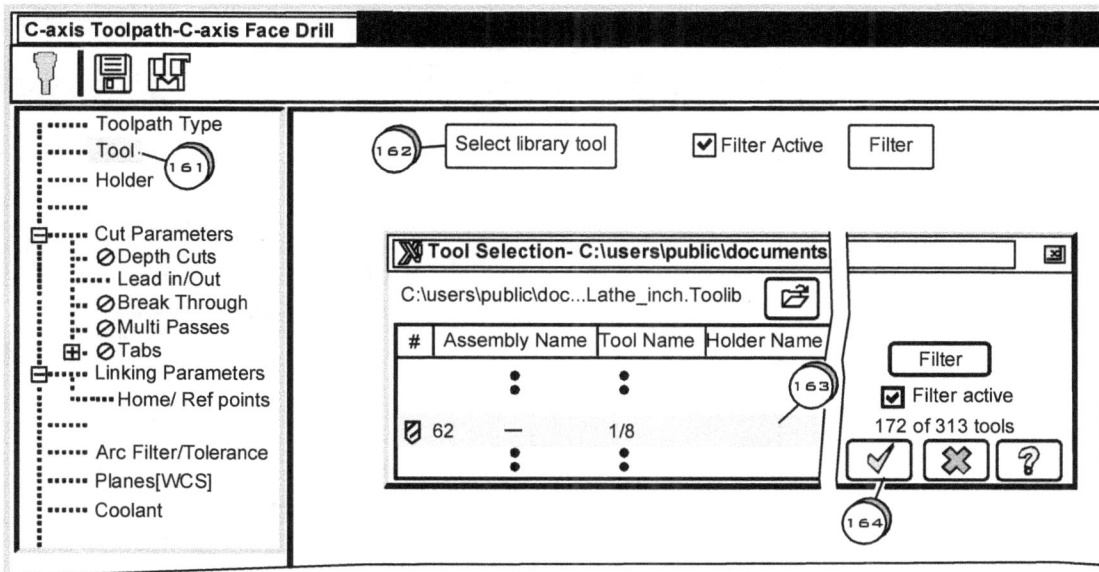

C-axis Toolpath-C-axis Face Drill

| Toolpath Type |
| Tool (161) |
| Holder |
| |
| Cut Parameters |
| ⊘ Depth Cuts |
| Lead in/Out |
| ⊘ Break Through |
| ⊘ Multi Passes |
| ⊘ Tabs |
| Linking Parameters |
| Home/ Ref points |
| |
| Arc Filter/Tolerance |
| Planes[WCS] |
| Coolant |

(162) Select library tool   ☑ Filter Active   Filter

Tool Selection- C:\users\public\documents

C:\users\public\doc...Lathe_inch.Toolib

#	Assembly Name	Tool Name	Holder Name
	•	•	
62	—	1/8	
	•	•	

(163)

Filter

☑ Filter active

172 of 313 tools

✓   ✗   ?

(164)

> Click (161) the Tool page

> Click (162) the | Select library tool | button

> Click (163) the 1/8 Drill

> Click (164) the OK button | ✓ | to bring the tool from the library into the part file for the operation.

♦ SPECIFY CROSS DRILL MACHINING PARAMETERS USING THE CURRENTLY SELECTED 1/8 DRILL TOOL

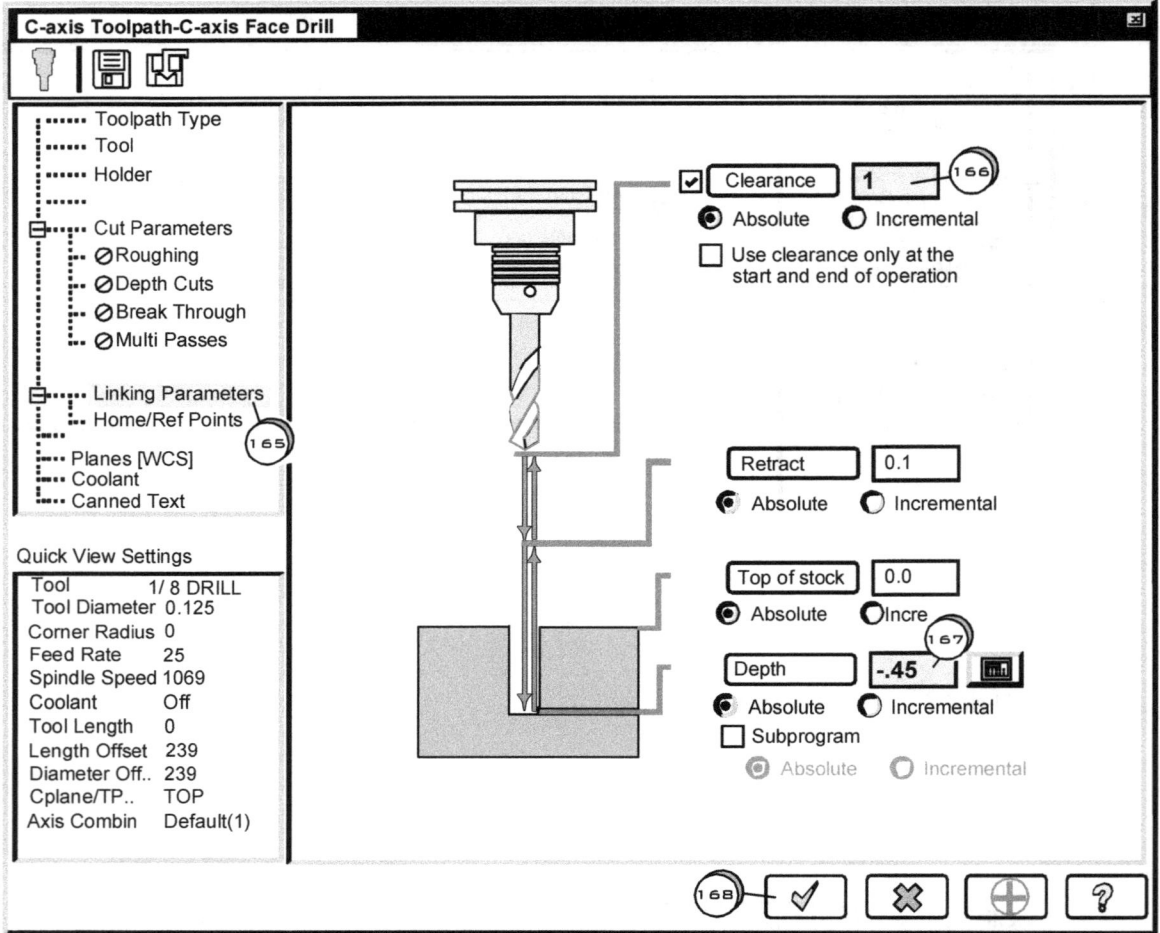

⯈ Click ⓰⓯ the Linking Parameters page.

⯈ Click ⓰⓰ ; enter ⟦ 1 ⟧ in the clearance box

⯈ Click ⓰⓱ in the Depth box and enter ⟦ -.45 ⟧

⯈ Click ⓰⓲ the OK button ✓ to *create* the operation

J) OPERATION#9 [Level: 1] - LATHE CUTOFF
◆ CREATE THE CUTOFF TOOLPATH

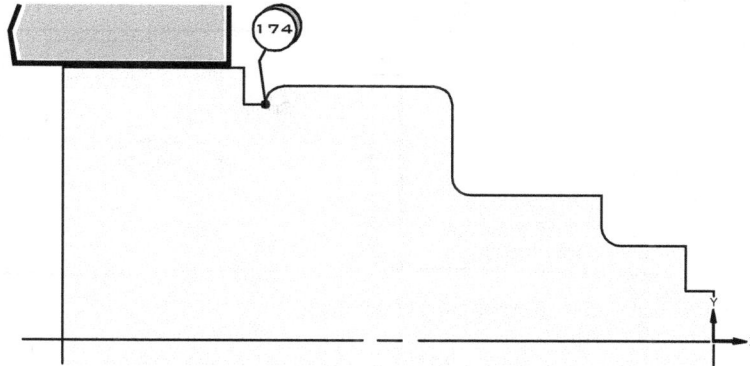

Change the view to Top.

➤ *Right* Click; Click (169) 🗇 Top   🗇 Top(WCS) (169)

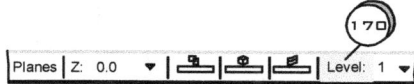

Planes | Z: 0.0 ▼ | 🖳 | 🖳 | 🖳 | Level: 1 ▼   (170)

➤ Click (170) the   Level: button

Level Manager

Number	Visible	Name
1	X	2D-PROFILE
2		C-FACE CON AT Z-1.75
3		C-CROSS CON AT Z-.75
4		C-CROSS DRILL(.109D)
5		C-CFACE DRILL(.125D)

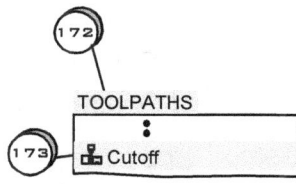

(171)

TOOLPATHS (172)

(173) 🔧 Cutoff

➤ Click (171) Level 1 as the current level

➤ Click the visibility (X) off for all the other levels

➤ Click (172) TOOLPATHS

➤ Click (173) 🔧 Cutoff

[Select cutoff boundary point]

➤ Click (174) near the end pt of the arc

◆ OBTAIN THE NEEDED R1/64 TNR, RIGHT CUTOFF TOOL

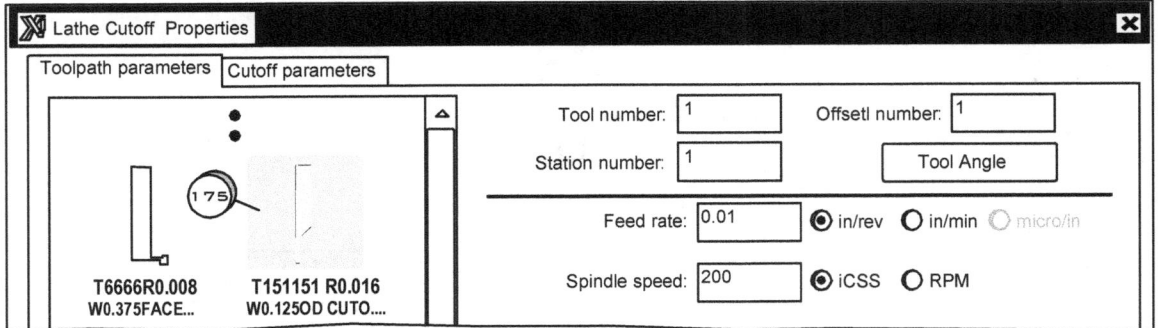

Lathe Cutoff Properties ☒

| Toolpath parameters | Cutoff parameters |

Tool number: 1   Offsetl number: 1

Station number: 1   Tool Angle

T6666R0.008   T151151 R0.016
W0.375FACE...   W0.1250D CUTO....

Feed rate: 0.01   ◉ in/rev   ○ in/min   ○ micro/in

Spindle speed: 200   ◉ iCSS   ○ RPM

➤ Click (175) the W0.125OD CUTOFF RIGHT tool

◆ ENTER THE LATHE CUTOFF MACHINING PARAMETERS

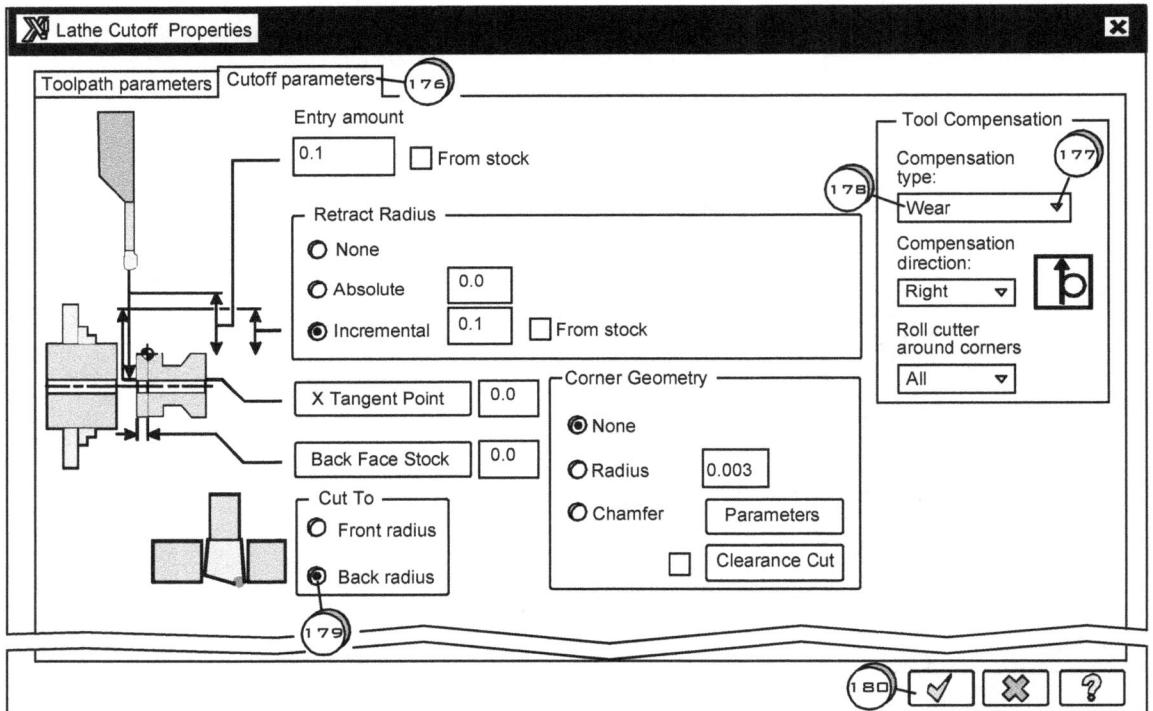

Lathe Cutoff Properties ☒

| Toolpath parameters | Cutoff parameters (176) |

Entry amount
0.1   ☐ From stock

Tool Compensation (177)

Compensation type:
(178) Wear

Compensation direction:
Right ▽

Roll cutter around corners
All ▽

Retract Radius
○ None
○ Absolute   0.0
◉ Incremental 0.1   ☐ From stock

X Tangent Point   0.0

Back Face Stock   0.0

Corner Geometry
◉ None
○ Radius   0.003
○ Chamfer   Parameters
☐ Clearance Cut

Cut To
○ Front radius
◉ Back radius
(179)

(180) ✓ ☒ ?

➤ Click (176)   Cutoff parameters tab

➤ Click (177) the Compensation type down button ▽

➤ Click (178) Wear

➤ Click (179) the Cut to Back radius radio button ◉

➤ Click (180)   the OK button ✓

♦ DIRECT *Mastercam* TO VERIFY ALL THE MACHINING OPERATIONS

⟫ Click (181) the Select all button ▷ₙ

⟫ Click (182) the Verify button 🗃

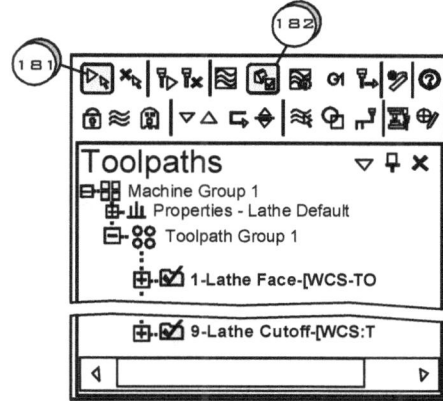

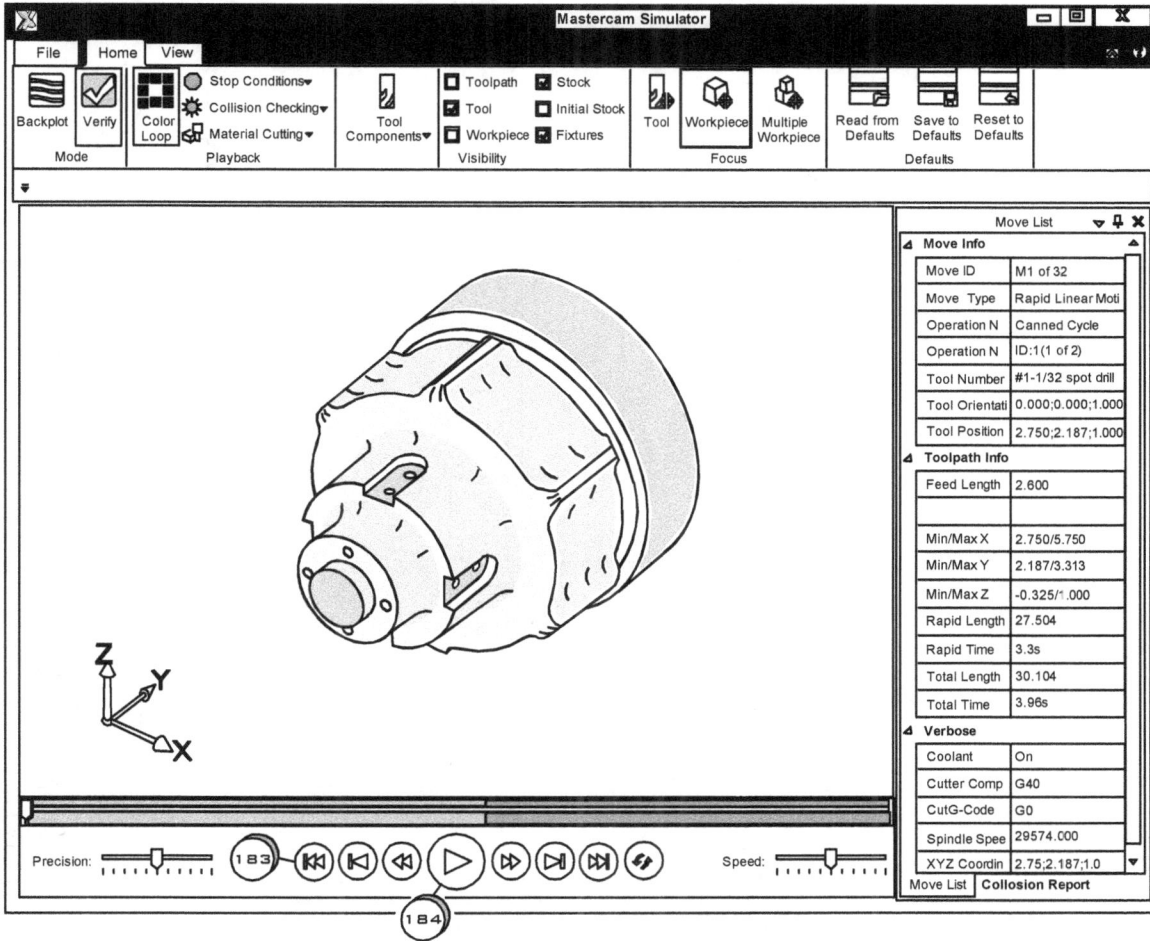

Toolpaths   ▽ ⚲ ✕

⊟ 🎛 Machine Group 1
  ⊞ 🏛 Properties - Lathe Default
    ⊟ 🎱 Toolpath Group 1

      ⊞ ☑ 1-Lathe Face-[WCS-TO

      ⊞ ☑ 9-Lathe Cutoff-[WCS:T

◁ ▭ ▷

⟫ Click (183) the Start (◄◄) button

⟫ Click (184) the Play (▷) button

EXERCISES

6-1) Get the CAD model file EX5-2XX. Using PROCESS PLAN 6P-1 generate the part program to produce the part shown in Figure 6p-1.

Material: 1030 Steel

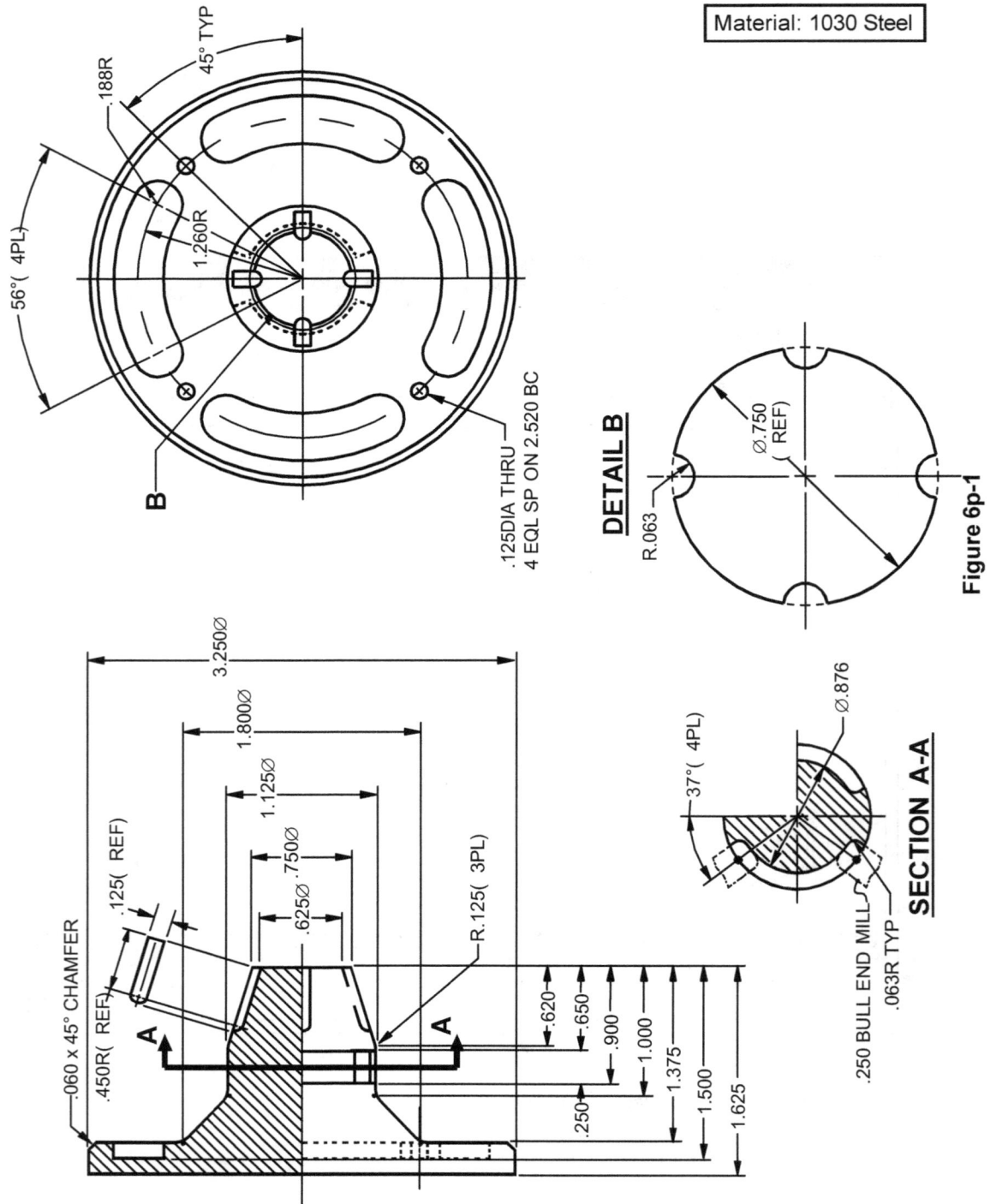

DETAIL B

SECTION A-A

Figure 6p-1

CHUCK

BAR STOCK

Part Origin

Ø.125 OD MARGIN

Ø3.25 OD

Ø.125

GRIP LENGTH ←.75→

LEFT MARGIN ←1→ ←1.625→ .10 RIGHT MARGIN

LENGTH

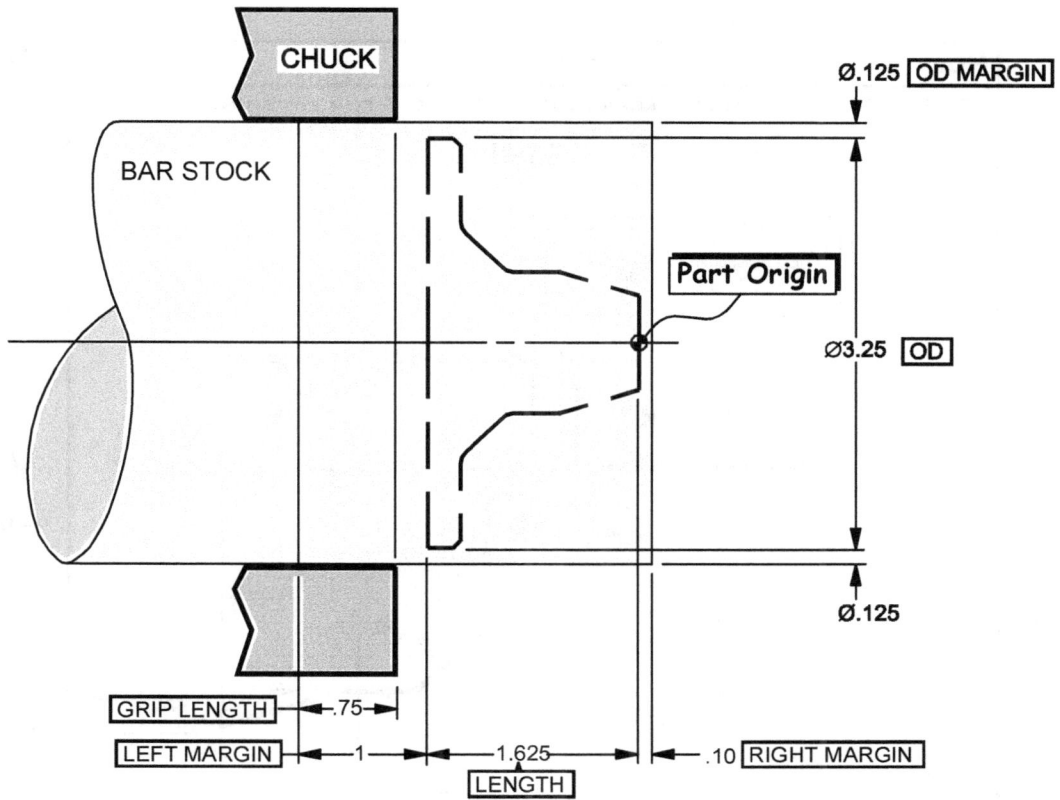

PROCESS PLAN 6P-1

No.	Operation	Tooling
1	Face End — Material Removed — .10	1/32 TNR, RH OD TURNING TOOL — R1/32

PROCESS PLAN 6P-1(*continued*)

No.	Operation	Tooling
2	**ROUGH TURN OD; LEAVE .01 IN X AND Z FOR FINISHING** ☑ Lead In/Out Level:1 Lead In │ Lead Out ┌ Adjust Contour ☑ Extend/shorten end of contour Ammount **.15** ⦿ Extend ○ Shorten .15 EXTEND END OF CONTOUR CHUCK Material Removed CHAIN 1 END PT ② CHAIN 1 START PT ① .01 LEFT FOR FINISH CUT	1/32 TNR, RH OD TURNING R1/32
3	**FINISH OD CONTOUR** ☑ Lead In/Out Level:1 Lead In │ Lead Out ┌ Adjust Contour ☑ Extend/shorten end of contour Ammount **.15** ⦿ Extend ○ Shorten .15 EXTEND END OF CONTOUR CHUCK CHAIN 1 END PT ② Material Removed CHAIN 1 START PT ① .01	1/64 TNR, RH OD FINISHING R1/64

PROCESS PLAN 6P-1(*continued*)

No.	Operation	Tooling
4	C-AXIS FACE CONTOUR x .125 DEEP	.375 DIA END MILL

Planes | Z: 0.0 ▼ | | | | Level: 1 ▼

Level Manager

Number	Visible	Name
1		2D-PROFILE
2	X	C-FACE CON
3		C-FACE DRILL
4		C-CONTOUR
5		C-CROSS CONTOUR

TOOLPATHS

C-axis   ○ Face Contour

PROCESS PLAN 6P-1(*continued*)

No.	Operation	Tooling
5	**C-AXIS FACE DRILL X .35 DEEP** Planes Z: 0.0 ▾ | [icons] | Level: 2 ▾ **Level Manager** | Number	Visible	Name
1		2D-PROFILE
2		C-FACE CON
3	X	C-FACE DRILL
4		C-CONTOUR
5		C-CROSS CONTOUR | TOOLPATHS ⋮ C-axis ◔ Face Drill Window Points ☑ ① ② 1.26R .35 controlled C-axis rotation	.125 1/8 DRILL
6	**C-AXIS CONTOUR X .125 DEEP** Planes Z: 0.0 ▾ | [icons] | Level: 3 ▾ **Level Manager** | Number	Visible	Name
1		2D-PROFILE
2		C-FACE CON
3		C-FACE DRILL
4	X	C-CONTOUR
5		C-CROSS CONTOUR | TOOLPATHS ⋮ C-axis ▦ C-Axis Contour ⟋ Single ☑ CHAIN 1 START PT CHAIN 1 START PT .438R .563R controlled C-axis rotation	1/4 DIA BULL END MILL .0625R

PROCESS PLAN 6P-1(*continued*)

No.	Operation	Tooling
7	C-AXIS CROSS CONTOUR x .063 DEEP	1/8 DIA BALL END MILL
8	LATHE CUTOFF	.016 TNR, RH OD CUTOFF TOOL

Operation 7:

Planes | Z: 0.0 ▼ | Level: 4 ▼

Level Manager

Number	Visible	Name
1		2D-PROFILE
2		C-FACE CON
3		C-FACE DRILL
4		C-CONTOUR
5	X	C-CROSS CONTOUR

TOOLPATHS

C-axis   ⬛ Cross Contour

Single

.063

controlled C-axis rotation

Operation 8:

Planes | Z: 0.0 ▼ | Level: 1 ▼

Material Removed

DETAIL A

1.8

A

R.016

6-2) Get the CAD model file EX5-2XX. Using PROCESS PLAN 6P-2
generate the part program to produce the part shown in Figure 6p-2.

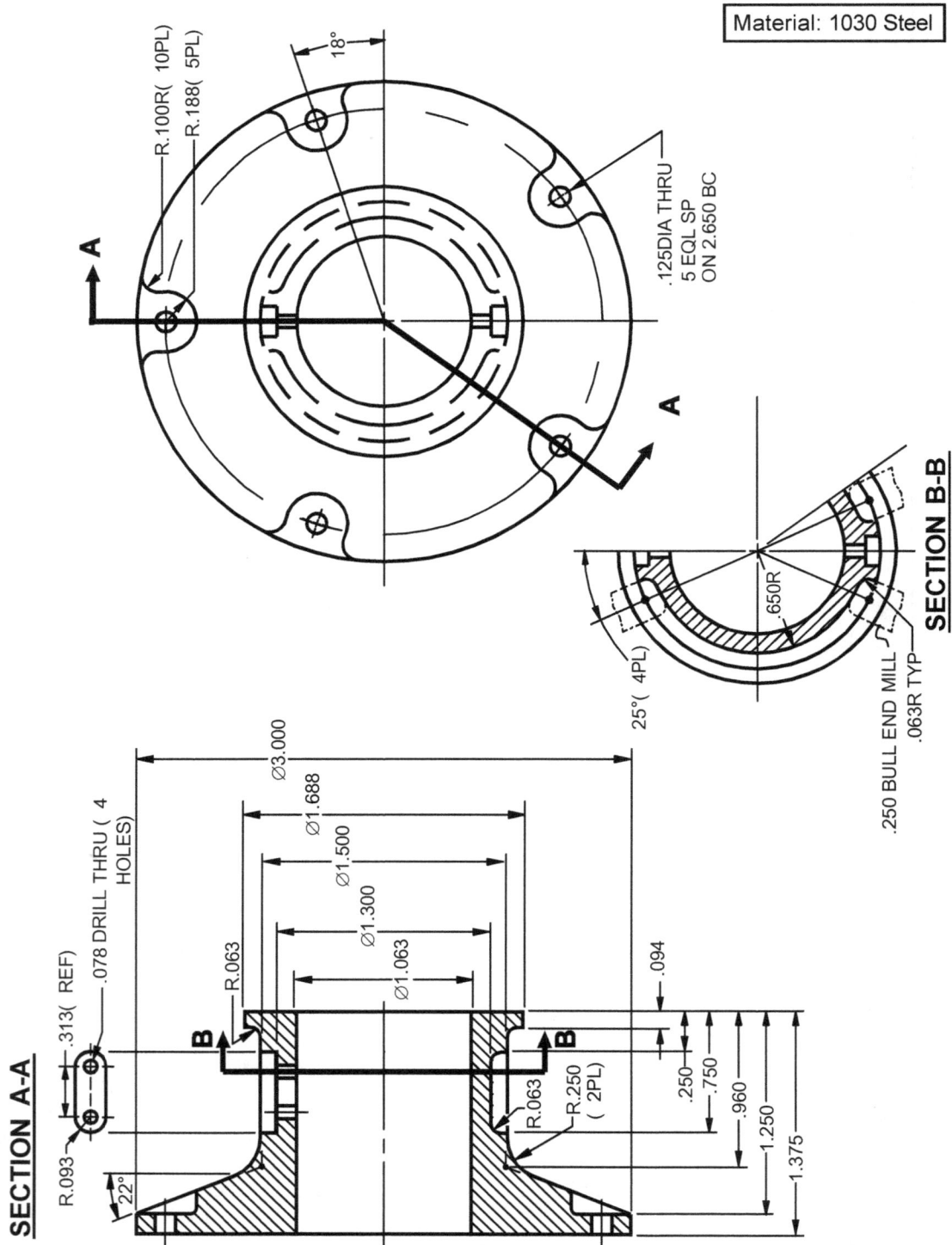

Material: 1030 Steel

R.100R(10PL)
R.188(5PL)

18°

.125DIA THRU
5 EQL SP
ON 2.650 BC

A

A

25°(4PL)

.650R

.250 BULL END MILL
.063R TYP

SECTION B-B

Figure 6p-2

Ø3.000
Ø1.688
Ø1.500
Ø1.300
Ø1.063

.078 DRILL THRU (4 HOLES)

.313(REF)

R.063

R.093

22°

R.063
R.250
(2PL)

.094

.250

.750

.960

1.250

1.375

B

B

SECTION A-A

CHUCK

BAR STOCK

Ø.125 OD MARGIN

Part Origin

Ø3.00 OD

Ø.125

GRIP LENGTH .75

LEFT MARGIN 1 1.375 .10 RIGHT MARGIN

LENGTH

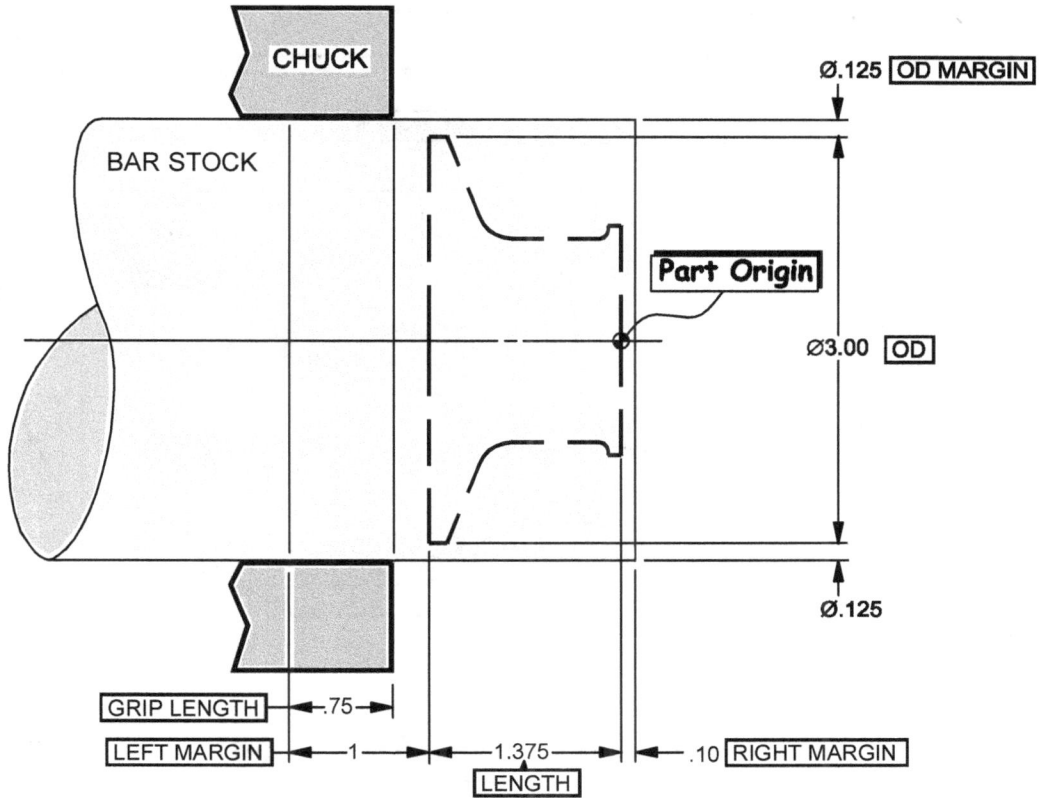

PROCESS PLAN 6P-2

No.	Operation	Tooling
1	Face End Material Removed .10	1/32 TNR, RH OD TURNING TOOL R1/32

PROCESS PLAN 6P-2(*continued*)

No.	Operation	Tooling
2	ROUGH TURN OD; LEAVE .01 IN X AND Z FOR FINISHING	1/32 TNR, RH OD TURNING
3	FINISH OD CONTOUR	1/64 TNR, RH OD FINISHING

Operation 2 panel:

Planes | Z: 0.0 ▼ | Level: 1 ▼   ☑ Lead In/Out

Lead In | Lead Out
Adjust Contour
☑ Extend/shorten end of contour
Ammount **.15**   ● Extend ○ Shorten

.15 EXTEND END OF CONTOUR

CHUCK

Material Removed

CHAIN 1 END PT ②

.01 LEFT FOR FINISH CUT

① CHAIN 1 START PT

R1/32

Operation 3 panel:

Planes | Z: 0.0 ▼ | Level: 1 ▼   ☑ Lead In/Out

Lead In | Lead Out
Adjust Contour
☑ Extend/shorten end of contour
Ammount **.15**   ● Extend ○ Shorten

.15 EXTEND END OF CONTOUR

CHUCK

Material Removed

CHAIN 1 END PT ②

.01

① CHAIN 1 START PT

R1/64

PROCESS PLAN 6P-2(*continued*)

No.	Operation	Tooling
4	ROUGH AND FINISH OD GROOVE ; LEAVE .01 IN X AND Z FOR FINISHING. Planes Z: 0.0 ▾ Level: 1 ▾ Groove Definition ○ 1 Point ○ 2 Points ○ 3 Lines ◉ Chain ○ Multiple chains CHUCK Material Removed CHAIN 1 END PT ② ① CHAIN 1 START PT	R.01, W .25 OD GROOVING RIGHT .75 ⟷ .15 ⟵ 1 ⟶ A DETAIL A ⟷ .144 .625 .250 .250 ⟷ R.01 LATHE(IN) GFG-250
5	LATHE DRILL THRU Planes Z: 0.0 ▾ Level: 1 ▾ Depth **-1.375** ◉ Absolute ○ Incremental Drill Cycle Parameters Cycle Peck drill ▾ ☑ Drill tip compensation Breakthrough ammount **.125** ⟶ ⟵ CHUCK Material Removed	1-1/16 DIA DRILL 6 10.5 1.063 ⟶ ⟵

PROCESS PLAN 6P-2(*continued*)

No.	Operation	Tooling
6	C-AXIS FACE CONTOUR X 0 DEEP Planes | Z: 0.0 ▼ | [icons] | Level: 1 ▼ **Level Manager** TOOLPATHS ⋮ C-axis ⊙ Face Contour 	.25 DIA END MILL

Level Manager

Number	Visible	Name
1		2D-PROFILE
2	X	C-FACE CON
3		C-FACE DRILL
4		C-CONTOUR
5		C-CROSS CONTOUR
6		C-CROSS DRILL

PROCESS PLAN 6P-2(*continued*)

No.	Operation	Tooling
7	**C-AXIS FACE DRILL X .225 DEEP** Planes Z: 0.0 ▼ Level: 2 ▼ **Level Manager** <table><tr><td>Number</td><td>Visible</td><td>Name</td></tr><tr><td>1</td><td></td><td>2D-PROFILE</td></tr><tr><td>2</td><td></td><td>C-FACE CON</td></tr><tr><td>3</td><td>X</td><td>C-FACE DRILL</td></tr><tr><td>4</td><td></td><td>C-CONTOUR</td></tr><tr><td>5</td><td></td><td>C-CROSS CONTOUR</td></tr><tr><td>6</td><td></td><td>C-CROSS DRILL</td></tr></table> TOOLPATHS C-axis ○ Face Drill Window Points	.125 1/8 DRILL
8	**C-AXIS CONTOUR X .1 DEEP** Planes Z: 0.0 ▼ Level: 3 ▼ **Level Manager** <table><tr><td>Number</td><td>Visible</td><td>Name</td></tr><tr><td>1</td><td></td><td>2D-PROFILE</td></tr><tr><td>2</td><td></td><td>C-FACE CON</td></tr><tr><td>3</td><td></td><td>C-FACE DRILL</td></tr><tr><td>4</td><td>X</td><td>C-CONTOUR</td></tr><tr><td>5</td><td></td><td>C-CROSS CONTOUR</td></tr><tr><td>6</td><td></td><td>C-CROSS DRILL</td></tr></table> TOOLPATHS C-axis C-Axis Contour Single	1/4 DIA BULL END MILL .0625R

PROCESS PLAN 6P-2(*continued*)

No.	Operation	Tooling
9	**C-AXIS CROSS CONTOUR X .1 DEEP** Planes | Z: 0.0 ▼ | [icons] | Level: 4 ▼ **Level Manager** | Number	Visible	Name
1		2D-PROFILE
2		C-FACE CON
3		C-FACE DRILL
4		C-CONTOUR
5	X	C-CROSS CONTOUR
6		C-CROSS DRILL | TOOLPATHS C-axis C-Axis Contour Single ✓	3/16 DIA END MILL
10	**C-AXIS CROSS DRILL X .5 DEEP** Planes | Z: 0.0 ▼ | [icons] | Level: 5 ▼ **Level Manager** | Number	Visible	Name
1		2D-PROFILE
2		C-FACE CON
3		C-FACE DRILL
4		C-CONTOUR
5		C-CROSS CONTOUR
6	X	C-CROSS DRILL | TOOLPATHS C-axis Cross Drill Window Points ✓	.0781 5/64 DRILL

PROCESS PLAN 6P-2(*continued*)

No.	Operation	Tooling
1 1	LATHE CUTOFF	.016 TNR, RH OD CUTOFF TOOL

Planes Z: 0.0 ▼ Level: 1 ▼

CHUCK

Material Removed

DETAIL A

1.8

A

R.016

6-3) Get the CAD model file EX5-3XX. Using PROCESS PLAN 6P-3 generate the part program to produce the part shown in Figure 6p-3.

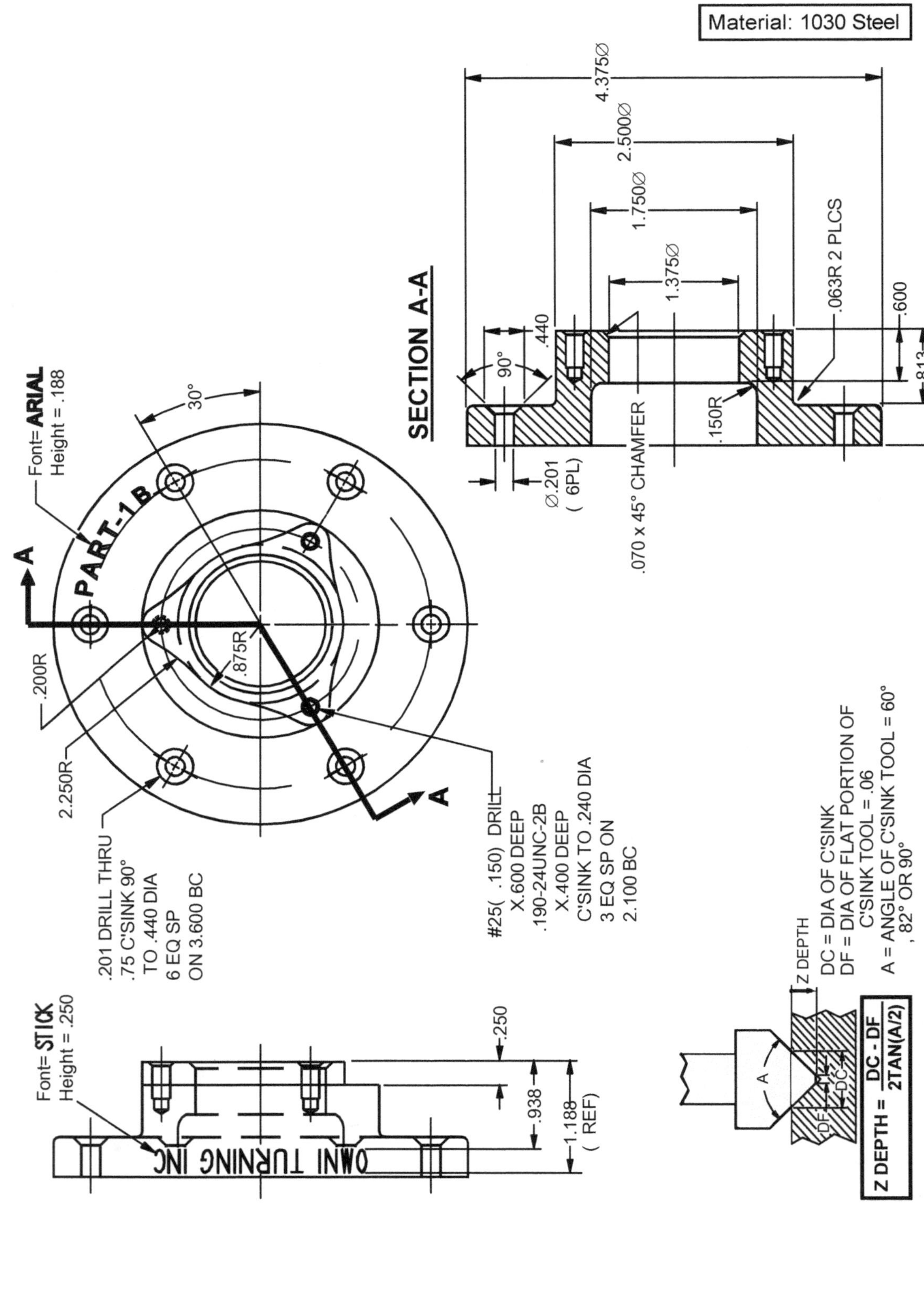

Material: 1030 Steel

SECTION A-A

4.375⌀
2.500⌀
1.750⌀
1.375⌀
.063R 2 PLCS
.600
.813
1.250
.440
90°
.070 x 45° CHAMFER
.150R
⌀.201 (6PL)

Font= **ARIAL** Height = .188

30°

PART-1B

A

.200R

.875R

2.250R

.201 DRILL THRU
.75 C'SINK 90°
TO .440 DIA
6 EQ SP
ON 3.600 BC

#25(.150) DRILL
X.600 DEEP
.190-24UNC-2B
X.400 DEEP
C'SINK TO .240 DIA
3 EQ SP ON
2.100 BC

Font= **STICK** Height = .250

OMNI TURNING INC

.250
.938
1.188 (REF)

Z DEPTH
DC = DIA OF C'SINK
DF = DIA OF FLAT PORTION OF
C'SINK TOOL = .06
A = ANGLE OF C'SINK TOOL = 60°
, 82° OR 90°

A

$$Z\ DEPTH = \frac{DC - DF}{2TAN(A/2)}$$

Figure 6p-3

CHUCK

BAR STOCK

Ø.125 OD MARGIN

Part Origin

Ø4.375 OD

Ø.125

GRIP LENGTH ├─.75─┤

LEFT MARGIN ├───1───┤──1.250──┤──.10 RIGHT MARGIN

LENGTH

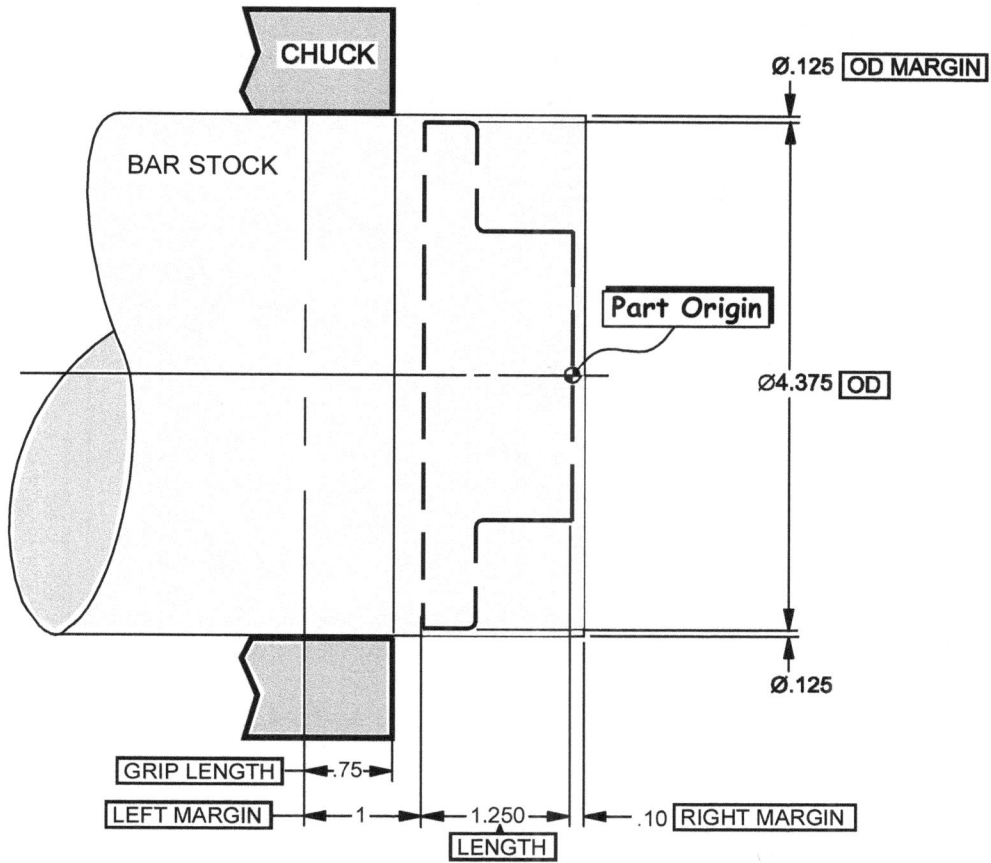

PROCESS PLAN 6P-3

No.	Operation	Tooling
1	FACE END • Material Removed • .10	1/32 TNR, RH OD TURNING TOOL • R1/32

PROCESS PLAN 6P-3(*continued*)

No.	Operation	Tooling
2	ROUGH TURN OD; LEAVE .01 IN X AND Z FOR FINISHING	1/32 TNR, RH OD TURNING
3	FINISH OD CONTOUR	1/64 TNR, RH OD FINISHING

PROCESS PLAN 6P-3(*continued*)

No.	Operation	Tooling
4	LATHE DRILL X .18 DEEP Planes	Z: 0.0 ▼	Level: 1 ▼ Depth　　-.18 ● Absolute ○ Incremental ☑ Drill tip compensation Material Removed	.25 DIA CENTER DRILL
5	LATHE DRILL THRU Planes	Z: 0.0 ▼	Level: 1 ▼ Depth　-1.125 ● Absolute ○ Incremental Drill Cycle Parameters Cycle Peck drill ▼ ☑ Drill tip compensation Breakthrough ammount　.125 CHUCK Material Removed	13/8 DIA DRILL

PROCESS PLAN 6P-3(*continued*)

No.	Operation	Tooling
4	ROUGH AND FINISH OD GROOVE ; LEAVE .01 IN X AND Z FOR FINISHING. Planes | Z: 0.0 ▼ | Level: 1 ▼ Groove Definition ○ 1 Point ○ 2 Points ○ 3 Lines ● Chain ○ Multiple chains ② CHAIN1 END PT — Material Removed — ① CHAIN1 START PT	R.01, W .125 ID GROOVING RIGHT .6 — A — .75 .093 8 DETAIL A .093 .313 .125 .125 — R.01 GFG-125CW
5	FINISH ID CHAMFER Planes | Z: 0.0 ▼ | Level: 1 ▼ ① CHAIN1 START PT ② CHAIN1 END PT — Material Removed	1/64 TNR, ID FINISH BORING TOOL ,.5 DIA R1/64 Ø.5

PROCESS PLAN 6P-3(*continued*)

No.	Operation	Tooling
6	C-AXIS FACE CONTOUR x .25 DEEP Planes | Z: 0.0 ▼ | ▣ ▣ ▤ | Level: 1 ▼ **Level Manager** 	Number	Visible	Name	 	1		2D-PROFILE	 	2	X	C-FACE CON	 	3		C-FACE DRILL	 	4		C-FACE DRILL	 	5		C-FACE CON	 	6		C-CON	 TOOLPATHS ⋮ C-axis ⊙ Face Contour ⟨ ∞ ⟩ ✓ 	.5 DIA END MILL

PROCESS PLAN 6P-3(*continued*)

No.	Operation	Tooling
7	C-AXIS FACE DRILL (PECK) x .6 DEEP 	#25(.1495) DRILL
8	C-AXIS FACE DRILL (TAP) x .4 DEEP 	.190 x 24 TAP RH

PROCESS PLAN 6P-3(*continued*)

No.	Operation	Tooling
9	C-AXIS FACE DRILL (NO PECK) x .09 DEEP 	3/4 CHAMFER MILL Z DEPTH = .5DC - .03 Z DEPTH = .5(.24) - .03 = .09
10	C-AXIS FACE DRILL (PECK) x .438 DEEP 	#7(.210) DRILL

PROCESS PLAN 6P-3(*continued*)

No.	Operation	Tooling
12	C-AXIS FACE DRILL (NO PECK) x .19 DEEP	3/4 CHAMFER MILL

For the tooling illustration:

Z DEPTH

90°

.06

DC

Z DEPTH
$$= .5DC - .03$$

Z DEPTH
$$= .5(.44) - .03$$
$$= .19$$

1.8R

controlled C-axis rotation

Planes Z: 0.0 ▼ | Level: 3 ▼

Level Manager

Number	Visible	Name
1		2D-PROFILE
2		C-FACE CON
3		C-FACE DRILL
4	X	C-FACE DRILL
5		C-FACE CON
6		C-CON

TOOLPATHS

C-axis   ⚙ Face Drill

Window Points

PROCESS PLAN 6P-3(*continued*)

No.	Operation	Tooling
13	C-AXIS FACE CONTOUR x .01 DEEP Planes | Z: 0.0 ▼ | [icons] | Level: 4 ▼ **Level Manager** | Number	Visible	Name | |---|---|---| | 1 | | 2D-PROFILE | | 2 | | C-FACE CON | | 3 | | C-FACE DRILL | | 4 | | C-FACE DRILL | | 5 | X | C-FACE CON | | 6 | | C-CON | TOOLPATHS ⋮ C-axis ● Face Contour ⬭⬭⬭ ☑ ② ① PART 1-B .01 1.8R controlled C-axis rotation	1/32 DIA BALL END MILL

PROCESS PLAN 6P-3(*continued*)

No.	Operation	Tooling
14	C-AXIS CONTOUR X .02 DEEP	1/16 DIA BALL END MILL

Planes | Z: 0.0 ▼ | | | | Level: 5 ▼

Level Manager

Number	Visible	Name
1		2D-PROFILE
2		C-FACE CON
3		C-FACE DRILL
4		C-FACE DRILL
5		C-FACE CON
6	X	C-CON

TOOLPATHS

C-axis   C-Axis Contour

Window

OMNI TURNING INC

1.8R

.938

controlled C-axis rotation

PROCESS PLAN 6P-3(*continued*)

No.	Operation	Tooling
1 5	LATHE CUTOFF Planes | Z: 0.0 ▼ | ⬓ | ⬙ | ⬗ | Level: 1 ▼ CHUCK Material Removed Cutoff parameters X Tangent point **.75** .75	.016 TNR, RH OD CUTOFF TOOL DETAIL A 1.5 A R.016

6-4) Get the CAD model file EX5-5XX. Using PROCESS PLAN 6P-4 generate the part program to produce the part shown in Figure 6p-4.

Material: Brass

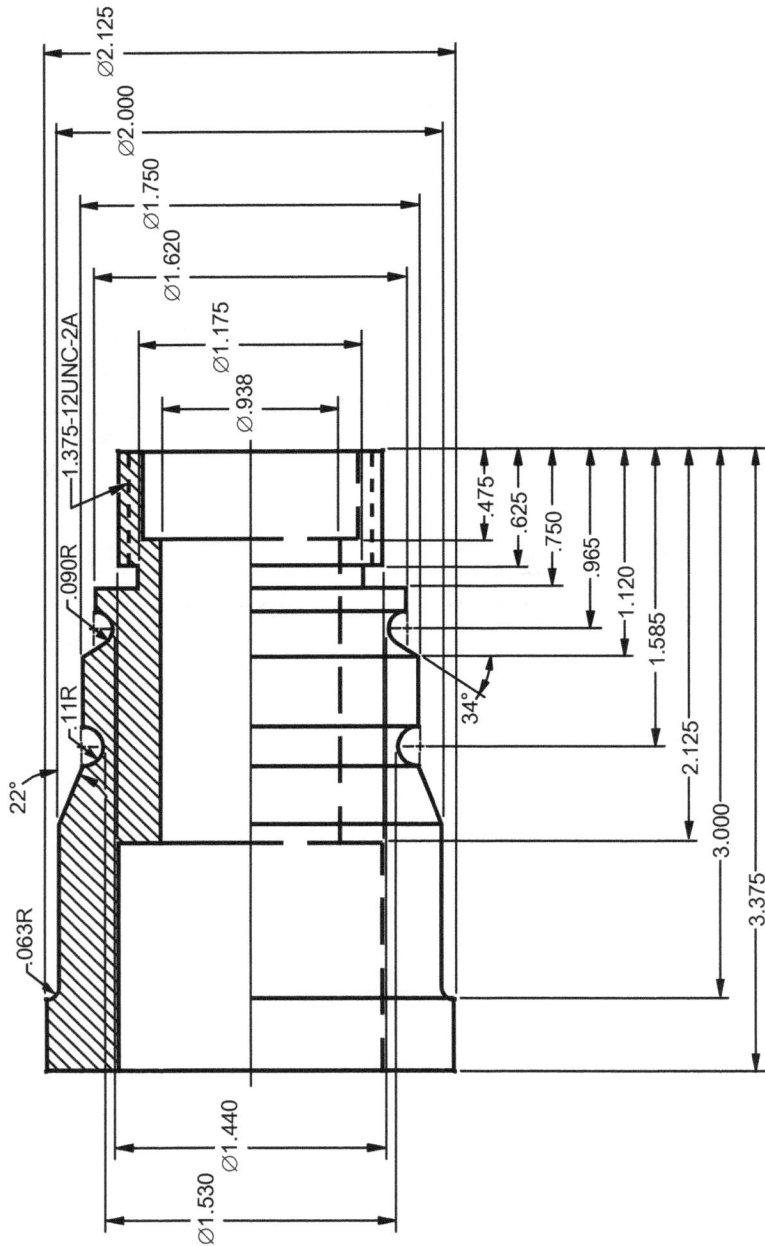

Figure 6p-4

TRUE SHAPE OF
UNROLLED TOOH

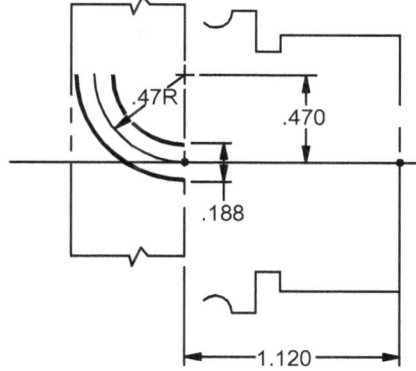

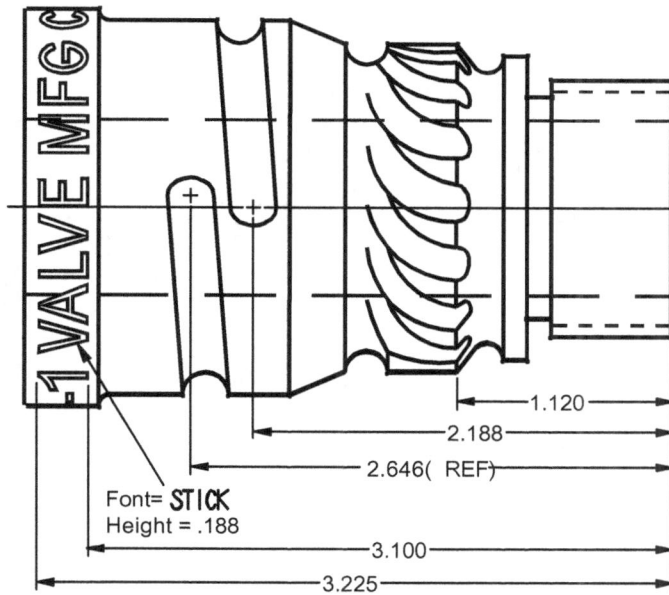

Font= STICK
Height = .188

SECTION B-B

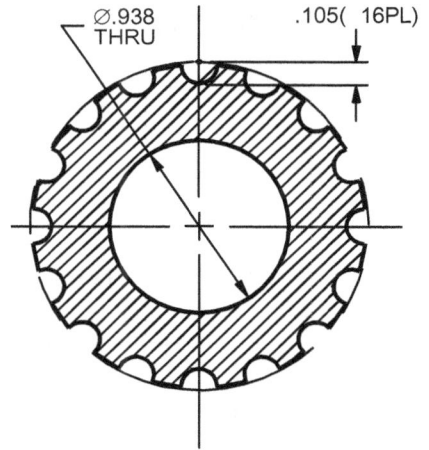

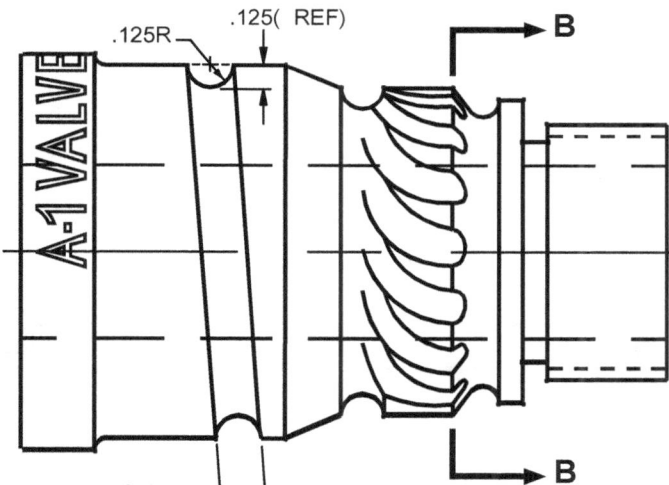

.125R

.125(REF)

B

B

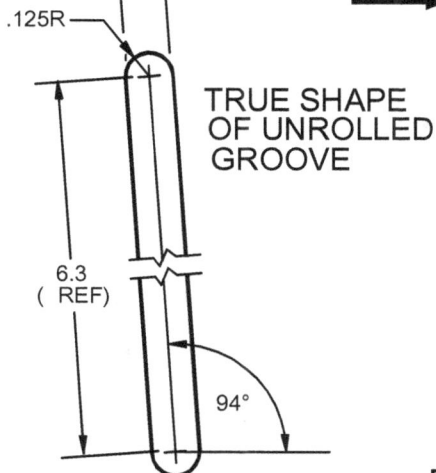

.125R

TRUE SHAPE
OF UNROLLED
GROOVE

6.3
(REF)

94°

Figure 6p-5

CHUCK

OD MARGIN Ø.0625

BAR STOCK

Part Origin

Ø2.125 OD

Ø.0625

GRIP LENGTH .75

LEFT MARGIN 1

3.375

.10 RIGHT MARGIN

LENGTH

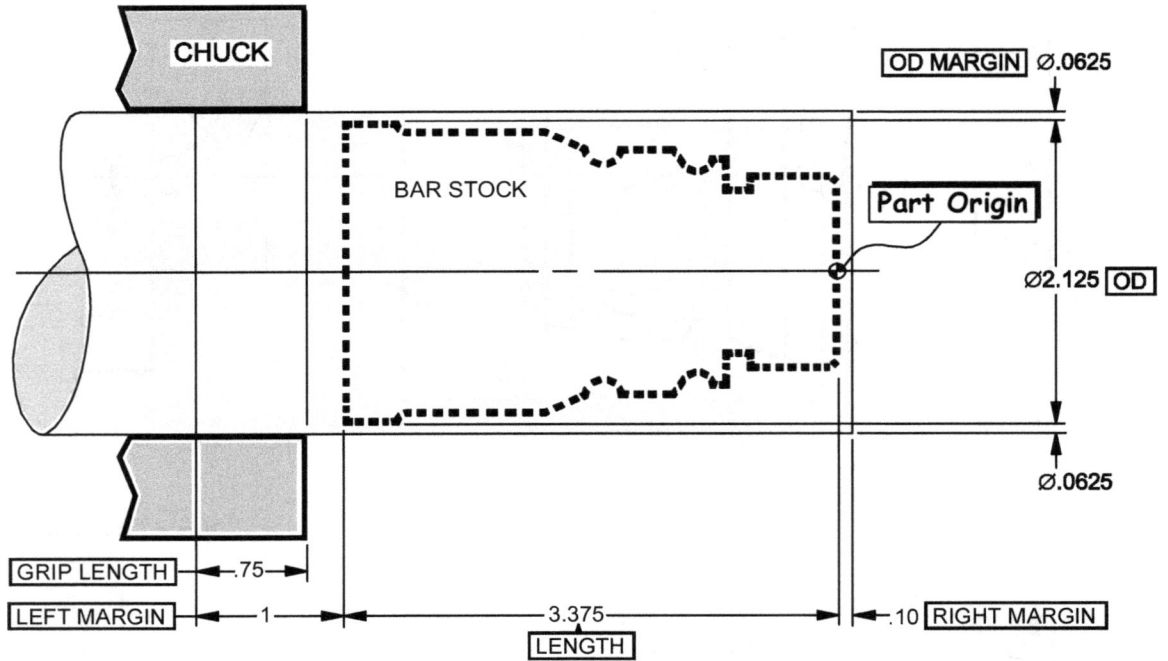

PROCESS PLAN 6P-4

No.	Operation	Tooling
1	FACE END	1/32 TNR, RH OD TURNING TOOL

Material Removed

.10

R1/32

PROCESS PLAN 6P-4(*continued*)

No.	Operation	Tooling
2	ROUGH TURN OD; LEAVE .01 IN X AND Z FOR FINISHING	1/32 TNR, RH OD TURNING

Planes | Z: 0.0 ▼ | | | | Level: 1 ▼   ☑ Lead In/Out

Lead In | Lead Out

┌─ Adjust Contour ─
☑ Extend/shorten end of contour

Ammount [.15]   ◉ Extend   ○ Shorten

.15 EXTEND END OF CONTOUR

CHUCK

Material Removed

CHAIN 1 END PT ②

.01 LEFT FOR FINISH CUT

CHAIN 1 START PT ①

R1/32

No.	Operation	Tooling
3	FINISH OD CONTOUR	1/64 TNR, RH OD FINISHING

Planes | Z: 0.0 ▼ | | | | Level: 1 ▼   ☑ Lead In/Out

Lead In | Lead Out

┌─ Adjust Contour ─
☑ Extend/shorten end of contour

Ammount [.15]   ◉ Extend   ○ Shorten

.15 EXTEND END OF CONTOUR

CHUCK

Material Removed

CHAIN 1 END PT ②

.01

CHAIN 1 START PT ①

R1/64

PROCESS PLAN 6P-4(*continued*)

No.	Operation	Tooling
4	ROUGH OD GROOVE IN A SINGLE PASS	R.01, W .25 OD GROOVING RIGHT

Planes | Z: 0.0 ▼ | | | Level: 1 ▼

Grooving Options ☒

Groove Definition
- ○ 1 Point
- ● 2 Points
- ○ 3 Lines
- ○ Chain
- ○ Multiple chains

Point Selection
- ● Manual
- ○ Window

✓ ✗ ?

Groove rough parameters

Groove Walls
- ● Steps
- ○ Smooth

Rough step:

Percent of tool width ▼

100

Stock to leave in X: **0**

Stock to leave in Z: **0**

Groove finish parameters

☐ Finish groove

NO CHECK APPEARS

Material Removed

① ②

1

5

.10

A

DETAIL A

.093

.5

.125

.125

R.01

GC-4125

PROCESS PLAN 6P-4(*continued*)

No.	Operation	Tooling
5	ROUGH AND FINISH OD GROOVE; LEAVE .01 IN X AND Z FOR FINISHING	R.024, W.255 OD GROOVING RIGHT

Planes Z: 0.0 ▼ ⬚ ⬚ ⬚ Level: 1 ▼

Groove Definition

○ 1 Point
○ 2 Points
○ 3 Lines
◉ Chain
○ Multiple chains

Groove Walls

○ Steps
◉ Smooth

☑ Lead In

Entry Vector
☑ Use Entry Vector

Fixed Direction
○ None
◉ Tangent
○ Perpendicular

Material Removed

CHAIN 1 END PT   ②   ①   CHAIN 1 START PT

1

5

.25

A

DETAIL A

R.024
.255
.25
.75

VIPV .255E .024

PROCESS PLAN 6P-4(*continued*)

No.	Operation	Tooling
6	ROUGH AND FINISH OD GROOVE; LEAVE .01 IN X AND Z FOR FINISHING Planes Z: 0.0 ▼ Level: 1 ▼ **Groove Definition** ○ 1 Point ○ 2 Points ○ 3 Lines ● Chain ○ Multiple chains **Groove Walls** ○ Steps ● Smooth ☑ Lead In **Entry Vector** ☑ Use Entry Vector **Fixed Direction** ○ None ● Tangent ○ Perpendicular Material Removed CHAIN 1 END PT ② ① CHAIN 1 START PT	R.024, W.255 OD GROOVING RIGHT 1 5 .25 A DETAIL A R.024 .255 .25 .75 VALENITE VIPV .255E .024

PROCESS PLAN 6P-4(*continued*)

No.	Operation	Tooling
7	CUT 1.375-12 UNC OD THREAD	60°V,RH,OD THREADING TOOL

Planes | Z: 0.0 ▼ | | | | Level: 1 ▼

Thread Form
- Select from table
- Compute from formula
- Draw Thread

Thread Table

Thread Form | Unified - UN, UNC, UNF, UNEF ▼

Common diameter/lead combinations up to 4 inches

Basic major	Lead	Major diameter	Minor diameter	Comment	▼
⋮	⋮	⋮	⋮	⋮	
1.3750	12.0000	1.3750	1.2728	1-3/8,12	
⋮	⋮	⋮	⋮	⋮	

Thread shape parameters

Thread cut parameters

NC code format: Canned ▼

Overcut
.06

End Position
-.625

StartPosition
0.0

.06 → | ← .875 →

Material Removed

DETAIL A

.195 →
← .117
1.0
R.0123

LT16ER 12 UN

PROCESS PLAN 6P-4(*continued*)

No.	Operation	Tooling
8	LATHE DRILL X .18 DEEP Planes | Z: 0.0 ▼ | Level: 1 ▼ Depth **-.18** ⦿ Absolute ◯ Incremental Material Removed ☐ Drill tip compensation	.25 DIA CENTER DRILL
9	LATHE DRILL THRU Planes | Z: 0.0 ▼ | Level: 1 ▼ Depth **-3.375** ⦿ Absolute ◯ Incremental Drill Cycle Parameters Cycle Peck drill ▼ ☑ Drill tip compensation Breakthrough ammount **.125** CHUCK Material Removed	15/16 DIA DRILL

PROCESS PLAN 6P-4(*continued*)

No.	Operation	Tooling
10	C-AXIS CROSS CONTOUR X .1 DEEP	3/16 DIA BALL END MILL

Planes | Z: 0.0 ▼ | | | | Level: 1 ▼

Level Manager

Number	Visible	Name
1		2D-PROFILE
2	X	C-CROSS CON
3		C-CON
4		C-CON

TOOLPATHS

C-axis   Cross Contour

Window

☑ Lead in/out

Lead in/out

☑ Entry
  Line
  ○ Perpendicular ● Tangent
  Length: 0   % 0

☑ Exit
  Line
  ○ Perpendicular ● Tangent
  Length: 0   % 0

② .016

.875R

1.12

controlled C-axis rotation

①

PROCESS PLAN 6P-4(*continued*)

No.	Operation	Tooling
11	C-AXIS CONTOUR x .125 DEEP	1/4 DIA BALL END MILL

Planes | Z: 0.0 ▼ | ⊡ ⊕ ▤ | Level: 2 ▼

Level Manager

Number	Visible	Name
1		2D-PROFILE
2		C-CROSS CON
3	X	C-CON
4		C-CON

TOOLPATHS

C-axis   ▦ C-Axis Contour

Single

Lead in/out

Lead in/out

Entry	Exit
Line	Line
○ Perpendicular ● Tangent	○ Perpendicular ● Tangent
Length: [0] % [0]	Length: [0] % [0]

.125

2.188

controlled
C-axis
rotation

PROCESS PLAN 6P-4(*continued*)

No.	Operation	Tooling
12	C-AXIS CONTOUR X .02 DEEP 	1/32 DIA BALL END MILL

PROCESS PLAN 6P-4(*continued*)

No.	Operation	Tooling
1 3	LATHE CUTOFF Planes | Z: 0.0 ▾ | 🔲 | 🔲 | 🔲 | Level: 1 ▾ Cutoff parameters X Tangent point **.375** **CHUCK** Material Removed .375	.016 TNR, RH OD CUTOFF TOOL 1.5 A DETAIL A R.016

INDEX

Welcome to Mastercam® X8 Demo/Home Learning Edition

Mastercam X8 Demo/Home Learning Edition is your window into Mastercam, the most widely used CAD/CAM software in the world.

With Mastercam X8 Demo/HLE, you can explore Mastercam thoroughly and at your own pace. Whatever your machining needs—2-axis machining, multiaxis milling and turning, wire EDM, router applications, artistic modeling and cutting, 3D modeling and more—there is a Mastercam product for your budget and application. Once you experience Mastercam X8, contact your Mastercam Reseller (http://www.mastercam.com/en-us/Communities/Find-A-Reseller) to find out which product is right for you.

Product Restrictions

Mastercam X8 Demo/HLE allows you to use many of the powerful Mastercam features with a few restrictions:

- Files are saved in EMCX format. This format cannot be read into industrial versions of Mastercam. Demo/HLE can read both MCX and EMCX formats. Once saved as EMCX there is no method to revert a file back to MCX for industrial use.
- Add-ons are not supported.
- Post processors are not available. Similarly, the .SET Setup Sheet function, which is based on a post processor, is not available.
- Mastercam X8 Demo/HLE cannot be upgraded and Maintenance is not available.
- Mastercam X8 Demo/HLE is intended for home use and may not be used in a classroom or in an industrial situation.
- We do not recommend that you install Mastercam Demo/HLE on the same system as Mastercam X8.
- Mastercam X8 Demo/HLE will expire permanently July 31, 2017.

Additional Information

Contact your Mastercam Reseller for more information about obtaining a full copy of Mastercam X8. For additional information about Mastercam X8, see the Mastercam X8 documentation after you have installed the product. The documentation is installed in the C:\Program Files\mcamDemoHLEx8\Documentation folder. Additional information is available from:

- **Mastercam Help** — Access Help by selecting *Help, Contents* from Mastercam's menu or press [Alt+H] on your keyboard. Most dialog boxes and ribbon bars feature a Help button that opens Mastercam Help directly.

- **Online help** — You can find a wealth of information, including many videos, at www.mastercam.com and www.mastercamedu.com.

- **Mastercam Reseller** — Your local Mastercam Reseller can help with most questions about Mastercam.

Thank you, and enjoy Mastercam!
CNC Software, Inc.

Mastercam

www.ingramcontent.com/pod-product-compliance
Lightning Source LLC
Chambersburg PA
CBHW080132220326
41598CB00032B/5044